Nanotechnology in Agriculture

Nanotechnology in Agriculture

Editors

K.S. Subramanian
K. Gunasekaran
N. Natarajan
C.R. Chinnamuthu
A. Lakshmanan
S.K. Rajkishore

Department of Nano Science & Technology
Tamil Nadu Agricultural University
Coimbatore – 641 003

NEW INDIA PUBLISHING AGENCY
New Delhi – 110 034

NEW INDIA PUBLISHING AGENCY
101, Vikas Surya Plaza, CU Block, LSC Market
Pitam Pura, New Delhi 110 034, India
Phone: + 91 (11)27 34 17 17 Fax: + 91(11) 27 34 16 16
Email: info@nipabooks.com
Web: www.nipabooks.com

Feedback at feedbacks@nipabooks.com

ISBN: 978-93-83305-20-9

Composed, Designed and Printed in India

Preface

Indian Agriculture is facing a wide spectrum of constraints such as burgeoning population, shrinking farm land, restricted water availability, imbalanced fertilization, low soil organic carbon, besides experiencing the fatigue of green revolution and vagaries of climate change. To address all the challenges ahead, we should think of an alternate technology such as "nanotechnology" to precisely detect and deliver the correct quantity of nutrients or other inputs required by crops that promote productivity while ensuring environmental safety. The word "nano" refers to the size of one-billionth of a metre or one-millionth of a millimeter in any one of the dimension. We are aware that all materials are made up of atoms which are the smallest units and for example, ten hydrogen atoms put together will measure one nanometer. As the size gets reduced, the surface area gets increased by several folds. For instance, a 5 cm^3 has a surface area of 120 cm^2 when the material is at its regular scale, if the same cube is divided 24 times to reach "nano" dimension, the fractionated cube can be spread over a surface area equivalent to a football stadium. Such a phenomenal increase in surface to mass ratio makes nanotechnology very unique and powerful. Nanotechnology is a fascinating field of science which manipulates atom by atom and thus processes and products evolved from nano science are the most précised ones that are impossible to achieve by the conventional systems. Despite the fact that nanotechnology is being exploited in electronics, energy and health sectors, and the field of agricultural science is just beginning to scratch the surface. Nanotechnology is visualized as a rapidly evolving field that has potential to revolutionize agriculture and food systems and improve the conditions of the poor. The word ***"nano agriculture"*** refers to the infusion of nanotechnology concepts and principles in agricultural sciences so as to develop processes and products that precisely deliver inputs and promote productivity without associated environmental harm. Nano Agriculture is quite appropriate in India in the context of changing scenarios in agricultural production systems which in the verge of transformation towards precision agriculture.

We take the opportunity of thanking the contributing authors for their commitment and patience, and the sense of friendship with which they worked with us on this assignment. We also thanks the publishers for their understanding and efficiency of handling this project.

Despite all our efforts to make this book error-free, there is every possibility of printing mistake due to oversight. We look forward to overcoming all the shortfalls, if any, in the book. Therefore, we request and expect from all our valuable readers to inform us/bring the shortfalls to our notice, which will help us to make the next edition of this book better and error-free. We also expect valuable suggestions to make the next edition of this book far better.

Editors

Contents

Section 1: General

Section 2: Synthesis

Section 3: Characterization

Section 4: Applications

Section 5: Biosafety

Contributors

A. Lakshmanan
Agro Climate Research Centre
Tamil Nadu Agricultural University
Coimbatore – 641 003
Tamil Nadu

B. Nalini
Professor (Physics)
Avinashilingam University
Coimbatore, Tamil Nadu

C.R. Chinnamuthu
Department of Nano Science and Technology
Tamil Nadu Agricultural University
Coimbatore – 641 003
Tamil Nadu

C. Sharmila Rahale
Department of Nano Science & Technology
Tamil Nadu Agricultural University
Coimbatore – 641 003
Tamil Nadu

C. Udayasoorian
School of Post Graduate Studies,
Tamil Nadu Agricultural University
Coimbatore – 641 003
Tamil Nadu

D.B.S. Sethi
Defence Research & Development
Organisation (DRDO), India

D. Nataraj
Associate Professor (Physics)
Thin Films & Nanomaterials Laboratory
Department of Physics, Bharathiar University
Coimbatore – 641 046, Tamil Nadu

Girija
Assistant Professor (Biotechnology)
Bharathiyar University
Coimbatore – 641 046, Tamil Nadu

K.S. Subramanian
Professor (SS & AC)
Dept. of Nano Science & Technology
Tamil Nadu Agricultural University
Coimbatore – 641 003, Tamil Nadu

K. Muraleedharan
Defence Research & Development
Organisation (DRDO), India

K. Pandian
Associate Professor (Chemistry)
University of Madras
Chennai – 600 025
Tamil Nadu

K. Srinivasan
Associate Professor (Physics)
Bharathiyar University
Coimbatore – 641 046
Tamil Nadu

K. Brindha
Department of Nano Science and Technology
Tamil Nadu Agricultural University
Coimbatore – 641 003, Tamil Nadu

K. Gunasekaran
Department of Nano Science and Technology
Tamil Nadu Agricultural University
Coimbatore – 641 003, Tamil Nadu

K. Ramaesh
Senior Scientist (Entomology)
NBPGR Regional Station
Hyderabad – 500030, Andhra Pradesh

K. Subaharan
Senior Scientist
CPCRI, Kasaragode – 671124, Kerala

K. Vinoth Kumar
School of Post Graduate Studies
Tamil Nadu Agricultural University
Coimbatore – 641 003, Tamil Nadu

K. Thangavel
Professor (F & APE)
AEC & RI
Tamil Nadu Agricultural University
Coimbatore – 641 003, Tamil Nadu

K. Ilamurugu
Professor (Microbiology)
Tamil Nadu Agricultural University
Coimbatore – 641 003, Tamil Nadu

M. L. Sharma
Consultant, ICON Analytical
Mumbai, Maharashtra

M. Kannan
Department of Nano Science & Technology
Tamil Nadu Agricultural University
Coimbatore – 641 003, Tamil Nadu

M. Praghadeesh
Senior Research Fellow
Department of Nano Science & Technology
Tamil Nadu Agricultural University
Coimbatore – 641 003, Tamil Nadu

M. R. Manikantan
Scientist (AS &PE)
Central Institute of Post-Harvest Engineering and Technology (CIPHET)
Ludhiana – 141004, Punjab

N. Sunil Kumar
Department of Nano Science and Technology
Tamil Nadu Agricultural University
Coimbatore – 641 003, Tamil Nadu

N. Varadharaju
Professor and Head, PHTC, AEC&RI
Tamil Nadu Agricultural University
Coimbatore – 641 003, Tamil Nadu

N.B. Nandakumar
Research Associates
Department of Nano Science & Technology
Tamil Nadu Agricultural University
Coimbatore – 641 003, Tamil Nadu

N. Natarajan
Department of Nano Science & Technology
Tamil Nadu Agricultural University
Coimbatore – 641 003
Tamil Nadu

P. Meenakshisundaram
Department of Nano Science & Technology
Tamil Nadu Agricultural University
Coimbatore – 641 003
Tamil Nadu

Rajeev Varshney
Defence Research & Development Organisation (DRDO), India

R.T. Rajendra Kumar
Associate Professor (Physics)
Bharathiyar University
Coimbatore – 641 046
Tamil Nadu

R. Sunitha
Department of Nano Science & Technology
Tamil Nadu Agricultural University
Coimbatore - 641 003
Tamil Nadu

R. Selvakumar
Assistant Professor in Nanobiotechnology,
Nanobiotechnology Laboratory
PSG Institute of Advanced Studies
Coimbatore, Tamil Nadu

R. Manimekalai
Senior Scientist (Biotechnology)
Central Plantation Crops Research Institute
Kasaragode – 671 124, Kerala

R.M. Jayabalakrishnan
Horticultural Research Station
Tamil Nadu Agricultural University
Ooty, Tamil Nadu

R. Selvarajan
Senior Scientist (Plant Virology)
NRC for Banana, ICAR
Trichy, Kerala

R. Velazhahan
Professor
Department of Plant Pathology
Tamil Nadu Agricultural University
Coimbatore – 641 003, Tamil Nadu

S. Thirunavukkarasu
Department of Nano Sci. & Tech.
Tamil Nadu Agricultural University
Coimbatore – 641 003, Tamil Nadu

S. Marimuthu
Department of Nano Science & Technology
Tamil Nadu Agricultural University
Coimbatore – 641 003, Tamil Nadu

S.K. Rajkishore
Research Associates
Department of Nano Science & Technology
Tamil Nadu Agricultural University
Coimbatore – 641 003, Tamil Nadu

S. Kalaivani
Department of Nano Science & Technology
Tamil Nadu Agricultural University
Coimbatore – 641 003, Tamil Nadu

S. Senthil Kumar
Department of Nano Science & Technology
Tamil Nadu Agricultural University
Coimbatore – 641 003, Tamil Nadu

S. Manikandan
Ph.D Scholar
Department of Nano Science & Technology
Tamil Nadu Agricultural University
Coimbatore – 641 003, Tamil Nadu

S. Manisankar
Professor & Head
Department of Industrial Chemistry
Alagappa University
Karaikudi – 630003, Tamil Nadu

S. Sithanantham
Sun Agro Biotech Research Centre
3/340, Madanandapuram Main Road
Mugalivakkam, Porur,
Chennai – 600125, Tamil Nadu

S.K. Rajkishore
Department of Nano Science & Technology
Tamil Nadu Agricultural University
Coimbatore – 641003, Tamil Nadu

V. Jayakumar
Senior Scientist (Plant Pathology)
Sugarcane Breeding Institute
Coimbatore, Tamil Nadu

Venkita Subbulakshmi
Chromous Biotech Private Limited
Bangalore, Karnataka

W. Selvamurthy
Defence Research & Development
Organisation (DRDO), India

Section 1: General

1

Nanotechnology for Precision Agriculture

K.S. Subramanian

Indian Agriculture is facing a wide spectrum of constraints such as burgeoning population, shrinking farm land, restricted water availability, imbalanced fertilization, low soil organic carbon, besides experiencing the fatigue of green revolution and vagaries of climate change. To address all the challenges ahead, we should think of an alternate technology such as "nanotechnology" to precisely detect and deliver the correct quantity of nutrients or other inputs required by crops that promote productivity while ensuring environmental safety. The word "nano" refers to the size of one-billionth of a metre or one-millionth of a millimeter in any one of the dimension. We are aware that all materials are made up of atoms which are the smallest units and for example, ten hydrogen atoms put together will measure one nanometer. As the size gets reduced, the surface area gets increased by several folds. For instance, a 5 cm^3 has a surface area of 120 cm^2 when the material is at its regular scale, if the same cube is divided 24 times to reach "nano" dimension, the fractionated cube can be spread over a surface area equivalent to a football stadium. Such a phenomenal increase in surface to mass ratio makes nanotechnology very unique and powerful. Nanotechnology is a fascinating field of science which manipulates atom by atom and thus processes and products evolved from nano science are the most précised ones that are impossible to achieve by the conventional systems. Despite the fact that nanotechnology is being exploited in electronics, energy and health sectors, and the field of agricultural science is just beginning to scratch the surface. Nanotechnology is visualized as a rapidly evolving field that has potential to revolutionize agriculture and food systems and improve the conditions of the poor. The word ***"nano agriculture"*** refers to the infusion of nanotechnology concepts and principles in agricultural sciences so as to develop processes and products that precisely deliver inputs and promote productivity without associated environmental harm. Nano Agriculture is quite appropriate in India in the context of changing scenarios in agricultural production systems which in the verge of transformation towards precision agriculture.

Global scenario

The potentials of nanotechnology has triggered interests of both developed and developing nations across the globe. The US National Science Foundation (NSF) has listed nanotechnology as one of the six priority areas and started investing on this cross-cutting theme areas. The recent statistics suggests that about 90% of the nano-based patents and products have come from just seven countries such as US, China, Germany, France, Japan, Switzerland and South Korea in the world while India's investments and progress is far from surfacing. Globally, there has been exponential increase in both investments and development of nano-based processes and products by private and public sectors. Total global investment in nanotechnology on research and development in 2007 was 13.5 billion USD which was 30 billion USD in 2010 and increased to 1000 billion USD by 2020. Such phenomenal increase will coincide with the production of plenty of nano-based products. The major industries that exploit nanotechnology include material science (coatings, automobiles, building materials), electronics (displays and batteries) and health sciences (pharmaceutical applications). The use of engineered nano-particles was just 2000 tonnes in 2004 and expected to be increased to 58,000 tonnes by the year 2020 (Maynard, 2006). The intensive use is likely to cause environmental hazard associated with engineered nano-particles (Subramanian et al., 2013).

Indian scenario

The Indian government is looking towards nanotechnology as a means of boosting agricultural productivity in the country. Recently, the Planning Commission of India recommended nanotechnology research and development (R&D) as one of the six areas for investment. In order to harness the benefits of nanotechnology, and transform Indian agriculture, an exclusive National Institute of Nanotechnology in Agriculture (NINA) has to be established under the NARS (National Agricultural Research System). The report says nanotechnology such as nano-sensors and nano-based smart delivery systems could ensure effective utilization of natural resources such as water, nutrients and chemicals in agriculture. Nano-barcodes and nano-processing could also help monitor the quality of agricultural produce. The report proposes a national consortium on nanotechnology R&D, to include the proposed national institute and Indian institutions that are already actively researching nanotechnology. It also recommended that Indian universities and institutions develop suitable graduate and postgraduate programs to train young scientists in nanotechnology. In order to gain the fruition of such development, more and more countries join the club of nanotechnology. To take the advantage of the fascinating field of nanoscience, the Government of India has invested Rs. 1000 crores through the Nano mission project during the 11th Five Year Plan and the investment is likely to be several folds higher during the 12th Plan period. The DST Nano Mission Project is headed by Dr. C.N.R. Rao. This investment primarily benefitted IITs, IISc, conventional universities, engineering colleges etc. Indeed, 19 Centres of Excellence were established across the country. The research investment on "Nano Agriculture" was mere Rs. 95 lakhs which is just 0.095% of the investment made to one of the State Agricultural Universities "TNAU". This opens up greater opportunity to harness nanotechnology in the field of agricultural sciences. Indeed, enhancement of agricultural productivity has been identified as a critical area of nanotechnology application for attaining the millennium development goals.

Status of nano agriculture in India

The Nano Mission in Agriculture will focus on undertaking research and development activities in a set of identified researchable areas of prime importance utilizing basic principles and applications of nanotechnology concepts in order to achieve the desirable goals besides ensuring environmental safety. Nano Agriculture is intended to infuse nanotechnology into agricultural sciences in order to transform traditional farming practices to precision agriculture that ensure evergreen revolution and food security of the country (Subramanian and Tarafdar, 2011). This encompasses early detection of pests and diseases, improving input use efficiencies, enhancing the shelf life of perishables (fruits, vegetables, flowers, meat and fish) and nano-toxicological studies. The skeleton of the proposed "Nano Mission in Agriculture" closely coincided with the reports of International Food Research Institute, Washington (IFRI, 2011). Further, the researchable areas identified are quite appropriate to augment research in Nano Agriculture as stipulated by United States Department of Agriculture (USDA, 2003). Despite there are indications of inclusion of nanoscience and technology in agricultural sciences, very scantly literature is available within the country or abroad. Scientists from ICAR Institutes and SAUs particularly TNAU, PAU and GBPNUAT besides national and international organizations have bestowed interest to involve themselves in nanotechnology research in agricultural sciences. The lessons learned from our experience and acquired from western laboratories abroad have served as a strong base to orchestrate this tailor-made document with a proposed action plan for 12th Five Year Plan.

The Indian Council of Agricultural Research (ICAR) has opened up an exclusive platform to target the field of Nanotechnology Applications in Agriculture and bestowed interest to invest about Rs. 200-250 Crores. The ICAR conducted five rounds of brainstorming sessions in various places such as CIFE, Mumbai (October, 2011), TNAU, Coimbatore (November, 2011), twice in NASC, New Delhi (March, 2012 & May, 2012) and CPRI, Shimla (September, 2012). These meetings helped the proposal to get refined and redesigned to suit the current scenarios in Indian agriculture. The ICAR - Nanotechnology Platform encompasses six major theme areas such as synthesis of nano-particles for agricultural sciences, quick diagnostic kits for early detection of pests and diseases, nano pheromones for effective pest control, nano agri inputs for enhanced use efficiencies, nano food systems and bio-safety besides establishing the policy frame work for the country.

Nano agriculture – principles and practices

Early detection of diseases & nutrient deficiencies using diagnostic kits

The pests, diseases and nutrient deficiencies constitute major loss to the tune of 40-65% of any agricultural or horticultural crops. Early detection is utmost essential to protect the crops from infection and prevent yield and quality losses. Conventionally, pesticides are sprayed only after symptoms are obvious based on visual diagnosis. When spraying is performed it may be too late to protect the crops. Nanotechnological approaches are widely used for early detection of diseases particularly cancer in humans. Similar diagnostic approaches and devices can be exploited in agricultural production systems. In the past two decades, viral diseases in several crops can be detected using ELISA (Enzyme Linked Immunosorbent Assay) tests. This method is based on antigen – antibody reaction that is very specific and accurately detects the disease. Despite this technique is highly useful, it takes a couple of weeks to get the ELISA tests done in nearby plant pathological laboratories by that time extensive damages could have already been done and it becomes too late to undertake control measures. The ICAR has taken assiduous efforts to undertake research on dip-stick method wherein proteins or nucleic acids serve as reference molecules. The plant extract is allowed to react with the stick and the detection is done within couple of minutes in the field itself. This technique has been proved effective in detecting viral diseases in potato. The precision and validity of the method can be further improved using nano-particles. Already two diagnostic kits for detecting banana bunchy top (Dr. Selvaraj, NRCB, Trichy) and potato curl virus (Dr. Chakroborty, CPRI, Shimla) are in the stage of validation and expected hit the farm gate in due course (Selvarajan, 2013). Nutrient deficiencies can be detected using spectral signature of the foliage which can be captured and developed as a reference kit. The development of diagnostic kits for detection of diseases, pests and nutrient deficiencies are quite appropriate to enable quick reaction time to take corrective measures.

Protection of soil and plant health requires rapid, sensitive detection of pollutants and pathogens with molecular precision. Biosensors can be developed in order to accurately measure the moisture, nutrient, pathogenicity and pest incidence so as to take up timely corrective measures. Sensors can be developed for the early detection of major pests (eg. Eriophid mite, mealy bugs, cotton weevil etc.), diseases (eg. red rot in sugarcane, downy mildew in grapes) and abiotic stresses (drought, salinity, Zn deficiency). In addition to the detection of pests and diseases, GM crops can be discriminated using a nano-based sensor. The development of "Diagnostic Kits" is one of priority areas in nano-agriculture for quick detection of plant diseases (caused by viruses, bacteria, fungi, phytoplasma), dairy animals, poultry and fisheries besides sensor development for assessing anti-nutritional factor (phytic acid) in poultry feed and nutritional disorders in crops.

Nano-pheromone for effective pest monitoring

Nanotechnological approaches can be exploited to protect the crops from devastating pests by effective monitoring which vociferously supports "Prevention is better than cure". One of the immediate possibilities of using nanoscience in pest control is the pheromones. Pheromones' are highly volatile and their release pattern can be regulated through nano-formulations. In this technology, pheromone lure compound is kept in a trap to attract female insects and effectively kill them for the effective control of pests. Research has been done for decades for the isolation, identification and synthesis of insect pheromones which is an effective tool for management of pest as one the IPM strategy but still the technology needs refinement for use in the applications of mass trapping and mating disruption that ensures formulations possessing, less photo-oxidation, least thermo degradation and prolonged shelf life. This study focuses on the synthesis of nanopherosensor (nanosensor used to sense pheromone) to detect pheromone molecules and provide concentration present in field or vineyard. After recognition of pheromone molecules this should act as trigger for device and the device should release optimum level of pheromones.

In India, pheromone lures are used for trapping fruit flies, stem borers in sugarcane, boll worms *Heliothis armigera* and *Spodoptera litura*, coconut red palm weevil, (*Rhychophorus ferrugineus*) and rhinoceros beetle (*Oryctes rhinoceros*). In certain cases like the banana stem weevil, *Odoiporous longicollis* and sweet potato weevil, *Cylas formicarius*. They are yet to be used on wide scale in field level due to non availability of a robust delivery system. For pests like cashew stem borer, citrus leaf miner and oilpalm psychid, *Metesia plana*, the compounds that cause physiological and behavioural response needs to be identified. Subhakaran (2013) reported that a lure has been developed at the CPCRI, Kasaragode, by fortifying ethyl 4 methyl octonate in a nano-matrix to attract rhinoceros beetles that possesses sustained release besides effective monitoring. Further, a lure has been developed by National Bureau of Agriculturally Important Insects (NBAII, Bangalore) to monitor fruit flies in guava orchards. The robustness of the chemo-ecological approach for pest management can be achieved by studying the interactions between organisms belonging to different trophic levels at a broader level. The results of such studies will pave way for environmentally sound practices of pest monitoring and control, as the plant derived compounds and species specific pheromones may interact synergistically in attracting targeted insect pest. The success of the pheromone/ kairomone technology depends of development of an effective delivery system. Development of a nano-matrix for release of pheromone / kairomone to be taken up under network mode will help to develop a clean technology for management of key pests of crops.

Enhanced input use efficiency

Nano-seed science in rainfed agriculture

Seed is a basic input deciding the fate of productivity of any crop. Conventionally, seeds are analyzed for their germination and distributed to farmers for sowing. Despite the fact that the germination percentage registered in the seed testing laboratory is about 80-90%, it hardly happens in the field due to the inadequacy or non-availability of sufficient moisture under rainfed system. In India, more than 60% of the net area sown is under rainfed system, it is quite appropriate to develop technologies for rainfed agriculture. Zinc oxide nano-particles are known to improve germination of blackgram or tomato seeds from 10-15% to 80-85% (Natarajan et al., 2013). During the storage, seeds produce reactive oxygen species (ROS) that results in lipid peroxidation causing serious reduction in germination. ZnO nano-particles donate electrons that quenche the ROS which eventually resulted in higher seed germination. Khodakovskaya (2009) and her team in 2009 at the University of Arkansas, USA, have reported the use of carbon nano-tube for improving the germination of tomato seeds through permeation of moisture. Their data have vividly shown that carbon nanotubes (CNTs) serve as new pores for water permeation by penetration of seed coat and act as a gate to channelize the water from the substrate in to the seeds. These processes facilitate germination which can be exploited rainfed agriculture system.

Nano-fertilizers for balanced crop nutrition

Nano-fertilizers are nutrient carriers of nano-dimensions ranging from 30-40 nm and are capable of holding bountiful of nutrient ions due to their high surface area and release it slowly and steadily that commensurate with crop demand. Subramanian (2008) reported that nano-fertilizers and nanocomposites can be used to control the release of nutrients from the fertilizer granules so as to improve the nutrient use efficiency while preventing the nutrient ions either get fixed or lost to the environment. Nano-fertilizers have high use efficiency and can be delivered in a timely manner to a rhizospheric target. There are slow-release and super sorbent nitrogenous and phosphatic fertilizers. Some new-generation fertilizers have applications to crop production on long-duration human missions to space exploration (Lal, 2008).

According to the report of Iranian Nanotechnology Initiative Council (2009), Iranian researchers have produced the first nano-organic iron-chelated fertilizer in the world. Nano fertilizers have unique features like ultra high absorption, increase of 20 % to 200 % in production, rise in photosynthesis by 3.5 times and a 70% expansion in the leaves' surface area, Iranian Nanotechnology Initiative Council reported. While foreign samples cause an increase of up to 30 % in photosynthesis, the Iranian nano fertilizers are able to cause a 350% increase. Moreover, these nano fertilizers are environmentally sustainable due to their organic base which makes them more suitable than foreign fertilizers that are hormone based.

Subramanian and Tarafdar (2011) suggested that clay particles are adsorptive sites carrying reservoir of nutrient ions. Major portion of nutrient fixation occurs in the broken edges of the clay particles. Zero valence nano-particles can absorb on to the clay lattice thereby preventing fixation of nutrient ions. Further, nano-particles prevent the freely mobile nutrient ions to get precipitated. These two processes assist in promoting the labile pool of nutrients that can be readily utilized by plants. Fertilizer particles can be coated with nano-membranes that facilitate in slow and steady release of nutrients. This process helps to reduce loss of nutrients while improving fertilizer use efficiency of crops.

The naturally occurring clay minerals and zeolites are reduced to the size of nano dimensions using top down approach and nutrient at desirable proportion have been fortified in the clays after surface modification (Bansiwal, 2006). The nanofertiliser formulations before and after loading with nutrients have been characterized using High Resolution Microscopes and Spectroscopy

(Liu et al., 2006). Nano zeolites are capable of retaining nutrients due to its extensive surface area and release slowly and steadily for an extended period of 1000-1200 hrs while conventional fertilizers could release for about 300-400 hours (Subramanian and Rahale, 2010). Recently, Subramanian and Rahale (2013) have monitored the nutrient release pattern of nano-fertiliser formulations carrying nitrogen. The data have shown that nano-clay based fertilizer formulations (zeolite and montmorillonite with a dimension of 30-40 nm) are capable of releasing the nutrients for a longer period of time (> 1000 hrs) than conventional fertilizers (< 500 hrs). These literatures strongly suggest that nano technological applications improve the NUE and productivity of crops without associated environmental hazard.

Nano-fertilizer technology is very innovative and scanty reported literature is available in the scientific journals. However, some of the reports and patented products strongly suggest that there is a vast scope for the formulation of nano-fertilizers. The Tamil Nadu Agricultural University is one of the pioneering institutes that initiated research in nano-fertilizers and the preliminary data are quite encouraging and shown to improve nutrient use efficiencies (Subramanian and Sharmila, 2012). Currently, research is underway to develop nano-composite that supply all the required essential nutrients in suitable proportion through smart delivery system. These literatures suggest that balanced fertilization may be achieved through nanotechnology. The impact of nano-fertilizer products on physiological, biochemical, nutritional and morphological changes in plants and the fate of nano-products in soil and plant systems have to be studied. In addition, the effects of nano-fertilizer products on rhizosphere microorganisms and biogeocycling of nutrients have to be explored under natural field conditions.

Nano-herbicide for effective weed control

Weeds are menace in agricultural production system. Since two-third of Indian agriculture is rainfed where usage of herbicide is very limited, weeds have the potential to jeopardize the total harvest in the delicate agro-ecosystems. Herbicides available in the market are designed to control or kill the growing above ground part of the weed plants. None of the herbicides inhibits activity of viable belowground plant parts like rhizomes or tubers, which act as a source for new weeds in the ensuing season. With a view to design and fabricate a nano-herbicide that is protected under natural environment and act only when there is a spell of rainfall that truly mimic the rainfed system, encapsulated nano-herbicides are relevant. The Tamil Nadu Agricultural University, Coimbatore, has successfully encapsulated the herbicide molecules (Pentimethalin and metalachlor) with polymers such as PSS (poly styrene sulphonate) and PAH (poly alylamine hydrochloride) that breaks open only the soil moisture is present. Encapsulated herbicides possess thermal and hydro stability besides sustained release of active ingredients that ensure effective weed control (Chinnamuthu et al., 2012). These nano-herbicides are to be tested under open environment before commercialization.

Nano insecticides

Persistence of insecticides in the initial stage of crop growth helps in bringing down the pest population below economic threshold level and to have an effective control for a longer period. Hence, the use of residues in the applied surface remains one of the most cost-effective and versatile means of controlling insect pests. In order to protect the active ingredient from the environmental conditions and to promote persistence, a nanotechnology approach "nano-encapsulation" can be used to improve the insecticidal value. Nano-encapsulation comprises nano-sized particles of the active ingredients being sealed by a thin-walled sac or shell (protective coating). In Tamil Nadu agricultural University, neem-based microemulsion (198 nm) developed has been found effective in controlling sucking pests such as thrips, aphids and mites (Gunasekaran, 2011). Recently, several research papers have been published on the encapsulation of insecticides. Nano-encapsuation of insecticides, fungicides or nematicides will help in producing a formulation which offers effective control of pests while preventing residues in soil.

Smart delivery system

Nanoscale devices are envisioned that would have the capability to detect and treat diseases, nutrient deficiencies or any other maladies in crops long before symptoms were visually exhibited. "Smart Delivery Systems" for agriculture can possess timely controlled, spatially targeted, self-regulated, remotely regulated, pre-programmed, or multi-functional characteristics to avoid biological barriers to successful targeting. Smart delivery systems can monitor the effects of delivery of nutrients or bioactive molecules or any pesticide molecules. This is widely used in health sciences wherein nanoparticles are exploited to deliver required quantities of medicine to the place of need in human system. In the smart delivery system, a small sealed package carries the drug which opens up only when the desirable location or infection site of the human or animal system is reached. This would allow judicious use of antibiotics than otherwise would be possible.

Nanodevices for identity preservation (IP) and tracking

One of the major constraints in Indian agriculture is the quality maintenance of agricultural produce. Proper monitoring of production system through nanotechnology will be very appropriate to promote quality and make clear distinction with organic products. Identity Preservation (IP) is a system that creates increased value by providing customers with information about practices and activities used to produce a particular crop or other agricultural products. Certifying inspectors can take advantage of IP as a more way of recording, verifying, and certifying agricultural practices. Through IP, it is possible to provide stakeholders and consumers with access to information, records and supplier protocols. Quality assurance of agricultural products safety and security could be significantly improved through IP at the nano-scale. Nano-scale IP holds a possibility of the continuous tracking and recording of the history which a particular agricultural product experiences. The nano-scale monitors linked to recording and tracking devices to improve identity preservation of food and agricultural products. The IP system is highly useful to discriminate organic versus conventional agricultural products.

Nanobiotechnology

Nanobiotechnology has the potential to increase the efficiency and quality of agricultural production and food storage, to enhance the safety of food supplies for the protection of consumers and producers and to introduce new functionality (value added products) for food, fiber and agricultural commodities. Nanobiotechnology will pave ways for new researchable areas and applications such as DNA chip, protein identification and manipulation, Novel nucleic acid engineering based films, smart delivery of DNA using gold nanoparticles. Biological tests measuring the presence or activity of selected substances become quicker, more sensitive and more flexible when nano-particles are put to work as tags or labels. Magnetic nanoparticles, bound to suitable antibody, are used to label specific molecules, structures or microorganisms. For example, gold nanoparticles tagged with short segments of DNA can be used for detection of genetic sequence in a sample. Multicolor optical coding for biological assays has been achieved by embedding different sized quantum dots into polymeric microbeads. Nanopore technology for analysis of nucleic acids converts strings of nucleotides directly into electronic signatures.

Nano-food industry

During last three years, food industries have witnessed that the nanotechnoloy has been really integrated in a number of food and food packaging products. There are now more than 300 nanofood products available on the market worldwide. These exciting achievements have encouraged a large increase of R&D investments in nanofood. Today, the nanotechnology is no longer an empty buzzword, but an indispensable reality in the food industry. Any food company who wants to keep its leadership in food industries must begin to work with nanotechnology right now. The impact of nanotechnology is huge, ranging from basic food to food processing, from nutrition delivery to intelligent packaging. It is estimated that the nanotechnology and nano-bio-info convergence will influence over 40% of the food industries up to 2015.

In the Post Harvest Technology Center (PHTC) of Tamil Nadu Agricultural University, Coimbatore, several food packaging materials have been developed by integrating nano-clays and chitosans. These are among the first nano-composites to emerge on the market as improved materials for packaging (including food packaging). Recently, the IDRC (International Development Research Center), Canada, sanctioned a project on "Enhanced Preservation of Fruits in South Asia using Nano-film" to TNAU to extend the shelf life of mango fruits. The budget outlay of the project is 2.5 million CAD. The institutes involved in this project include TNAU, India, University of Guelph, Canada, and ITI (Industrial Technology Institute), Sri Lanka.

Nanotechnology for environmental safety

One of the major constraints in Indian agriculture is irrigation water and the availability is shrinking alarmingly. It has been estimated that the per capita availability of water has reduced nearly 50% in the past 40 years. This necessitates for remediation of contaminated waters and to use them for agriculture besides domestic purposes. Nanotechnology is a powerful tool to remediate aquatic and soil pollutants. The scientists of Banaras Hindu University in India have devised a simple method to produce carbon nanotube filters that efficiently remove micro to nano-scale contaminants from water and heavy hydrocarbons from petroleum. Magnetic nanoparticles offer an effective and reliable method to remove heavy metal contaminations from waste water by making use of magnetic separation technique. Nanotechnology can introduce new methods for the treatments and purification of water from pollutants, as well as new techniques for wastewater management and water desalinization. In TNAU, efforts are being undertaken to use Fe^0 nanoparticle to decontaminate soil and aquatic systems (Udayasoorian et al., 2009).

Precision farming

Precision farming has been a long-desired goal to maximize output (i.e. crop yields) while minimizing input (i.e. fertilizers, pesticides, herbicides etc.) through monitoring environmental variables and applying targeted action. Precision farming makes use of computers, global satellite positioning systems, and remote sensing devices to measure highly localized environmental conditions thus determining whether crops are growing at maximum efficiency or precisely identifying the nature and location of problems. By using centralized data to determine soil conditions and plant development, seeding, fertilizer, chemical and water use can be fine-tuned to lower production costs and potentially increase production – all benefiting the farmers. Precision farming can also help to reduce agricultural waste and thus keep environmental pollution to a minimum. Although not fully implemented yet, tiny sensors and monitoring systems enabled by nanotechnology will have a large impact on future precision farming methodologies.

ICAR nanotechnology platform

The Indian Council of Agricultural Research (ICAR) has opened up a dedicated Platform on "Application of Nanotechnology in Agriculture" with a major focus on diagnostics, pheromones, enhanced input use efficiencies, nano-food system besides biosafety. The platform carries the following specific tasks:

- Early detection of pests, diseases and nutrient deficiencies using nano-based quick diagnostic kits
- Nanotechnological approaches to evolve sustained release of pheromones and insecticide molecules for effective insect pests management
- Design, fabricate and develop smart delivery systems to enhance use efficiencies of nano-agricultural inputs (seeds, fertilizers, herbicides, insecticides, biofertilizers etc.)
- Develop smart food packaging to extend shelf life of fruits, vegetables, dairy and poultry products besides quality control and value addition of agricultural products
- Evolve standard operational protocols for assessing the nanotoxicity, biosafety and policy framework to ensure that nano-based agricultural products are safer to the consumers and environment
- Nano-biotechnological approaches to gain insights into the mechanisms involving interactions between nano-particles and macro and micromolecules at the organelle levels
- Infrastructure development and capacity building of Indian agricultural scientists to train large number of scientists in the fascinating field of nanoscience to augment farm productivity and food security of the country without associated environmental harm.

Benefits of nanotechnology

- Nanotechnology today is regarded as a revolutionary technology. Worldwide, there has been an increasing interest in nanotechnology as evident from the rising trends in investment and policy initiatives directed towards this end.
- Nanotechnology interventions could enable the successful development of renewable energy solutions and reduce our dependence of fossil fuels. Enhancement of agricultural productivity has been identified as the second most critical area of application of nanotechnology for attaining the Millennium Development Goals.
- Nanotechnology is believed to enhance agricultural productivity through genetic improvement and make crops more resistant to heat and water logging. Water treatment and remediation has been cited as the third most critical area where nanotechnology includes water purification, detection of contaminants and waste water treatment.

Risks in nanotechnology

- Nanotechnology risks can be best understood in conjunction with its benefits. The complexity of the technology, the breadth of nanomaterials and applications, coupled with the possibility of its wide dissemination in the globalised world renders the technology unpredictable in many senses. The risks are heterogeneous as the field of nanotechnology itself and include environmental, health, occupational and socio economic risks.
- The unusual properties of nanomaterials that can enable rewarding applications for society might post unknown or unforeseen environment, health and safety challenges. The pro-technology stance taken in several developed and developing countries at the cost of risk related research has led to an information gap around the impacts of nanomaterials. These aspects together with the commercialization and pervasiveness of nanoproducts serve to heighten the risk from this emergent technology.
- By virtue of their size, nanomaterials like other tiny particles might be able to enter the human body and those of other species imperceptibly through various pathways-inhalation, ingestion, dermal contact, etc. Early research also indicates that nanoparticles could reach various parts of the body where they may exert adverse effects. Nanoparticles, it is believed might be able to disrupt cellular, enzymatic and other organ related functions posing health hazards.

- On the other hand, nanoparticles might also be non-biodegradable and on disposal, these disposed materials might form a new class on non-biodegradable pollutant and pose a new threat to the environment (air, water, soil) and health. In light of these facts it appears that the greatest current risk is to the occupational health of the workers involved in the production, packaging or transport of the nanomaterials. With the increase in the application of nanomaterials in various products, the risk of the exposure of the consumers and the general public will also increase. Therefore it is crucial to examine and estimate the risk for regulating the production, use, consumption and disposal of these materials.
- The nano-based processes and products are ubiquitous in market but relatively little is known about the environmental or industrial health and safety of nanoparticles (EPA, 2005). The EPA insists that the following questions to be addressed during the evolution of nano-products. They include (1) are manufactured nanomaterials used in commerce sufficiently different from their macroscale chemicals or products to warrant special regulatory attention; (2) do we know enough about the hazards and environmental fate of these products to adequately assess risk; and (3) how much knowledge about potential risk should governments require before products are brought to market. The nanotoxicity studies in agriculture is very limited, there are literature suggesting potential risks to plants, microbes, animals and even humans.

Challenges

- Producing the nanomaterials in large enough volumes, with consistent quality, and at acceptable costs.
- Supplying the nanomaterials in a form (such as proper particle size, surface chemistry, dispersion capability, compatibility with various media, etc.) that would allow integration into the process
- Engineering and customizing the nano-based system to local requirements.
- Addressing environmental, health and safety concerns in the use and disposal of nano products.
- One of the biggest challenges has been in terms of the interdisciplinary nature of nanotechnology *per se* and the scope of its applications. This has lead to significant overlaps in the areas for R & D support identified by different agencies.
- The gap between basic research and application is another challenge in nanotechnology, like several other technologies. There is poor lab-firm

integration which is compounded by the paucity of skilled manpower that could provide linkages between the technology and commercial domains.

- Being cost and risk intensive, and being dependent upon sophisticated and complex equipment, technical know-how and capacity, financial constraints often act as impediment in this regard.
- The main challenges faced by regulatory institutions currently relate to the regulatory capacity, information asymmetry and absence of inter-agency coordination.

Action plan

The Nano Science and technology interventions in agriculture is a fairy recent and several strategic action plans have been evolved in order to infuse nanotechnology in agricultural sciences to gain solution to unresolved problems besides ensuring food security in the country.

- The Indian Council of Agricultural Research (ICAR), New Delhi, has opened up an exclusive platform on "Applications of Nanotechnology in Agriculture" to be introduced in the current 12th five year plan (2012-2017). The platform encompasses five major theme areas namely, nano – based diagnostic kits for the detection of diseases, pests and nutrient deficiencies, nano-pheromones for effective pest control, nano-agri inputs (fertilizers, herbicides, insecticides, growth hormones etc.,) for enhanced used efficiency, nano-food systems (nano-packaging and encapsulation of functional foods) and biosafety and risk assessment with a budget outlay of Rs.200 crores (40 million USD). This is one of the major initiatives of Government of India to augment nanotechnology research in the sphere of agricultural sciences.
- Nano-agriculture is at the nascent stage and the agricultural scientists, students and development officials have to be exposed through periodical training in the field of nano science and technology. The SAU / ICAR institutes that have some expertise may take a lead role in offering training. Indeed, ICAR has given a task to TNAU to give training to scientists through NAIP Program.
- The nanotechnology research and development activities require sophisticated equipment facility and thus ICAR bestowed interest to initiate a network of three national level nodal centers in order to cater to the needs and requirements of the scientists involved in nanotechnology research. In addition to the infrastructure, at least five faculties from basic sciences such as Physics, Chemistry, Nanotechnology,

Nanobiotechnology and Electronics are required to gain insights into the exciting field besides transforming the processes into products quickly that flow into Indian agriculture. Further, for the better function and upkeep of the laboratories at least 5 technicians are utmost essential.

- It is appropriate to start a new PG degree program in Nanotechnology in Agriculture to take advantage of youth being trained in this field and later they become strong human resource for the country. Indeed, the TNAU is one of the early birds to initiate M.Tech. Agri in Nano Science & Technology since 2011.

References

Bansiwal, A.K., Sadhana, R., Labhasetwar, S.S., Juwarkar, A.A. and Devotta, S. 2006. Surfactant-Modified Zeolite as a Slow Release Fertilizer for Phosphorus. *Journal of Agriculture Food Chemistry*, **54**: 4773 – 4779.

Chinnamuthu, C.R., Kanimozhi, V. and Dhinesh, V. 2012. Synthesis and Characterization of MnCO3 core-shell nanoparticles for action loading of herbicides. *Biomedical Applications of nano Structured materials,* 411-416.

IFRI, 2011. Leveraging Agriculture for Improving Nutrition and Health", to be held February 10-12. Independent Impact Assessment Report of 2020. pp. 1-75.

Khodakovskaya, M., Dervishi, E., Mahmood, M., Xu, Y., Li, Z., Watanabe F. and Biris, A.S. 2009. Carbon nanotubes are able to penetrate plant seed coat and dramatically affect seed germination and plant growth. *American Chemical Society*, **10(3):** 3221-3227.

Lal, R. 2008. Soils and India's Food Security. *Journal of Indian Society of Soil Science,* 56:129-138

Liu, F., Wen, L.X., Li, Z.Z., Yu, W., Sun, H.Y. and Chen, J.F. 2006. Porous hollow silica nanoparticles as controlled delivery system for water-soluble pesticide. *Material Research Bulletin*, **41(12):** 2268–2275.

Maynard, A. D.2006. Nanotechnology: Managing the risks. *Nano Today*. **1:** 22-33.

Selvarajan, R. 2013. Nanotechnological Approaches for Early Detection of Plant Pathogens. *In*: Subramanian, K.S., K. Gunasekaran, N. Natarajan, S.K. Rajkishore, R. Sunitha and N.B. Nandakumar. Nanotechnology and Plant Disease Management. Pp. 129-136.

Subaharan, K. 2013. Nano-Pheromones – Frontier Areas in Nano Technology. *In*: Subramanian, K.S., K. Gunasekaran, N. Natarajan, S.K. Rajkishore, R. Sunitha and N.B. Nandakumar. Nanotechnology and Plant Disease Management. Pp. 211-214.

Subramanian, K.S. 2008. Nanotechnological Approaches for Soil and Crop Management. Paper presented at the National Trainings on Seed Quality Regulation and Seed Health Testing and Varietal Identification through In-vitro techniques, National Seed Research and Training Center, Government of India, Varanasi, Feb.18-22, 2008.

Subramanian, K.S. and Sharmila Rahale, C. (2012) Slow Release of Potassium through Nano-fertilizers formulations. *International Journal of Nanotechnology and Applications*. 3(1): 15-29.

Subramanian, K.S. and Sharmila Rahale, C. 2010. Synthesis of Nano-Fertiliser for Regulated Release of Nutrients. Book Chapter in the "Biomedical Applications of Nanostructured Materials" Ed: V.Rajendran, B.Hillebrands, P.Prabu and K.E.Geckeler. Published by Macmillan Publishers India Ltd. pp58-61.

Subramanian, K.S. and Sharmila Rahale, C. 2013. Nano Fertilizers- Synthesis, characterization and Applications. Ed. Tapan Adhikari, S. Kundu and A. Subba Rao. Nanotechnology in Soil Science and Plant Nutrition. Published by New India Publishing Agency. Pp. 263-276.

Subramanian, K.S. and Tarafdar, J.C. (2011) Prospects of Nanotechnology in Indian Farming. *Indian Journal of Agricultural Sciences,* **81(10):** 887-893.

Subramanian, K.S., and K. Gunasekaran, 2013. Nanotechnological Applications in Agriculture. *In:* Subramanian, K.S., K. Gunasekaran, N. Natarajan, S.K. Rajkishore, R. Sunitha and N.B. Nandakumar. Nanotechnology and Plant Disease Management. Pp. 1-6.

Subramanian, K.S., S.K. Rajkishore and A. Lakshmanan. 2012. Nanotoxicity: A Challenge to Nano Agriculture. *Nano Digest,* **3(9):** 36-38.

Udayasoorian, C., E. Parameshwari, S. Paul Sebastin and R.M. Jayabalakrishnan. 2009. Decolorization of water soluble Azo-Dye by nano scale zero valent Iron particles. *Research Journal of Chemistry and Environment*, **1:** 60-70.

USDA (United States Department of Agriculture) 2003. *Ochroma pyramidale* syn. *O. lagopus*. Wood Technical Fact Sheet. Forest Products Laboratory: Madison Wisconsin. Pp. 2.

2

Nanotechnology Interventions in Agriculture

W. Selvamurthy, K. Muraleedharan, D.B.S. Sethi and Rajeev Varshney

Introduction

Nature has been performing wonderful feats for millions of years. All organisms from microbes to humans are powered by highly evolved molecular and cellular machines that operate at nano scale. The capability to manipulate, control, assemble, functionalise, produce and manufacture things at atomic precision provides an opportunity to understand this new and exciting field of science. Nanotechnology is application of science to control the properties and structure of matter at nanometric scale.

Late Nobel Laureate Richard Feynman would have been amazed to see how his brain child is being put to use in a wide range of applications such as diagnostics, energy, health, food, resource management and environment sustainability. In India, immense potential is available in agriculture as 60 % of the nation's population depends on this primary occupation. Agricultural productivity can be enhanced through better investigation of pests and diseases, targeted treatment through precision farming, automated weather forecasting and advance post harvest technologies. In all these priority areas, nanotechnology interventions have a key role to play for taking Indian agriculture to new heights.

In addition, the scope of nanotechnology is further extended to protect against the threat of agro bioterrorism. DRDO has started research to utilize the database of IDBD (Infectious Disease Biomarker Database) and screen for biomarkers in potential bioterrorism threats for agriculture in bordering areas.

Challenges in the India's agriculture sector

About three-fourths of the poor in developing countries, including the poorest of the poor, live in rural areas, and most depend on agriculture for their livelihood. In this context, agriculture is widely viewed as a strategic area to reduce poverty and foster development. Advances in nanoscale sciences may present a new opportunity to help address these issues and improve the livelihood of the poor. New challenges like declining farm profitability, depletion of natural resources, resurgence of new pests and diseases, besides effects of climate change on plant physiology need to be addressed in a short span of time. New S&T interventions that are cost and time effective are needed. A number of projects around the world are exploring the use of nano-particles/ nanotechnology on the farm for enhanced photosynthesis to better germination and soil management.

In agricultural sector, nanotech research and development is likely to facilitate and frame the next stage of development of genetically modified crops, animal production inputs, chemical pesticides and precision farming techniques. These applications are largely intended to address some of the limitations and challenges facing large-scale, chemical and capital intensive farming systems. This includes the fine-tuning and more precise micro-management of soils; more efficient and targeted use of inputs; new toxin formulations for pest control; new crop and animal traits; diversification and differentiation of farming practices and products within the context of large-scale and highly uniform systems of production.

Issues in agriculture that can be addressed through nanotechnology:

a. Lower agricultural input efficiency

- Nanofertilizers for slow release and efficient use of water and fertilizers by plants
- Nanocides – pesticides encapsulated in nanoparticles for controlled release,and nanoemulsions for greater efficiency
- Delivery of nutrients and drugs for livestock and fisheries

b. Unsustainable farm management

- Nanoparticles for soil conservation
- Nanosensors for soil quality and plant health monitoring and by precise control of the appropriate environment at every stage of agriculture.
- Nanobrushes and membranes for soil and water purification
- Nanomagnets for removal of soil contaminants

c. Food Shortage and Agricultural productivity enhancement

- Genetic improvements of plants and animals
- Delivery of genes and drug molecules to specific sites at cellular levels in plants and animals
- Nano-array based gene technologies for gene expressions in plants and animals under stress conditions
- Water delivery systems for farming

d. Vulnerabilities to climate change

- Nano based gene technologies in crops to withstand heat, salt, submergence or water logging.

e. Food Security and Safety

- Nanopackaging for enhanced protection of foodstuffs
- Nano-sensors for detection of food spoilage and
- Nanobarcodes to track identity and quality of packaged foods.

Examples of Nanotechnology Interventions

1. Farming

a. ***Plant growth treatment:*** PrimoMaxx, Syngenta has developed 100nm particle size emulsion. Using this nano-sized particles, the potency of active ingredients can be greatly increased besides reducing the quantity to be applied.

b. ***Nanoparticle Farming- Grow Nanoparticles in Genetically Engineered Crops:*** In the future, industrial nanoparticles may not be produced in a laboratory, but produced in fields by growing genetically engineered crops - what might be called "particle farming." It's been known for some time that plants can use their roots to extract nutrients and minerals from the soil but research from the University of Texas-El Paso confirms that plants can also soak up nanoparticles that could be industrially harvested. For industrial-scale production, the researchers speculate that the plants can be grown indoors in gold-enriched soils, or they can be farmed nearby abandoned gold mines. Nanobiotech researchers at the National Chemical Laboratory in Pune, India, have been carrying out similar work with geranium leaves immersed in a gold-rich solution. After 3-4 hours, the leaves produce 10 nm-sized particles shaped as rods, spheres and pyramids which, appear to be shaped according to the aromatic compounds in the leaves.

c. ***Plant disease Diagnostics & Treatment***: Control of powdery mildew was controlled within 0-7 days after nanosized silica-silver application.

d. ***Seeding Iron***: To improve the germination of tomato seeds by spraying a solution of iron nanoparticles on the fields, the Russian Academy of Sciences reported improved germination.

e. ***Soil Binder:*** A nanotech-based soil binder, called SoilSet developed by Sequoia Pacific Research of Utah (USA), is a quick-setting mulch which relies on chemical reactions on the nanoscale to bind the soil together.

f. ***Soil Clean-Up Using Iron Nanoparticles:*** A nano clean-up method of injecting nano-scale iron into a site contaminated by heavy metals and PCBs has been tested. The particles flow along with the groundwater and decontaminate *en route*, which is much less expensive than digging out the soil to treat it. The tests with nano-scale iron show significantly lower contaminant levels within a day or two. The tests also show that the nano-scale iron will remain active in the soil for six to eight weeks, after which time it dissolves in the groundwater and becomes indistinguishable from naturally occurring iron.

g. ***Agrochemicals:*** Pakistan-US Science and Technology Cooperative Program is working on Super combined fertilizer and pesticide using Nano-clay capsule that contains growth stimulants and biocontrol agents that can be designed for slow release of active ingredients and such treatment requires only one application over the life of the crop. Herbicide being developed by Tamil Nadu Agricultural University (India) and Technologico de Monterry (Mexico) has Nano-formulation designed to attack the seed coating of weeds that destroys soil weed seed banks and prevent weed germination. Pesticides, including herbicides are developed by Australian Commonwealth Scientific and Industrial Research Organization that has Nano-encapsulated active ingredients. Very small size of nanocapsules increases their potency and may enable targeted release of active ingredients.

2. Food industry

Nanotechnology applications in food industry are expected to improve the nutritional security of the country.

- ***Nutritional supplement:*** Nanoceuticals 'mycrohydrin' powder by RBC Lifesciences has Molecular cages 1-5 nm diameter made from silicamineral hydride. Nano-sized mycrohydrinhas showed increased potency and bioavailability. Exposure to moisture releases H- ions and acts as a powerful antioxidant.

- ***Nutritional drink:*** Oat Chocolate Nutritional Drink Mix for Toddler Health has 300nm particles of iron (SunActive Fe) Nano-sized iron particles have increased reactivity and bioavailability.
- ***Food additive:*** Aquasol preservative by AquaNova has Nanoscale micelle (capsule) of lipophilic or water insoluble substances. Surrounding active ingredients within soluble nanocapsules increases absorption within the body (including individual cell).

Experience of DRDO in food systems

Nano encapsulation of functional foods and bioactive compounds has thrown a new challenge in nutrition. A feasibility study was undertaken by DFRL Lab of DRDO for the development of micronutrients; iron and vitamins incorporated liposomes using nanotechnology. Milk has been fortified with vitamins and the bioavailability of nanoemulsion is well achieved. The following areas are being explored to take advantage of nanotechnology.

a) Nano-encapsulations of micronutrients and their incorporation in different food matrices.

b) Carotenoids to increase their oxidative stability.

c) Vitamin A, C, D and E for food fortification;

d) L-carnitine for increased fatty acids catabolism and branched chain amino acids for decreased protein catabolism and together to combat fatigue.

e) Essential fatty acids like omega-3 and 6 fatty acids for performance enhancement.

f) Herbal extracts/active components to develop 'phytosomes' and their targeted delivery for therapeutic applications.

g) Bitter bioactive principles from varied natural sources as a taste modifier.

h) Development of nanoceuticals/ nano-functional foods by incorporation of nano-emulsions.

DFRL, Mysore has also developed high barrier nanocomposite films for food packaging applications by incorporating nanoclays at various concentrations. The results were fruitful that gas transmission rate decreased, mechanical and thermal properties improved. Further, they have incorporated Silver and Gold nanoparticles into PVA matrix to form nanocomposites and silver nanoparticles imparted antimicrobial properties.

3. Food Packaging

a. ***Electrospinning Nanofibres Can Turn Waste Into New Products:*** To produce low cost, high-value, high-strength fiber from a biodegradable and renewable waste using the recently perfected technique of electrospinning is being employed to spin nanofibers from cellulose.

b. ***Clay Nanoparticles to Improve Plastic Packaging for Food Products:*** Clay nano-particles when dispersed throughout the plastic film can block oxygen, carbon dioxide and moisture from reaching fresh meat or other foods stuffs. Chemical giant Bayer produces a transparent plastic film (called Durethan) containing clay nanoparticles. The nanoclay also makes the plastic lighter, stronger and more heat-resistant. Creating a molecular barrier by embedding nanocrystals in plastics can greatly improve packaging. Until recently, industry's quest to package beer in plastic bottles (for cheaper transport) was unsuccessful because of spoilage and flavour problems. Today, Nanocor, a subsidiary of Amcol International Corp., is producing nanocomposites for use in plastics beer bottles that give the brew a six-month shelf-life. By embedding nanocrystals in plastics, researchers have created a molecular barrier that helps prevent the escape of oxygen. Nanocor and Southern Clay Products are now working on a plastic beer bottle that may increase shelf-life to 18 months.

c. Adhesive for McDonald's burger containers has Ecosynthetix 50-150nm starch nanospheres. These nanoparticles have 400 times the surface area of natural starch particles. When used as an adhesive they require less water and thus less time and energy to dry.

4. Nanosensor Development

(i) *Networks to monitor farm produce & track Livestock*

Advances in nanotechnology enabled sensors will have a large impact in the area of nanobiotechnology. Many of the advances in sensing biological systems (e.g., proteins and micro-organisms) and biomolecular phenomena (e.g., DNA identification and biological process) are enabled by improvements in nanoscale sensing. Dendrimers based nanoparticles for specific targeting can be tagged. The nanosensors enable new tests that were previously difficult, time consuming, or non-existent. Improvements in sensor technology with nano concepts will allow devices that are portable and would be able to perform biological detection in the field with minimal human interaction. Injectable microchips are already used in a variety of ways with the aim of improving animal welfare and safety besides studying the animal behaviour in the wild, to track meat products back to their source. In the nanotech era, however, retrofitting farm animals with sensors, drug chips and nanocapsules will further extend the vision of animals as industrial production units.

(ii) *Embedded Sensors in Food Packaging and 'Electronic Tongue' Technology*

Embedded sensors can detect substances even at a concentration of parts per trillion and would trigger a colour change in the packaging to alert the consumer if a food has become contaminated or if it has begun to spoil. Using a Nanotech Bioswitch in 'Release on Command' food packaging researchers in the Netherlands are going one further to develop intelligent packaging that will release a preservative if the food within the container begins to spoil. This "release on command" preservative packaging is operated by means of a bioswitch developed through nanotechnology.

(iii) *Linking Nanosensors to Drug Delivery Systems for livestock*

Research in nanobiotechnology not only has the potential to improve biological detection methods but also other aspects of biology, including drug discovery, drug delivery, surgical methods, biocompatibility, diagnosis, implants, and even prosthetics. Other active areas of nanobiotechnology research include spectroscopic tools for characterizing biological processes, biomimetic artificial nanostructures, materials for biomolecular sorting and sensing, and functionalized nanostructures for controlled drug delivery. Research into nanobiotechnology is also resulting in the emergence of a new field of computation, by contributing to the design of DNA and molecular-based computers. These computational devices, whose computational power is derived from the ability to explore a very large number of combinations of molecules in a relatively short time, are potentially a very powerful tool for investigation.

A smart treatment delivery system could be a miniature device implanted in an animal that samples saliva on a regular basis. Long before a fever develops, the integrated sensing, monitoring and controlling system could detect the presence of disease and notify the farmer and activate a targeted treatment delivery system. This is a vision of precision agriculture where the long-term aim is not merely to monitor, but also to automatically and autonomously intervene with pharmaceuticals using small drug delivery devices that can be implanted into the animal in advance of illness. The notion of linking in-built sensors to in-built smart delivery systems has been called "the fuel injection principle" since it mimics the way modern cars use sensors to time fuel-delivery to the engine. One of the current barriers to implantable medical devices is that their composite materials (e.g., metal or plastics) are often incompatible with living tissue. New materials, engineered at the nano-scale with biocompatible property can address this problem.

5. Nano-Aquaculture

The world's fastest growing area of animal production is the farming of fish, crustaceans and molluscs, particularly in Asia. With a strong history of adopting new technologies, the highly integrated fish farming industry may be among the first to incorporate and commercialise nanotech products.

a. ***Cleaning Fish Ponds with Nanotechnology Devices:*** Nevada-based Altair Nanotechnologies makes a water cleaning product for swimming pools and fishponds called 'NanoCheck.' It uses 40 nm sized particles of a lanthanum-based compound which absorbs phosphates from the water and prevents algae growth. There may be potentially large demand for NanoCheck for use in thousands of commercial fish farms worldwide where algae removal and prevention is costly at present.

b. ***DNA Nano-Vaccines Using Nanocapsules and Ultrasound Methods:*** The USDA is completing trials on a system for mass vaccination of fish using ultrasound. Nanocapsules containing short strands of DNA are added to a fishpond where they are absorbed into the cells of the fish. Ultrasound is then used to rupture the capsules, releasing the DNA and eliciting an immune response from the fish. This technology has so far been tested on rainbow trout by Clear Springs Foods (Idaho, US).

c. ***Using Iron Nanoparticles to Speed Up the Growth of Fish:*** Scientists from the Russian Academy of Sciences have reported that young carp and sturgeon exhibited a faster rate of growth (30% and 24% respectively) when they were fed with iron nanoparticles.

Societal Implications

It may become increasingly important to consider the life cycle assessment, also known as cradle-to-grave analysis, for products containing nanomaterials. The use of nanotechnology and nanomaterials in commercial products applicable to agriculture is expected to grow for the foreseeable future. Consequently, human beings and livestock equally will be exposed through the food chain and the whole environment will be increasingly exposed to manufactured nanostructured materials. Yet there is still no nanotechnology specific regulation or safety testing required before manufactured nanomaterials can be used in food, food packaging or agricultural products. Hence, research on biosafety of engineered nanoparticles has to go hand in hand in order to protect from undesirable consequences if any.

Conclusion

Nanotechnology applications for agriculture can be explored in multiple ways to achieve evergreen revolution. Nano principles and concepts can be extended to soil management, crop improvement, water management, plant diseases and diagnostics so as to improve the food and nutritional security. Nevertheless, there is a science to explore to bring out benefits that humanity can derive from the mother earth by taking advantage out of this emerging nanotechnology.

Section 2: Synthesis

3

Physical and Mechanical Methods of Synthesis of Nano-materials

D. Nataraj

Introduction to Nano-materials

Nano-sized materials are naturally present from forest fires and volcanoes, but are also generated unintentionally from anthropogenic sources as a by-product of combustion and deliberately as manufactured nanomaterials. Nanoparticles have been used for centuries. The coloured glass that we see in many old cathedrals from the middle age was made of gold nanosized clusters that created different colour depending on the size of the nanoparticles. The most prominent example of engineered nanoparticulate material is carbon black which has been around us for decades in applications like printing inks, toners, coatings, plastics, paper, tires and building products.

Two of the major factors why nanoparticles have different properties (optical, electrical, magnetic, chemical and mechanical) than bulk material are because in this size-range, the quantum effects start to predominate and the surface area to volume ratio is increased.

Effect of quantum confinement in nanostructures

$$\lambda = \frac{h}{p} = \frac{h}{\sqrt{3m^{*}_{e(h)}kT}} \qquad (1.1)$$

Quantum size effects arise when the size of a nanocrystal is comparable to the length parameters i.e., the de Broglie wavelength ë and exciton Bohr radius a_B of the quasi particles (electrons, holes and excitons).

In the above equation m_e^*, m_h^* is the effective mass of electron and hole respectively and T is the temperature. At room temperature, ë is around 10 nm. For most common semiconductors a_B is in the range of 1-10 nm which indicates that quantum size effects are pronounced even for a particle with dimension ten to hundred times larger than the lattice constant. In this size range we can consider a nanocrystal as a macroscopic crystal with respect to the lattice constant but should be treated as a quantum box for quasi particles. Sufficient deep confining potential induced by the small size of the nanoparticle localize the carriers in all the three dimensions.

The consequence is a breakdown of the classical band structure of a solid with a continuous dispersion of energy as a function of momentum to atom like discrete levels. For a particle in an infinite periodic potential, the energy and crystal momentum ($\hbar k$) may both be precisely defined, while the position is not. In a nanocrystal, due to spatial confinement, the uncertainty in the position of the quasi particle decreases. But momentum is no longer be well defined. The wide range of the momentum of the carrier is translated to higher average energies. As a result the electronic states in a crystal shifted to higher energy with respect to bulk.

Confinement Regimes

Efros Al. L. and Efros A. L. [Efros 1982] classified the extent of confinement in a low-dimensional structure with respect to the exciton Bohr radius (a_B). The confinement regimes are classified into three limiting cases.

Weak Confinement Regime ($R>>a_B$)

This corresponds to the case when the quantum dot (QD) radius R is a few times larger than a_B. In this case the quantization of the exciton's translational centre-of-mass motion can be described as a single, uncharged particle with mass $M=m_e^*+m_h^*$ in a spherical potential. In the weak confinement regime, since the binding energy of an exciton is larger than the quantization energy of both the electron and the hole, the character of the exciton as a quasi particle is conserved. The exciton energy levels can be expressed in the form of eqn (1.2).

$$E_{nml} = E_g - \frac{R_y^*}{n^2} + \frac{\hbar^2 \chi_{ml}^2}{2MR^2} \tag{1.2}$$

Where, $\div_{ml}$ - is the roots of Bessal function, R_y* - is the exciton Rydberg energy, å - represent the dielectric constant of the crystal and nml - represents quantum numbers.

$$R_y^* = \frac{e^2}{2\varepsilon a_B} \qquad (1.3)$$

2.1.2 Intermediate confinement Regime ($a_e > R > a_h$)

Intermediate confinement regime in a QD corresponds to be the situation, where the Bohr radius of the electron $\left(a_e = \frac{\hbar^2 \varepsilon}{e^2 m_e^*}\right)$ is larger than that of hole $\left(a_h = \frac{\hbar^2 \varepsilon}{e^2 m_h^*}\right)$. This is due to the difference in effective masses of electron and hole.

2.1.3 Strong Confinement Regime ($R << a_B$)

In this case the particle size is smaller than a_B. As a result, the kinetic energy of electron and hole are much higher than the exciton binding energy R_y*. Therefore, confined electrons and holes have no bound state like an exciton. Electron and hole are separately confined. Therefore, the absorption spectrum reduces to a set of discrete bands peaking at the energies

$$E_{nml} = E_g - \frac{R_y^*}{n^2} + \frac{\hbar^2 \chi_{ml}^2}{2MR^2}$$

Where, $\mu = \frac{m_e^* m_h^*}{m_e^* + m_h^*}$ - is the reduced mass.

More quantitatively, such discretization of energy levels and widening of the band gap can be explained in terms of density of states (DOS).

Nano-material synthesis methods

The synthesis of nanoscale materials is an essential part of nanotechnology which is concerned with the development and utilization of structures and devices scaling between individual molecules and features below 100 nm. Nanomaterial synthesis can be classified into two different approaches.

1. Top down approach
2. Bottom up approach

In general, most of the physical and mechanical methods available for nanomaterial synthesis are under top-down approach, while the chemical methods are bottom-up approaches.

Ball milling method

It is a ball milling process where a powder mixture placed in the ball mill is subjected to high-energy collision from the balls. This process was developed by Benjamin and his coworkers at the International Nickel Company in the late of 1960 and it is one of the simple physical method to produce nanoparticles.

The alloying process can be carried out using different apparatus, namely, attritor, planetary mill or a horizontal ball mill. However, the principles of these operations are same for all the techniques. Since the powders are cold welded and fractured during mechanical alloying, it is critical to establish a balance between the two processes in order to alloy successfully. Planetary ball mill is a most frequently used system for mechanical alloying since only a very small amount of powder is required. Therefore, the system is particularly suitable for research purpose in the laboratory. The ball mill system consists of one turn disc (turn table) and two or four bowls. The turn disc rotates in one direction while the bowls rotate in the opposite direction. The centrifugal forces, created by the rotation of the bowl around its own axis together with the rotation of the turn disc, are applied to the powder mixture and milling balls in the bowl. The powder mixture is fractured and cold welded under high energy impact.

During the high-energy ball milling process, the powder particles are subjected to high energetic impact. Microstructurally, the mechanical alloying process can be divided into four stages: (a) initial stage, (b) intermediate stage, (c) final stage, and (d) completion stage.

a. At the initial stage of ball milling, the powder particles are flattened by the compressive forces due to the collision of the balls. Micro-forging leads to changes in the shapes of individual particles, or cluster of particles being impacted repeatedly by the milling balls with high kinetic energy.

b. At the intermediate stage of the mechanical alloying process, significant changes occur in comparison with those in the initial stage. Cold welding is now significant. The intimate mixture of the powder constituents decreases the diffusion distance to the micrometer range. Fracturing and cold welding are the dominant milling processes at this stage.

c. At the final stage of the mechanical alloying process, considerable refinement and reduction in particle size is evident. The microstructure of the particle also appears to be more homogenous in microscopic scale than those at the initial and intermediate stages.

d At the completion stage of the mechanical alloying process, the powder particles possess an extremely deformed metastable structure. At this stage, the lamellae are no longer resolvable by optical microscopy.

Laser ablation method

In the ablation, the energy of the laser beam is transferred to the target material. There are three identifying pathways by which the laser energy gives rise to ablation and removes material from the target. These pathways are called: (1) photothermal ablation, (2) plasma-assisted ablation and (3) photochemical ablation. The primary factors to determine which pathway is favored are the surface nature, laser power, wavelength and pulse duration. In the thermal process, the energy of the photon in the laser beam is transferred to the target *via* the Inverse Bremsstrahlung effect, where the photons of the beam are absorbed by the free or bound electrons on the target surface, resulting in the electron oscillation.

This process is favored at lower laser power (100–500 MW/Cm^2) and longer pulse length (> 10 ns). The vibrating electrons cause forward radiation seen as a reflection of the incoming laser beam from the irradiated surface. At the same time the excited electrons relax through the interaction with other electrons and lattice photons, giving rise to a temperature increase of the material. Collisional interactions among electrons and phonons govern the heat transfer and the material temperature increases. Due to the large amount of energy transferred to the material in a very short pulse time, the energy is restricted from transferring to the interior of the material by thermal diffusion and is contained in the first few atomic layers of the structure. The surface atoms of the target are allowed to cool *via* evaporative cooling while the subsurface atoms cannot dissipate their energy and their respective volume increases rapidly. Subsurface heated atoms are then forcibly ejected along with the cooler surface atoms due to this rapid volume expansion. In this case, threshold laser energy has been observed for laser ablation to occur. Below the threshold, the absorbed laser energy can be dissipated to the bulk of the material via diffusion and hence, no material removal is observed. Above the threshold, ablation is observed.

In the second pathway, plasma-assisted ablation, high peak power in the laser pulse produces a temporary plasma above the target. The plasma is an ionized gas, which absorbs energy of the laser pulse. High-energy ions bombard the target surface and giving up their energy. The target atoms are then heated and are thermally ejected from the surface. Plasma processes are favored at high laser power (500 MW/cm^2).

Generally, there is a smooth transition from thermal to plasma ablation. The third pathway, photochemical or photolytic ablation, is a result of quantum phenomenon. If the laser photon energy is of the same order of the bonding energy of the molecules, the energy is absorbed and accumulated by the bond, leading to dissociation of the bond, and causing the ejection of material. Polymers are usually subjected to the photochemical ablation. Photothermal and plasma-assisted ablation leave behind small droplets and liquid expulsion on the target. Also there may be a heat-affected zone remaining after the ablation due to some of the target material heating, but not to the point of ablating or evaporating off the surface. In the photochemical ablation, only target material irradiated by the laser beam is ablated off the surface, and the remaining material is left unaffected by the process.

By all of the above said pathways one can get nanoscale films (two dimensional films). If one wants one dimensional nanostructure, the one has to deposit film on lithographically patterned substrates.

Chemical vapor deposition (CVD) method

In chemical vapor deposition (CVD), the vaporized precursors are introduced into a CVD reactor, where the precursor molecules adsorb onto a substrate held at an elevated temperature. These adsorbed molecules will be either thermally decomposed or reacted with other gases/vapors to form a solid film on the substrate. Such a gas–solid chemical reaction at the surface of a substrate is called the heterogeneous reaction. Because a number of chemical reactions may occur in the CVD process, CVD is considered to be a process of potentially great complexity as well as one of great versatility and flexibility.

Principle of Chemical Vapor Deposition

Thermodynamically, the chemical vapor deposition requires high temperature and low pressure for most of the systems. Under these conditions, a chemical system will rapidly fall to the minimum Gibbs free energy, leading to formation of solid reaction products. The application of thermodynamics to the CVD is normally the first step to understand the CVD process, and is also the underlying science for vapor transport, reaction kinetics, nucleation, and growth of deposited materials. It can also be extremely valuable in selecting an entire materials system. The chemical vapor deposition process is considered to consist of three steps.

1. Mass transport of reactants to the growth surface through a boundary layer by diffusion.
2. Chemical reactions on the growth surface, incorporating the new material into the growth front.
3. Removal of the gas–phase reaction by–products from the growth surface.

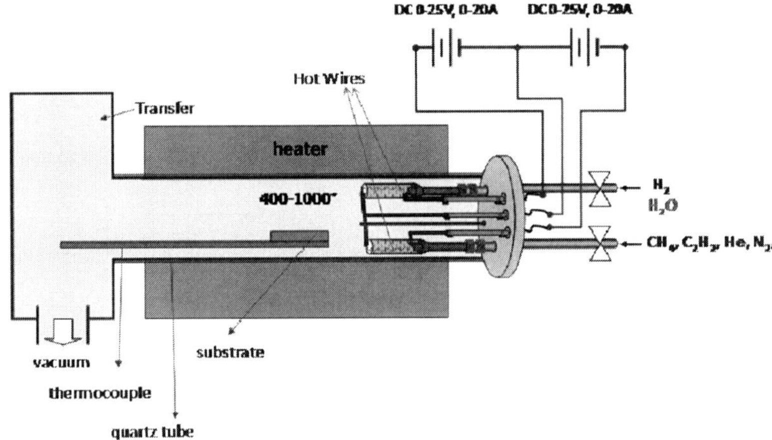

Fig. 1: Schematic diagram of chemical vapour deposition

The first and the third steps are generally not independent, since the diffusion rate of reactants affects the diffusion rate of the by–products. These two steps are coupled by the stoichiometry of the reaction, and both affect the chemical reaction rate. The second step is normally exceedingly complex and may involve surface and/or gaseous reaction as well as simultaneous adsorption–desorption (both chemisorption and physical adsorption) and nucleation processes. The slowest step in this sequence will determine the growth rate. If the substrate temperature is held at a value substantially higher than the pyrolysis temperature of the reactants, ensuring their rapid decomposition at the growth surface, the growth rate is then limited by the mass transport of the reactants to the growth surface through the boundary layer. In general, if the mass–transfer coefficient is very large compared to the product of the kinetic rate constant and the system pressure, the reaction is said to be kinetically controlled. If the opposite is true, the process is said to be diffusionally controlled.

In this CVD method, one can get a well controlled two dimensional epitaxial films. We can also get one dimensional wire like structure by growing the film on lithographically patterned substrates. Using catalyst assisted growth, one can even get vertically aligned one dimensional nanorods as well. This method is mostly used for the growth of multilayer structures and singlen devices.

Molecular beam epitaxy method

In the process of epitaxy, a thin layer of material is grown on a substrate. With respect to crystal growth it applies to the process of growing thin crystalline layers on a crystal substrate. In epitaxial growth, there is a precise crystal orientation of the film in relation to the substrate. For electronic devices, the substrate is a single crystal (usually Si or GaAs) and therefore so is the epitaxial layer (epilayer). In the most basic form of molecular beam epitaxy (MBE), the substrate is placed in ultra high vacuum (UHV) and the source materials for the film are evaporated from elemental sources.

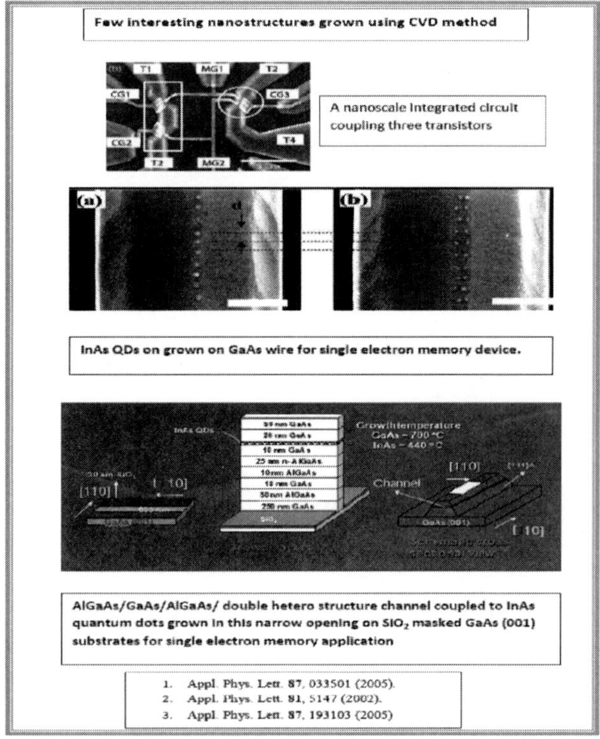

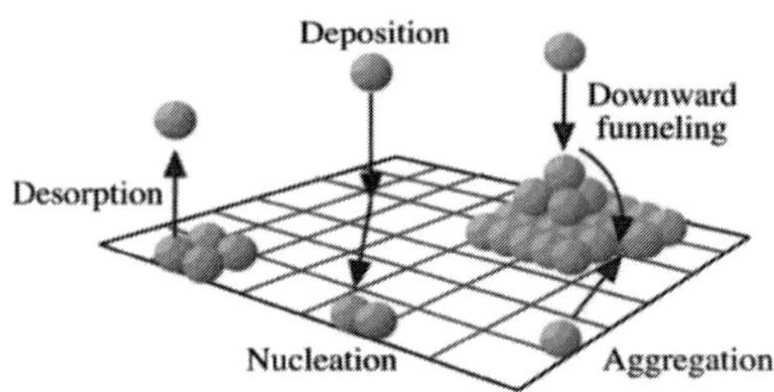

Fig. 2: Schematic illustration of processes on growing surface during MBE

The evaporated molecules or atoms flow as a beam, striking the substrate, where they are adsorbed on the surface. Once on the surface, the atoms move by surface diffusion until they reach a thermodynamically favorable location to bond to the substrate. Molecules will dissociate to atomic form during diffusion or at a favorable site. Figure 2 illustrates the processes that can occur on the surface. Because the atoms require time for surface diffusion, the quality of the film will be better with slower growth. Typically growth rates of about 1 monolayer per second provide sufficiently high quality.

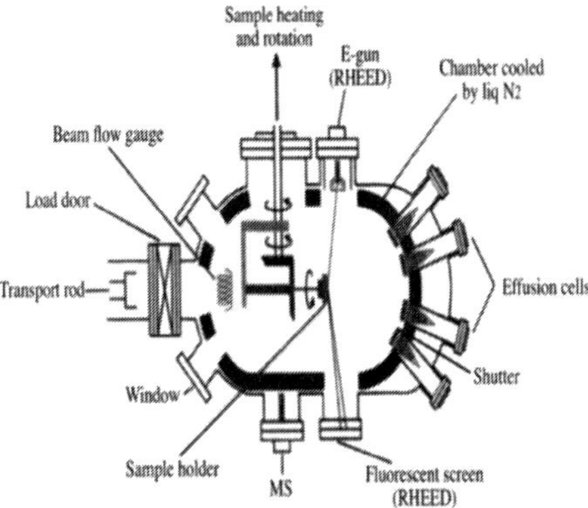

Fig. 3: The MBE growth chamber

A typical MBE chamber is shown in figure 3. The substrate is chemically washed and then put into a loading chamber where it is further cleaned using argon ion bombardment followed by annealing. This removes the top layers of the substrate, which is usually an undesired oxide which grew in air and contains impurities. The annealing heals any damage caused by the bombardment. The substrate then enters the growth chamber *via* the sample exchange load lock. It is secured on a molybdenum holder either mechanically or with melted indium or gallium which hold the substrate by surface tension.

Each effusion cell (Figure 3) is a source of one element in the film. The effusion cell, also called a Knudsen cell, contains the elemental form in very high purity (greater than 99.99999% for Ga and As). The cell is heated to encourage evaporation. For GaAs growth, the temperature is typically controlled for a vapor pressure of 10^{-2} to 10^{-3} Torr inside the effusion cell, which results in a transport of about 10^{15} molecules/cm^2 to the substrate when the shutter for that cell is opened. The shape and size of the opening in the cell is optimized for an even distribution of particles on the substrate. Due to the relatively low concentration of molecules, they typically do not interact with other molecules in the beam during the 5-30 cm journey to the substrate. The substrate is usually rotated, at a few rpm, to further even the distribution.

Using MBE also, one can get nanoscale films/rods, etc. GaAs, AlGaAs, silicon based devices and their integrated circuits are usually grown by this method. But this method is high cost involving one and need ultra high vacuum as well.

Sputtering method

The radio frequency (rf) sputtering, allowing the direct deposition of insulators, and magnetron sputtering, which enables much higher deposition rates with less substrate damage, have evolved more recently. These two developments have allowed sputtering to compete effectively with other physical vapor deposition processes such as electron beam and thermal evaporation for the deposition of high quality metal, alloy, and simple inorganic compound coatings, and to establish its position as one of the more important thin film deposition techniques. Today sputtering is widely applied to surface cleaning and etching, thin film deposition, surface and surface layer analysis etc.

Sputtering Basics

Sputtering is a relatively simple process in which an energetic particle bombards a target surface with sufficient energy to result in the ejection of one or more atoms from the target. The ejected or sputtered atoms can be condensed on a substrate to form a thin film. The sputtering yield is just the ratio of the number of emitted particles to the number of bombarding ones. Physical sputtering can result from bombardment with a variety of incident species. The most commonly used species is an inert gas ion (e.g. Ar^+, Kr^+). A sputtering system consists of the target (made up of the material to be deposited), substrate holder (on which the substrate or the sample is placed), sputtering gas, vacuum chamber with pumps and the power supplies. Sputtering is similar to evaporation in that reduced pressures are required for each.

The principal difference is that while thermal energy is used in evaporating the coating material, ion bombardment of the material, causing ejection of atoms, is used for sputtering. Thus, thin films of refractory materials may be deposited by sputtering without high source temperatures such as required by evaporation. The ions are formed when a high electric field is applied to a low-pressure gas such as argon, creating a glow-discharge in the deposition chamber. The positively charged argon ions are accelerated through the field to strike a cathode, made of the material to be sputtered. The atoms of a cathode surface are torn loose by this ion bombardment and condense on the surroundings. In addition to neutral (sputtered) material liberated from the bombarded target, other effects include secondary electron emission, secondary ion emission, reflection of incident particles, emission of radiation, adsorptions of gases, chemical dissociation etc.

When compared to CVD and MBE methods, this method is cost effective, but can produce polycrystalline form of nanostructured films. Even we can use this method to grow seed crystals (nanoscale dimension) and upon these seed crystals the nanowires can be grown later.

References

Al. L. Efros and A. L. Efros, *Sov. 6. Phys. Semicond.*, 1982, 16, 772.

Andrew R. Barron, http://cnx.org/content/m25712/latest/, 13.06.2013.

Ganesh Skandan and Amit Singhal, Nanomaterials handbook, CRC Press, Taylor & Francis Group, London.

P. Holister, J. –W, Weener, C. R. Vas, and T. Harper, Nanoparticles; Technology White Papers nr. 3. Cientifica, 2003.

W. Cao, http://www.understandingnano.com/nanomaterial-synthesis-ball-milling.html. 08.06.2013.

4

Chemical Method of Synthesis of Nano-materials

K. Pandian

Introduction

The metal particles in nanometer scale shows exciting optical and electronic characteristics when compared with respective bulk metals. Because of the inherent electronic properties of the metal and metal oxide nanoparticles these nanostructure materials are applied in various disciplines in recent days. Mostly these nanoparticles can be synthesized either top down or bottom up approaches. The conversion of bigger size particle or matter into tiny particles is called top down approach where as smaller particles or atoms joined together forming particles in regular shape within nanometer scale dimension is called bottom up approach. The formation of nanoparticles by chemical reduction is an example for the later one and the reduction size with the help of mechanical crusing leads to the formation of fine nanoparticles is named as top down approach. Mostly nanoparticles can be synthesized by chemical method because of its easiness to synthesis particles with uniform size and high purity.

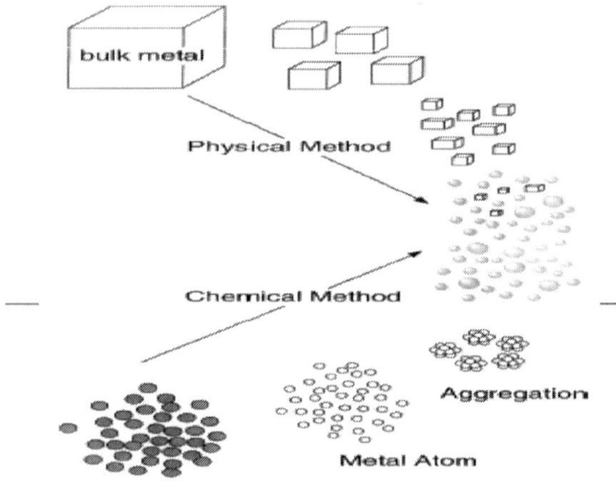

Fig. 1: Schematic illustration of top-down and bottom-up approaches for the preparation of metal nanoparticles.

Synthesis of metal nanoparticles

Nanostructured metal particles can be synthesized either by the so called "top down methods", *i.e.* by the mechanical grinding of bulk metals, or *via* "bottom-up methods" which rely on the wet chemical reduction of metal salts (nucleation and growth of metallic atoms) (Fig. 2). For the synthesis of nanoparticulate metal colloids, a large variety of stabilizers, e.g. donor ligands, polymers, and surfactants, are used to control the growth of the initially formed nanoclusters and to prevent them from agglomeration.

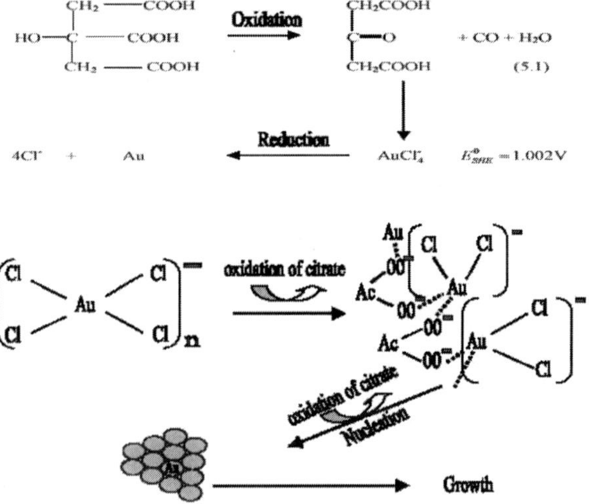

Fig. 2: Mechanism of gold nanoparticle formation by citrate reduction method.

A broad particle size distribution is obtained from the physical methods. Traditional colloids are typically larger (>10 nm) and not reproducibly prepared giving irreproducible catalytic activity. Chemical methods such as the reduction of transition metal salts are the most convenient ways to control the size of the particles. A key goal in the transition metal colloid area is the development of reproducible nanoparticles in opposition to traditional colloids. Generally nanoclusters should be or have at least (i) specific size (1-10 nm), (ii) well defined surface composition, (iii) reproducible synthesis and properties, and (iv) should be isolable and redispersible. Colloidal suspensions can be obtained by various methods leading to various particle size distributions. Nevertheless, whatever the method used, a stabilizing agent is always essential to prevent the colloids from aggregation. Four general synthetic methods are mainly used in the literature to synthesize transition metal colloids: (i) Chemical reduction of transition metal salts, (ii) Thermal, photochemical, or sonochemical decomposition, (iii) Ligand reduction and displacement from organometallics, and (iv) Electrochemical reduction.

The synthesis of various types of nanoparticles through solution processes has several advantages. Formation of nanoparticles dispersed in a solvent is the most common approach and prefers several advantages, which include easiness of:

1. Stabilization of nanoparticles from agglomeration,
2. Extraction of nanoparticles from solvent,
3. Surface modification and application,
4. Processing control, and
5. Mass production.

Reduction of metal complexes in dilute solutions is the general method in the synthesis of metal colloidal dispersions, and a variety of methods have been developed to initiate and control the reduction reactions. The formation of monosized metallic nanoparticles is achieved in most cases by a combination of a low concentration of solute and polymeric monolayer adhered onto the growth surfaces. Both a low concentration and a polymeric monolayer would hinder the diffusion of growth species from the surrounding solution to the growth surfaces, so that the diffusion process is likely to be the rate limiting step of subsequent growth of initial nuclei, resulting in the formation of uniformly sized nanoparticles.

Influence of reducing agents in the synthesis of nanoparticles

The size and size distribution of metallic colloids vary significantly with the types of reduction reagents used in the synthesis. In general, a strong reduction reaction promotes a fast reaction rate and favors the formation of smaller nanoparticles. However, a slow reaction may result in either wider or narrower size distribution. If the slow reaction leads to continuous formation of new nuclei or secondary nuclei, a wide size distribution would be obtained. On the other hand, if no further nucleation or secondary nucleation occurs, a slow reduction reaction would lead to diffusion-limited growth of the nuclei would be controlled by the availability of the zero valent atoms. Metal nanoparticles have a tendency to agglomerate, and therefore, it is necessary to protect them using surfactants or polymers, such as cyclodextrin, PVP , PVA , citrate or quaternary ammonium salts.

Using the same reduction reagent, nanoparticle size can be varied by changing the synthesis conditions. In addition, it was found that the reduction reagents have noticeable influences on the morphology of the gold colloidal particles. Gold particles with spherical shape were obtained using sodium citrate or hydrogen peroxide as reduction reagents, whereas faceted gold particles were formed within hydroxylamine hydrochloride and citric acid were used as reduction reagents. Furthermore, concentration of the reduction reagents and pH value of the reagents have noticeable influences on the morphology of the grown gold nanoparticles.

Synthesis of gold nanoparticles by citrate reduction method

Colloidal gold has been studied extensively for a long time. In 1857, Faraday published a comprehensive study on the preparation and properties of colloidal gold. A variety of methods have been developed for the synthesis of gold nanoparticles, and among them, sodium citrate reduction of chloroauric acid at 100 °C was developed more than 50 years ago and remains the most commonly used method.

Turkevich and coworkers investigated the nucleation and growth of Au nanoparticles prepared using citrate ions as reluctant wherein citrate ions not only act as reducing agent but also as an ionic stabilizer. To 1 mL of ~5.0 x 10^{-3}M $HAuCl_4$ solution, add another 18 mL of H_2O. Heat the 19 mL solution of $HAuCl_4$ until it begins to boil. Add 1 mL of 0.5% sodium citrate solution, as soon as boiling commences. The solution continuously heated until colour change is evident (pale purple). Remove the solution from the heating element and continue to stir until it has cooled to room temperature. The nanoparticles synthesized by this method will show an absorbance maximum at 520 nm.

The nanoparticles synthesized by citrate method are nearly monodisperse spheres of size controlled by initial reagent concentrations. They have a negative surface charge as a consequence of weakly bound citrate coating. Figure 3 shows the mechanism of synthesizing gold nanoparticles using citrate ions. Here, acetone dicarboxylic acid plays the role of precursor, reducing and nucleating agents. The first stage is the oxidation of citrate ions to acetone dicarboxylic followed by the reduction of gold ions into atoms. The acetone dicarboxylic acid molecule has two groups per unit, which may form a bond with the gold ion. In fact, the reduction of gold from the oxidation state +III to +I is not a direct process and an intermediate product is formed before further reduction.

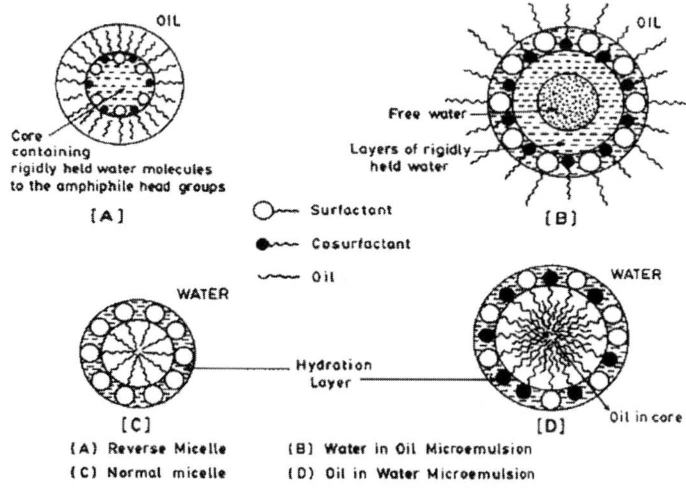

Fig. 3: Representations of reverse micelles and micro-emulsions.

Nanoparticles of other metals may also be prepared by citrate reduction, such as silver nanoparticles from $AgNO_3$, palladium from H_2PdCl_4, and platinum from H_2PtCl_6, and iridium nanoparticles from $IrCl_4$. However citrate anion is a non-innocent ligand and has one significant disadvantage because of the simultaneously formation of the intermediate acetone dicarboxylic acid. The silver nanoparticles prepared by this method were greenish yellow and had an absorption maximum at 420 nm.

Synthesis of silver nanoparticles by EDTA reduction method

Silver nanoparticles can be prepared by a variant of the standard EDTA reduction procedure. To 100 ml of boiling water were added 1.0 ml of 0.1 M EDTA and 4.0 ml of 0.1 M NaOH followed by 1.3 ml of 0.1 M $AgNO_3$. After boiling for about 60 seconds, 0.3 ml of 0.1 M HCl was added. The mixture was then kept at boiling for 2-3 minutes until an intense yellow-brown colour developed. The silver nanoparticles prepared by this technique had an absorption maximum at 414 nm.

Synthesis of silver nanoparticles by sodium borohydride reduction method

A solution of 5 x 10^{-3} M $AgNO_3$ (100 ml) was added portion-wise to 300 ml of vigorously stirred ice-cold $NaBH_4$. A solution of 1% PVA (Poly vinyl Alcohol) (50 ml) was added during the reduction. The mixture was then boiled for 1 hour to decompose any excess of $NaBH_4$. The final volume was adjusted to 500 ml. The silver nanoparticle prepared was brownish and had absorption maximum at 400 nm.

Synthesis of platinum nanoparticles

The colloidal platinum nanoparticles can be synthesized by Rampino and Nord (1942) method in which a solution of 1×10^{-4} M K_2PtCl_4 was prepared in 250 ml water to which 0.2 ml and 0.5 ml of 0.1 M sodium polyacrylic acid (average m.wt. 2,100) was added separately to the starting solution. Later, reduction of platinum ions was carried out by bubbling hydrogen gas vigorously into the solution for 10 min and then the reaction vessel was tightly sealed and left overnight, the solution turned light golden indicates the formation of colloidal platinum nanoparticles.

$$PtCl_4^{2-} + H_2O \rightarrow PtCl_3(H_2O)^- + Cl^-,$$

$$PtCl_3(H_2O)^- + H_2O \rightarrow PtCl_2(H_2O)_2 + Cl^-.$$

The rate of reduction of the aquated complexes by hydrogen was faster than that of freshly prepared K_2PtCl_4 solution.

Polyol method can be employed for the synthesis of relatively large Pt nanopowder (e.g. 5-13 nm) with high purity. The reduction is based on the decomposition of ethylene glycol and its conversion to diacetal. Toshima et al., have introduced alcohol reduction method in the field of nanoparticles synthesis. Alcohols such as methanol, ethanol or propanol work simultaneously as solvent and as reducing agents being oxidized to aldehydes or ketones.

$$\underset{OH}{CH_2}-\underset{OH}{CH_2} \xrightarrow{-H_2O} H_3C-\overset{H}{C}{=}O \xrightarrow{M(II)} H_3C-\underset{O}{\overset{\|}{C}}-\underset{O}{\overset{\|}{C}}-CH_3 + M_{(0)}$$

Refluxing metal salts or complexes such as H_2PtCl_6, $HAuCl_4$, $PdCl_2$, $RhCl_3$ in an alcohol/water (1/1, v/v) mixture yields nanocrystalline metal powders in the absence of stabilizers. The homogeneous dispersion of colloidal nanoparticles can be obtained in the presence of protective polymers such as polyvinylpyrrolidone (PVP).

Stability of metal nanoparticles

One of the main characteristics of colloidal particles is their small size. Because of the small size of the particles, they often tend to fall out of the solution. These metallic nanoparticles are unstable with respect to agglomeration to the bulk. In most cases, this aggregation leads to loss of the properties associated with the colloidal state of these metallic particles. For example, during catalysis the coagulation of colloidal particles used as catalyst leads to a significant loss of activity. Thus, the stabilization of metallic colloids and the means to preserve their finely dispersed state are the crucial aspects to be considered during their synthesis. In general the roles of the stabilizers are (i) prevention of uncontrollable growth of particles (ii) prevention of particle aggregation (iii) control of growth rate (iv) control of particle size and (v) allowing particle solubility in various solvents.

Stabilization by ligands or the solvent molecules

One of the most frequently employed stabilizers for metal nanoparticles is the ligand notably called *ligand stabilized metal nanoparticles*. Ligands such as phosphines, thiols, (Fig. 4) amines, or carbon monoxide have been invariably used for the protection of various metal nanoparticles. Such type of stabilization occurs through coordination between the ligand moiety and metal nanoparticles. It has recently been reported that solvent molecules can stabilize the nanoparticles more efficiently. Thus, Ti and Ru nanoparticles were synthesized in tetrahydrofuran or thioethers without adding steric or electrostatic stabilizers.

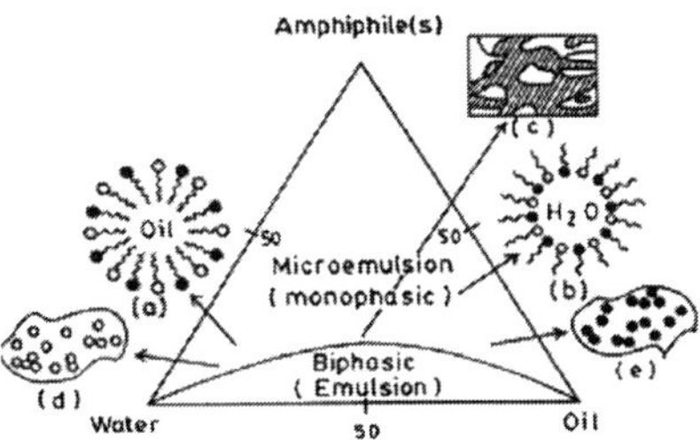

Fig. 4: A ternary phase diagram showing how various complexes depend on the different surfactants and intrinsic structures: **(a)** o/w micro-emulsion,**(b)**w/o micro-emulsion, **(c)** bicontinuous dispersion, **(d)** isolated and aggregated o/w dispersion, and **(e)** isolated and aggregated w/o dispersion.

Micro-emulsion method

Micro-emulsions are complex liquids with potentials for current and future application prospects. They consist of oil, water, surfactant and co-surfactant that form a clear solution. The amphiphilic nature of a surfactant, such as cetyltrimethylammonium bromide (CTAB) as described later, makes them miscible with water and oil. Micro emulsions can be either 'water in oil' (w/o) or 'oil in water' (o/w) stabilized by surfactant molecules with morphologies similar to those of micelles and reverse micelles where the interfacial tension is extremely low between the two phases. With these characteristics, micro-emulsions appear to be an ideal plat form to bring together the metal precursor (e.g. water-soluble) and the reactant (e.g.oil-soluble) to enable the reduction of the metal to occur. Generally, the size of micro emulsions is small and uniform. Hence, they can be regarded as a micro reactor for the synthesis of nanoparticles. As the water concentration alters, the system can change from a w/o to an o/w micro-emulsion, which is illustrated in Figures 5. These systems offer a wide range of possibility and flexibility to carry out the synthesis.

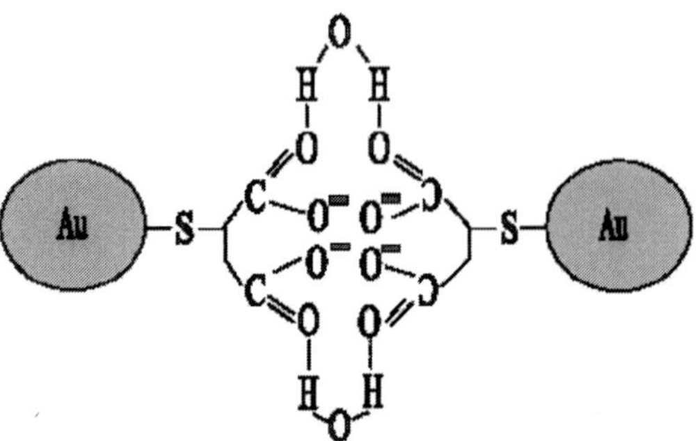

Fig. 5: Schematic representation showing the stabilization of Au nanoparticles by ligand (mercaptosuccinic acid).

Green synthesis of nano-materials

Nanotechnology is mainly concerned with synthesis of nanoparticles with variable sizes, shapes, chemical compositions and controlled dispersity and their potential use for human benefits . Ravindan et al have discussed green chemistry method for the synthesis of silver and gold nanoparticles using glucose as reducing agent in presence of water soluble starch as stabilizing agent. Since then many research groups are actively involved to synthesis metal nanoparticles in a green approach using plant extract, biomass, bacteria, fungus, biocompatible polymer, and microwave and ultrasound mediated synthesis method. Recently many review articles have been appeared in the literature on the important application of green chemical synthesis of metal nanoparticles. Although chemical and physical methods may successfully produce pure, well-defined nanoparticles, these are quite expensive and potentially dangerous to the environment. Use of biological organisms such as microorganisms, plant extract or plant biomass could be an alternative to chemical and physical methods for the production of nanoparticles in an eco-friendly manner.

Several plants have been successfully used for efficient and rapid extracellular synthesis of silver and gold nanoparticles. Leaf extracts of geranium (*Pelargonium graveolens*), lemongrass (*Cymbopogon flexuosus*), *Cinnamomum camphora*, neem (*Azadirachta indica*), *Aloe vera*, tamarind (*Tamarindus indica*) and fruit extract of *Emblica officinalis* have shown potential activity in reducing Au (III) ions to form gold nanoparticles Au (0) and silver nitrate to form silver nanoparticles Ag(0). Biomasses of wheat (*Triticum aestivum*) and oat (*Avena sativa*), alfalfa (*Medicago sativa*), native and chemically modified hop biomass and remnant water collected from soaked Bengal gram bean (*Cicer arietinum*) have also been used for gold nanoparticles synthesis. However, alfalfa (*Medicago sativa*), *Chilopsis linearis* and *Sesbania* seedlings showed synthesis of gold nanoparticles inside living plant parts. We have tabulated these various plants exploited for the synthesis of gold and silver nanoparticles (Table 1).

Table 1: Plants used in biogenic synthesis of silver and gold nanoparticles

S.No.	Plant	Nanoparticles	Applications
1.	Aloe vera	Au and Au	-
2.	Casicum	Au and Ag	-
3.	Tea leaf	Au, Ag and Fe	Dye removal
4.	*Tamarindus indica*		
5.	Coriander leaf		
6.	Algae	Au	-
7.	Cassia		
8.	*Azadirachta indica*		
9.	Cinnamomum camphora		
10.	*Arbutus Unedo*	Ag	-
11.	*Rosa damascene*	Ag and Au	electrochemistry
12.	*Rosa rugosa*	Ag and Au	-

In characterizing the components of living organisms involved in such nanoparticle synthesis, the role of reductases and reducing equivalents has been discovered. In one such example, nitrate reductase from a fungus (*Fusarium oxysporum*) has been documented to catalyze the reduction of silver nitrate to silver nanoparticles utilizing NADPH as reducing agent. Besides these extracellular enzymes, several naphthoquinones and anthraquinones with excellent redox properties have been reported in *F.oxysporum* that could act as an electron shuttle in metal reductions. These studies have demonstrated the involvement of proteins, polyphenols and carbohydrates in the synthesis of metal nanopartilces. All these constituents are present in plants and might be responsible for the synthesis of metal nanoparticles. However, in plants the involvement of such constituents in nanoparticle synthesis needs experimental proof. Isolated quercetin (natural plant pigment) and polysaccharides have been used for silver nanoparticle synthesis. The influence of different factors on the formation of nanoparticles in freshly brewed tea extracts has been investigated. The mean particle size and number of nanoparticles increase with decreasing temperature. In the presence of caffeine more particles were formed, while theophylline catalyzes synthesis of fewer nanoparticles. This suggests that these various compounds could be responsible for various kinds of nanoparticle synthesis through biogenic routes.

Nanoparticle application in catalysis, sensors and medicine depends critically on the size and composition of the nanoparticles. Thus different routes leading to the synthesis of nanoparticles of various shapes and sizes have extended the choice of properties that can be obtained. The optoelectronic and physiochemical properties of nanoscale matter are a strong function of particle size. Nanoparticle shape also contributes significantly to modulating their electronic properties. In the subsequent sections we have described the use of plants or their extracts in the synthesis of gold, silver and gold/silver nanoparticles in a controlled manner for various purposes.

Carbon nanoparticles and their important applications

Lee et al., have discovered carbon nanoparticles by fractionating candle shoot for the first time. Since then many research group have been concentrating on preparing carbon nanoparticles from various green sources from soyabean milk, chitosan, EDTA and glucose. These nanoparticles are brilliant fluorescent in nature. These fluorescence colors can be tuned by various chemical treatments. Such a kind of cheap fluorophore can be utilized as an alternate probe for imaging of cancer cell, diseases diagnosis and environmental applications.

Sol-gel method for synthesis of metal oxides

Controlled hydrolysis of metal alkoxides using either acid or base leads to the formation of uniform size metal oxide nanoparticles. The schematic diagram for the solvent process is shown below;

Synthesis of Zerovalent Iron (ZVI)

The zerovalent iron nanoparticles can be synthesized by chemical reduction of ferrous sultate using $NaBH_4$ in presence of suitable stabilizing agent like polyvinylpyrrolidone, PVP. The reaction medium should be free from dissolved oxygen because iron nanoparticles easily oxidized to nonmagnetic Fe_3O_4.

Recently, it has been reported that green chemistry approach of synthesis of iron nanoparticles green tea extract. The green nanoparticles can be directly used for the degradation of textile dyes and removal chloroorganic pollutants.

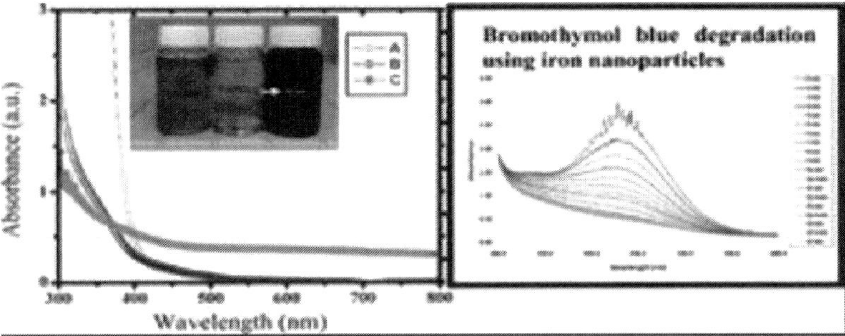

Fig. 6: UV-Visible spectrum of zerovalent iron and left side of the diagram depicted the degradation of bromothymol blue using tea extract protected zerovalent iron nanoparticles

Quantum dots

Semiconductors comprise the class of materials whose electrical conductivities are higher than those of insulators and lower than those of metals. It has an electrical conductivity between 10^{-9} ohm^{-1} cm^{-1} and 10^{2} ohm^{-1} cm^{-1}.

Binary semiconductors (e.g.CdS, CdSe, TiO_2) are the topic of much research in material physics and chemistry. The conductivity of semiconductors can be deliberately manipulated through the introduction of specific chemical impurities. This process is known as doping. A material that has been doped with donors is called a n-type semiconductor (e.g.TiO_2, ZnO) while one that has been doped with acceptors is called a p-type semiconductor (e.g.GaAs). These chemical impurities or dopants, either donate charges to the semiconductor or accept charges from the solid. The introduction of dopants can greatly modify the electrical properties of semiconductors. These nanoparticles are crystalline materials with strong inter atomic chemical bonding. Semiconductor cluster is focused on the properties of quantum dots fragments of semiconductor consisting of hundreds to many thousands of atoms.

The important characteristics of semiconductors are electronic and optical absorption properties. These properties arise from the delocalized electronic charges of these materials. To understand the origin of this electronic delocalization, we must consider the nature of the bonding within semiconductor crystals. Band Theory of solids strongly says that band gap determines the optical changes in semiconductors.

Synthesis of quantum dots

Many synthetic methods for semiconducting nanoparticles have been reported in the literature. An ideal route should result in pure, monodispersed and crystalline particles which are stable, (i.e) they have robust surface passivation. Most of the methods based on colloid chemistry, decomposition of organometallic compounds or growth in a restricted reaction space. This chapter is taking into account the various preparation methods of semiconducting quantum dots. Many researchers are interested in synthesizing of II-VI materials especially CdSe nanopaticles in different routes.

Murray et al., have been reported a simple route to prepare semiconducting nanoparticles with a high degree of monodispersity. The synthesis is based on the pyrolysis of organometallic reagents (Me_2Cd-TOPSe) by injection into a hot co- ordinating solvent (TOPO). This procedure provides temporally controlled growth of nanoscopic quantities of nanocrystallites. Size selective precipitation technique has been used for separation of narrow size selection of clusters. Purified nanocrystallites are dispersed in anhydrous 1-butanol forming an optically clear solution. Anhydrous methanol is then added drop wise to the dispersion until opalescence persists upon stirring or sonication. Separation of the supernatant as flocculate by centrifugation produces a precipitate enriched with the large crystallites in the sample.

$$(H_3C)_2Cd + \left[CH_3{-}(CH_2)_7\right]_3P{=}Se \xrightarrow{TOPO} CdSe + \left[CH_3{-}(CH_2)_7\right]_3P(CH_3)_2$$

Size selective precipitation can be carried out in a variety of solvent/non-solvent pairs, including pyridine / hexane and chloroform / methanol. The color photograph shows the as prepared CdSe – TOP/ TOPO nanoparticles with different sizes and TEM image has also shown.

Later, Alivisatos et al., have been prepared spherical semi-conducting nanoparticles of CdSe with slight modification procedure of Murray et al. In this method, dimethyl cadmium and selenium powder are co-dissolved in a tri-alkyl phosphine (butyl or octyl) and the solution was injected into hot (340-360°C) trioctyl phosphine oxide (TOPO), nucleation occurs rapidly followed by growth at 280-300°C. At high

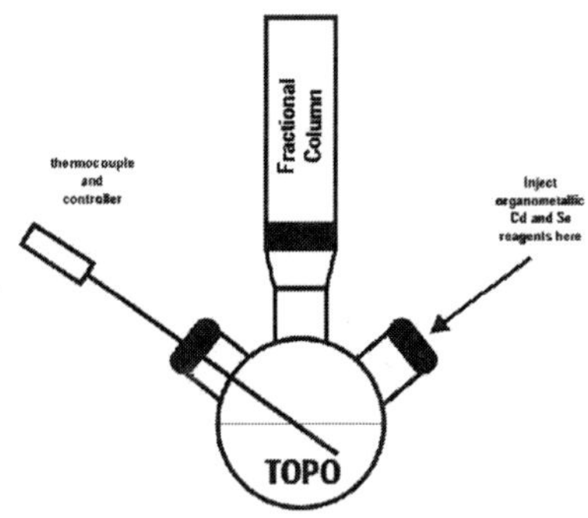

Fig. 7: Procedure for synthesizing monodisperse CdSe Nanocrystals.

temperature surfactants molecules adsorb and desorb rapidly from the nanocrystal surface enabling the addition as well as removal of atoms from the crystallites. The aggregation is suppressed by the presence of surfactant at the crystallite surface at higher monomer concentration, the smaller particles grow faster than the larger one and as a result the crystallites became mono dispersive in nature. Shape of the nanocrystals can be achieved by further manipulation of growth kinetics.

As an expansion to the wet chemical route for the preparation of thiol–capped CdSe nanoparticles has been synthesized in aqueous solution using mercaptoalcohols and mercaptoacids by Rogach et al. They have synthesized a series of thiols stabilized cadmium chalogenides nanocrystals by a wet chemical route in aqueous solution. Colloidal solution of CdSe nanoparticles have been synthesized through addition of freshly prepared oxygen free NaHSe solution of N_2 saturated Cd $(ClO_4)_2 \cdot 6H_2O$ solution at pH 11.2 in the presence of different thiols such as thioalcohols and thioacids. The particle size is controlled by the type of stabilizing agent and through size selective precipitation. Particle growth occurred while refluxing of the reaction mixture for 12 hours, which is due to Ostwald ripening. During heating of the thioacid stabilized CdSe solution, the particle growth proceeded about five times faster than for thioalcohol stabilized CdSe and larger CdSe nanoparticles was formed. A prolonged refluxing of the solution leads to red shift of the absorption edge upto 350-470 nm due to continuous growth of the particles.

Magnetic nanoparticles

Tailored design of colloidal MNPs plays a crucial role in determining the effectiveness of functional magnetic nanostructures for a given biomedical application. A number of key requirements related to the intrinsic properties of the magnetic material, to the occurrence of size and shape effects, to the nature of its surface, to the stability in water-rich environment, as well as to its non-toxicity need to be taken into account. While the MNP size and shape, surface coating and colloidal stability can be tuned through appropriate synthetic procedures, the choice of the magnetic material is rather restricted to iron based magnetic oxides which to date represent the best compromise among good magnetic properties (such as saturation magnetisation) on one hand and stability under oxidizing conditions and limited toxicity on the other. In particular, the US Food and Drug Administration (FDA) and the European Medicines Agency (EMA) have already approved the medical use of magnetic iron oxides MNPs which include magnetite (mixed Fe^{2+} and Fe^{3+} ions), and maghemite, which retains the same spinel structure as magnetite, but is fully oxidized to Fe^{3+}. As bulk materials, both are ferrimagnetic. As magnetite has a larger magnetisation its use should be preferred. In contrast, maghemite is usually more stable in aqueous media, ensuring therefore a more durable magnetic behaviour. Actually, the ease of formation of nonstoichiometric iron oxide nanostructures has been often overlooked due to the difficulty to distinguish magnetite and maghemite at the nanoscale by conventional characterization techniques and many of the reported results refer to a mixture of magnetite and maghemite. In general, MNPs are termed depending on the final size of the coated and functionalized particle. MNPs smaller than 50 nm are called ultra small superparamagnetic iron oxide nanoparticles (NPs), and superparamagnetic iron oxide NPs (SPIONs) when they are larger than 50 nm.

Wet chemical synthesis of magnetic nanoparticles

Wet chemistry routes for the preparation of ferrite MNPs can be conveniently grouped into hydrolytic and non- hydrolytic approaches, which exhibit distinctive advantages and suffer from specific drawbacks. Magnetic ferrites for biomedical use are most commonly prepared by hydrolytic synthesis, with particular reference to co-precipitation techniques. A widespread approach relies on the Massart method, where magnetite is obtained by alkaline coprecipitation of stoichiometric amounts of ferrous and ferric salts (usually chlorides). The experimental parameters affecting this process, which involves the formation of intermediate hydroxyl species, such as temperature, pH, concentration of the cations and nature of the base, have been studied in order to vary the average MNP size in the range from 3 to 20 nm. It is noteworthy that the pH is a critical parameter in affecting both the MNP size (by increasing the pH the repulsion among primary MNPs is induced and smaller magnetite MNPs are obtained) and the stability of the MNP dispersion. In particular, thanks to the surface electrostatic repulsion stable ionic ferroûuids can be obtained in a wide range of pH. The co-precipitation approach oûers a wide range of advantages including: the use of cheap chemicals and mild reaction conditions; the possibility to perform direct synthesis in water; the ease of scale-up; the production of highly concentrated ferroûuids thanks to the high density of surface hydroxyls. Most important, the synthetic route is extremely flexible when it comes to the modulation of the core and surface properties. Conversion to maghemite is easily obtained by chemical oxidation of the agnetite colloids, and substituted ferrites can be prepared by alkalinization of aqueous mixtures of a ferric salt and a salt of a divalent metal under boiling conditions. Likewise, thanks to the high density of reactive sites surface modiûcation can be easily performed by direct incorporation of additives, which is particularly useful for large scale production as it is the case of the carbonate method. A general limitation of the hydrolytic approach lies in the large number of parameters, which have to be carefully monitored in order to control the synthetic outcome, which deals with the complex aqueous chemistry and rich phase diagram of iron oxide phases.

References

Colloids and Clusters, Ed.G. Schmid, VCH Press, NY, 1995
Encyclopedia of nanotechnology, Nalwa series

5

Biological Synthesis of Nano-materials

Girija

The history of nanotechnology in the modern world dates back to 1959 when American physicist Richard Feynman inspired the scientific community in his lecture "There is a plenty of rooms at the bottom". Research on nanosized particles dramatically developed into an important field of modern research with potential effects in electronics to medicine and recently in agriculture and food industry.

Nanoparticles exhibit unique properties in terms of chemical, physical, photoelectrochemical and electronic properties when compared to their respective bulk materials. The applications of these nanosized particles are defined in a bigger ways in the field of Biotechnology. All these applications require custom based synthesis methods for specific applications.

The synthesis of nanoparticles involves usually the physical and chemical methods but those techniques have the dispute of being hazardous. In order to minimize the toxicity, the biological nanoparticle synthesis has been practiced to synthesize the nanoparticles of our interest, which is considered to be less toxic due to their environmentally being surface. Recently, integration of biological components in the formation of nanosized particles leading to the complete green synthesis of nanoparticles has emerged as novel method and gaining more importance among researchers. Biologically inorganic nanomaterials are produced either intra- or extracellular by employing microorganisms or plant derived materials. Various metal nanoparticles like Silver, Gold, Cadmium selenium, Magnetic, Greigite, Pallidium, Barium titanate and Titanium has been successfully synthesized by biological methods (Fig. 1). Among the various metal nanoparticles is synthesized using biological method; noble metal nanoparticles such as silver and gold have been studied extensively.

Microorganisms such as Bacteria, Actinomycetes, yeast and fungi are used for synthesis of various metal nanoparticles with different shape and size. Some well-known examples are *Pseudomonas stutzeri*, *Magnetospirillum magneticum*, *Pseudomonas aerugi-nosa*, *Thermomonospora* sp., *Fusarium oxysporum*, *Candida glabrata*, *Schizosaccharomyces prombe*, *Penicillium fellutanum* and *Cladosporium cladosporioides*. Towards elucidating mechanism of nanoparticle formation, most of the literatures suggest the role of enzyme NADH dependent reductase and also by the high molecular weight proteins released from the microorganisms. In synthesis of CdS nanocrystals, the formation was explained as: upon exposure to Cd as a stress, a series of biochemical reactions were triggered to overcome the toxic effect of the metals and enzyme phytochelatins synthase was activated to synthesis phytochelatins which aids in formation of high molecular PC-CdS 2-complex. Biosynthesis of nanoparticles using microorganisms has some disadvantages beyond its successful synthesis of various metal nanoparticles. The difficulty in maintaining the culture, exclusive medias used for culturing, maintain the optimum culture conditions, time period in formation of the nanoparticles and difficulty in nanoparticle product recovery are some of the reported disadvantages of the synthesis process using microorganisms. Overcoming all the above disadvantages, the synthesis of nanoparticles using plant materials offer numerous advantages. Simple synthesis procedures, mostly single step and one pot synthesis and easy product recovery from the final solutions. In addition, the synthesis of plant materials are also eco-friendly, compatible for pharmaceutical and biomedical applications, cost effective, economic viability, non toxic and easily scaled up for large scale synthesis and also the less time consuming synthesis method. The plants already exploited for the synthesis of nanoparticles and possible uses of plant resources as a nanofactories to synthesize less toxic metal nanoparticles. Green nanoparticles have been synthesized using various plant species such as *Avena sativa*, *Azadirachta indica*, *Aloe vera*, *Alfalfa*, papaya fruit extract, lemongrass, *Sesbania drummondii*, latex of *Jatropha curcas*, etc.

The biomolecules present in these plants are acting as a reducing agent and also as capping agent which favours the synthesis of size controlled nanoparticles through the biological synthesis method. The biomolecules like reducing sugars, phenolic compounds and protein molecules are reported to aid in reduction and proteins in capping the formed nanoparticles. Further detail studies on the mechanism of nanoparticle synthesis in plants in the cellular level and the possible extracellular synthesis of nanoparticles using plant extract in large scale would revolutionize the field of Plant Biotechnology in Nanoscience. This technology untaps the utilization of nanomaterial for agricultural purposes.

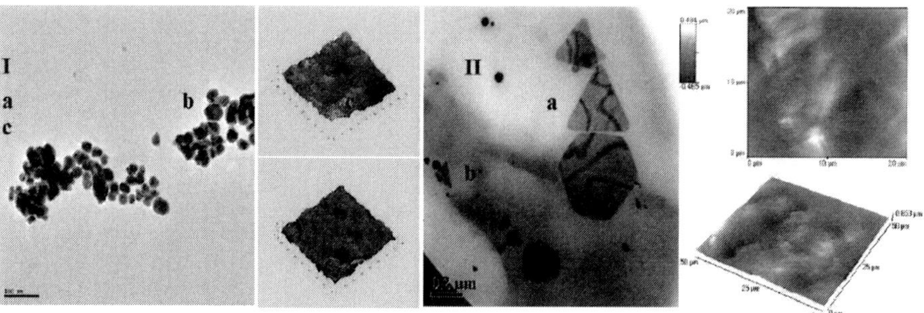

Fig. 1: Silver and gold nanoparticles using plant leaf extract. I) TEM image (a) and AFM images (b,c) of silver nanoparticles. II) TEM image (a) and AFM images (b,c) of gold nanoparticles.

These nanoparticles synthesized by various techniques have received special attention because they have found potential application in many fields such as catalysis, sensors, drug delivery systems etc. The green synthesized nanoparticles can have a wide application in not only medical field like artificial implantations, Diagnostic tools and biological research, it also applied in Cosmetics, Agriculture, Environmental protection and Food industries. Using different plant sources such as *Azadirachta indica*, *Aloe vera* etc. for synthesizing nanoparticles will create a tremendous application in agricultural practices such as to increase the potentiality of fertilizers and insecticides and also in food preservation.

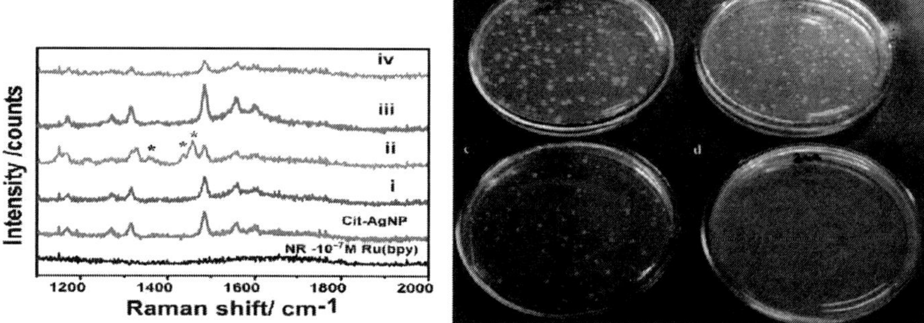

Fig. 2: (I) SERS and (II) anti-bacterial properties of silver nanoparticles synthesized used plant leaf extract.

It has been found that excellularly synthesized silver nanoparticles using *Fusarium oxysporium*, can be incorporated in several kinds of materials such as clothes and these cloths with silver are effective towards bacteria such as *S. aureus*. Metal sulphide microcrystallites formulated using *S. pome* can function as quantum semiconductor crystals and have properties like optical absorption, photosynthetic and electron transfer.

Recently, we have shown the optical enhancing properties using SERS of the silver nanoparticles synthesized with plant leaf extract of *O. sanctum* in different reaction pH (Fig. 2). The SERS signals were better than that of the commonly synthesized silver nanoparticles using citrate.

Biological systems have large potent for the fabrication of nanoparticles which can be utilized for wide application areas. Desirable nanoparticles can be obtained by using products or live single celled microorganism to complex eukaryotes. Though many biological systems are identified to synthesis nanoparticles, the identification of key bio-components involving reduction and stabilization will further help in the production of nanoparticles with desirable shape and size. However, the elucidation of the exact mechanism of nanoparticle formation in biological methods needs much more experimental observations, so that well-defined nanoparticles can be fabricated for application in Agriculture and food industry.

Section 3: Characterization

6

Properties of Nano-materials – I

B. Nalini

Nano-material possess unique properties such as dimension, shape, specific surface area, surface mass ratio etc., that are highly influenced by the mechanical and physical properties. It is pertinent to understand the basic physical properties of nano-materials in order to apply them in a wide array of functions. In this context, this chapter is devoted to provide basic information on physical, mechanical and photo-luminescence properties of nano-materials.

Mechanical properties of nanostructures

- Compressive strength
- Ductility
- Fatigue limit
- Hardness
- Poisson's ratio
- Shear modulus
- Shear strength
- Tensile strength
- Yield strength
- Young's modulus

Mechanical properties of Nanomaterials

Due to the nanometer size, many of the mechanical properties of the nanomaterials are modified to be different from the bulk materials including the hardness, elastic modulus, fracture toughness, scratch resistance and fatigue strength etc. An enhancement of mechanical properties of nanomaterials can result due to these modifications, which are generally resultant from structural perfection of the materials. The small size either renders them free of internal structural imperfections such as dislocations, micro twins, and impurity precipitates or the few defects or impurities present cannot multiply sufficiently to cause mechanical failure. The imperfections within the nano dimension are highly energetic and will migrate to the surface to relax themselves under annealing, purifying the material and leaving perfect material structures inside the nanomaterials. Moreover, the external surfaces of nanomaterials also have less or free of defects compared to bulk materials, serving to enhance the mechanical properties of nanomaterials. The enhanced mechanical properties of the nanomaterials could have many potential applications both in nano scale such as mechanical nano resonators, mass sensors, microscope probe tips and nanotweezers for nano scale object manipulation, and in macro scale applications structural reinforcement of polymer materials, light weight high strength materials, flexible conductive coatings, wear resistance coatings, tougher and harder cutting tools etc.

Among many of the novel mechanical properties of nanomaterials, high hardness has been discovered from many nanomaterials system. A variety of superhard nanocomposites can be made of nitrides, borides and carbides by plasma-induced chemical and physical vapor deposition. In the appropriately synthesized binary systems, the hardness of the Nanocomposite exceeds significantly that given by the rule of mixtures in bulk. For example, the hardness of nc-MnN/a–Si_3N_4 (M=Ti, W, V,...) nano-composites with the optimum content of Si_3N_4 close to the percolation threshold reaches 50 GPa although that of the individual nitrides does not exceed 21 GPa. These superhard nanocomposites will have promising potential in hard protective coatings.Superhardness also comes from pure nanoparticles. For example, Gerbericha report the superhardness from the nearly spherical, defect-free silicon nanospheres with diameters from 20 to 50 nm of up to 50 GPa, fully four times greater than the bulk silicon. Since their discovery, carbon nanotubes have stimulated intensive research interests. As the Smallest carbon fibers discovered, carbon nanotubes have been found to have excellent mechanical properties. The strength of the carbon fibers would increases with graphitization along the fiber axis.

Carbon nanotubes, which are formed of seamless cylindrical graphene layers, represent the ideal carbon fiber and should presumably have the best mechanical properties in the carbon fibers species, showing a high Young's modulus and high tensile strength. Theoretical research has predicted the high modulus of carbon nanotubes aside from the direct experimental measurement, which calculated the Young's moduli Y of single wall carbon nanotubes to be from 0.5-5.5 TPa, much higher than high-strength steel of ~200 GPA. The Young's moduli and the tensile strength have also been measured experimentally. The first experimental measurement of Young's modulus of Multiwall carbon nanotubes was obtained by measuring thermal vibrations of carbon nanotubes using transmission electron microscopy (TEM), yielding the Young's modulus of 1.8+/-0.9 TPa .In this method, the amplitude of the thermal vibrations of the free ends as a function of temperatures of anchored nanotubes was correlated with the young's modulous.

Figure shows the free standing multi wall carbon nanotubes with tip blurring due to thermal vibration. By using a similar technology, Krishnan measured the young's modulous of single wall carbonnanotubes, resulting in average value of Y=1.25 -0.35/+0.45 TPa.

Atomic force microscope (AFM)

AFM is employed to measure the young's modulus of the carbon nanotubes. This is realized by bending the anchored carbon nanotubes with AFM tip while simultaneously recording the force by the tube as a function of the displacement from its equilibrium position. The resultant Young's modulus was 1.28+/- 0.5 TPA. The values of Young's modulus measured from different ways were all in the range in theorical prediction, proving the existence the high elastic modulus of the carbon nanotubes.The tensile strength of carbon nanotubes has also been measured.

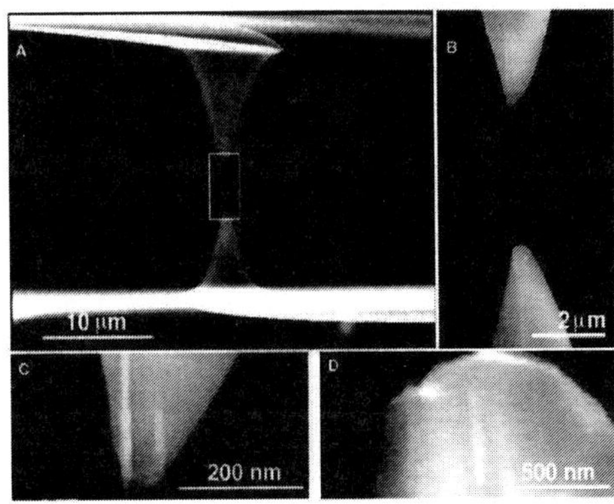

Figure A multi wall carbon nanotube was aligned between two AFM tips. The lower AFM tip is on a soft cantilever whose deflection is recorded to determine the force applied on the carbon nanotube.

An individual multi wall carbon nanotube was mounted between two AFM tips, one on rigid cantilever and the other on soft cantilever. By recording the whole tensile loading experiment, both the deflection of the soft cantilever from which the force applied on the nanotube and the length change of the nanotube were simultaneously obtained. The carbon nanotubes brokein the outermost layer ("sword-in-sheath" failure), and the tensile strength of this layer ranged from11 to 63 GPA and the measured strain at failure can be as high as 12%. For comparison, the tensile strength of high- strength steel is 1-2 GPA.The excellent mechanical properties of nanomaterials could lead to many potential applications inall the nano, micro and macro scales. High frequency electro-mechanical resonators have been made from carbon nanotubes and nanowires.

Figure 1 (a) SEM image showing the carbon nanotubes suspended between two electrodes (Top). The schematic device geometry is shown at the bottom left. Scale bar: 300 nm. (b) SEMimage of the resonator with a suspended nanowire (bottom right).

An individual carbon nanotube was contacted with two metal electrodes and was doubly clampedand suspended across a trench, which was realized by conventional lithography technique. The nanotube resonator was actuated and detected through electrostatic interaction using backside gate electrode underneath the tube in a vacuum. The guitar-string-like oscillation modes of doubly clamped nanotube was discover from the resonator with the resonant frequency as high as55 MHz. A similar nanoelectromechanical resonator can also be fabricated from platinum nanowires. The nanowire resonator bear a similar but a with side gate electrode to detect the actuation, which gave a resonant frequency of 105MHz. These NEMS oscillators could potentially used in ultra sensitive mass detection1, radio-frequency signal processing, and as a model system for exploring quantum phenomena in macroscopic systems.

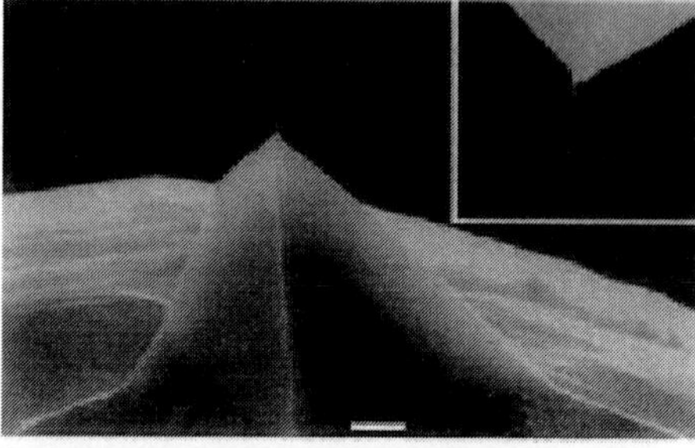

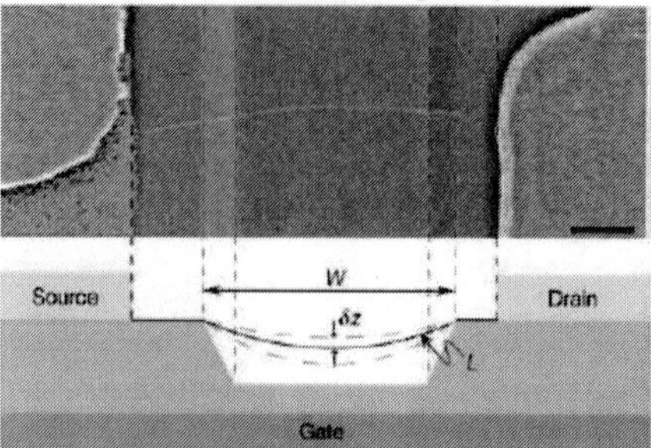

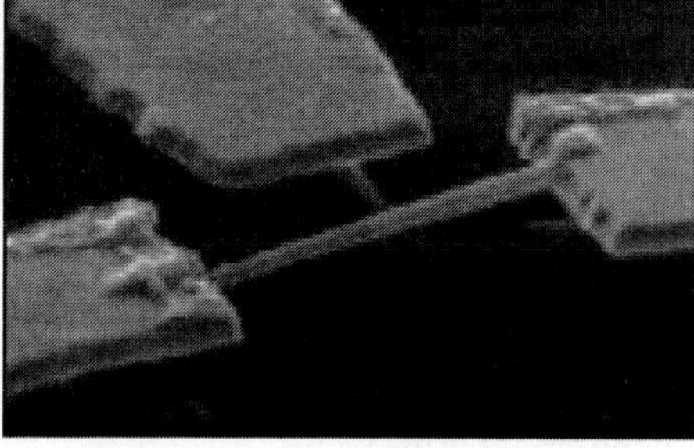

Fig. 1: SEM image of carbon nanotube attached on the silicon cantilever tip

Nano structured materials can also be used as nanoprobes or nanotwizzers to probe and manipulate nanomatierals in a nanometer range. Due to their high aspect ratio and small dimensions, one-dimension nano structures such as carbon nanotubes can also be used as nano probe tips. Nanotube to conventional pyramidal tip of a silicon cantilever for scanning force microscopy.

Table 1: The Comparison of the Resolution Obtained with Nanotube and Silicon Tips on Fibrils and Protofibrils.

Sample	Tip	width at half maximum (nm)a	depth between subumits (mn)b	tip radius (nm)c
Type-I fibril	MWNT	18.6 ± 2.2	3.2 ± 0.4	9.3
	SWNT	11.9 ± 0.7	2.7 ± 0.4	2.6
	Si-TESP	26.0 ± 0.9	2.7 ± 0.3	19.7
	Si-FESP	21.5 ± 1.8	2.5 ± 0.3	12.9
Protofibril	MWNT	11.4 ± 0.4	1.1 ± 0.1	9.7
	Si-TESP	14.4 ± 0.3	0.6 ± 0.1	15.9
	Si-FESP	14.8 ± 1.7	0.7 ± 0.2	16.9

Stanislaus S. Wong acquired high resolution AFM image for biological systems by using carbonnanotubes tips. Figure shows a typical nanoprobe tip made from carbon nanotube. While most conventional tips suffer from "tip crash", posse significant constraints on potential lateral resolution, and furthermore, the large probe size restricts the ability of these tips to access narrow and deep features.

Probe tips from carbon nanotubes could offer several advantages

The flexibility of carbon nanotubes makes the tip more resistant to damage from tip crashes; the high aspect ration of carbon nanotubes makes them suitable to image sharp recesses in surface topography; Tips from carbon nanotubes could significantly improved lateral resolution; carbon nanotubes tips are more robust and less prone to contamination. By investigating the amyloid-β1-40 (Aβ40) fibrils under AFM, Stanislaus compared the performance of nanotubes tips and the single crystal silicon tips.

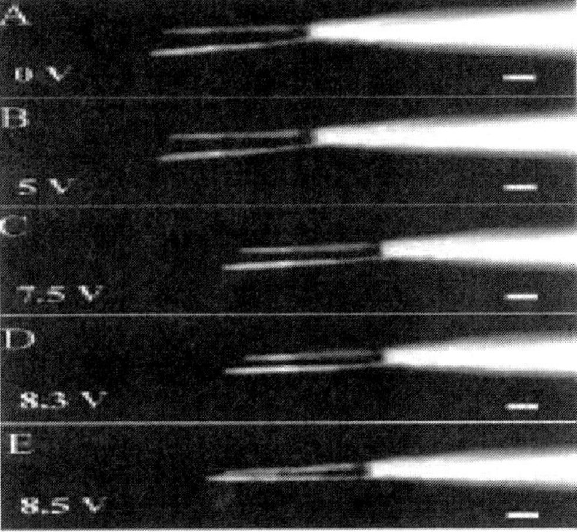

Fig. 2: Dark field optical images of the electro-actuated nanotube nanotwizzers.

An improvement of the image resolution (smaller fibrils and protofibrils were imaged using carbon nanotube tips) due to a reduction of the effective tip radii when imaging with nanotube tips was acquired.

Similar to the carbon nanotube nanoprobe fabrication, if on the supporting micro structure, two individual carbon nanotubes are attached instead of only one nanotube in case of nanoprobe, then a nanotube nano tweezers would result. Phillip Kim showed such a nanotube tweezers by attaching two carbon nanotubes on glass micropipettes.

Table 2: Enhancement of mechanical properties of polymers

Polymer	E'(40 °C, 1 Hz) (matrix) (MPa)	E'(40 °C) CNTs composite (MPa)	CNTs (wt%)	% Increase E'
PMMA	≅ 800	≅ 1600	26	100
PS	≅ 2400	≅ 3500	5	44
PSBA	≅ 0.681	≅ 1.584	7	132
PVA	≅ 5000	≅ 11200	60	124
MEMA	708	2340	1	230

This nano tweezers operated under electrical stimulus and was used to probe the electrical characteristics of nanostructures. It could be useful both in the nanostructure characterization and the manipulation. The nanotweezers could be used as a novel electromechanical sensor that can detect pressure or viscosity of media by measuring the change in resonance frequency and Q-factor of the device. They can also be explored into manipulation and modification of biological systems such as structures within a cell. The enhancements of mechanical properties of polymeric materials by nanofillers are also very active applications of nanomaterials. Micrometer size fillers were used in traditional polymer composites and showed improvements in their mechanical properties such as the modulous, yield strength and glass transition temperature. However, these performance enhancements will sacrifice the ductility and toughness of the materials and large amounts of filler were needed to achieve the desired properties. Comparably, polymer nanocomposites from nano size fillers could result in unique mechanical properties at very low filler weight fractions. Sumita found dramatic improvements in the yield stress (30%) and Young's modulus (170%) in polypropylene filled with with ultrafine SiO_2, compared to micrometer-filled polypropylene with 40 wt % 12 nm silica, compared to micrometer-sized filler reinforced polymer. Vinyl acetate, acrylic ester, synthetic rubber, and other polymer latexes have been used in coatings and adhesives; colloidal silica is used with these polymer emulsions in order to improve adhesion, durability, and abrasion resistance. The silica also serves to prevent stickiness and improves the washing resistance of the coatings. Besides nanoparticles, One-dimension nanostructures such as carbon nanotubes are also superior candidates for nano fillers because of their high aspect ratio and excellent mechanical properties. Nanocomposite materials from carbon nanotubes are expected to exhibit outstanding mechanical properties, such as high Young's modulus, stiffness and flexibility.

Comparative results (E′, % increase E′) of CNTs composites obtained in different Polymer matrices at 40 °C 1Hz of frequency.

Table 2 summarized some results from several research groups of the enhancement of mechanical properties of polymers nanocomposites containing carbon nanotubes fillers without the use of additives, generally more than 100% in "E" was acquired. Besides the improvement of mechanical properties in carbon nanotube- polymer nanocomposites, they can further be employed as multifunctional materials providing light weight, strong and tough structural materials as well as excellent electrical conducting and thermal conducting properties.

Technical Description

Applications of nanoscale multilayers in the near term include use as protective coatings, whether for thermal barriers, ultrahard coatings, or for wear resistance needs. Future uses might include structural applications. Within the next decade, physical vapor deposition (PVD) structures are expected to amount to as much as 2% of the estimated $25B high performance coatings industry in the U.S., as the dependence on environmentally hazardous electroplated hard chrome decreases. Nanolayered composites are being commercially applied presently, and advances in fabrication technology will allow the exceptional physical properties of this class of materials to be more extensively utilized in high performance and high value- added applications. Potential users of this technology include users of critical, high performance coatings: automotive engines, aircraft engines, cutting and machine tools, and ceramic parts manufacturing.

Technical Objectives

- Examine the fundamental mechanisms that operate within nanolayered materials that result in such extraordinary properties as ultrahardness, wear resistance, and high thermal conductivity.
- Evaluate how these properties can be controlled using the ability to prescribe the makeup of the microstructure at nearly the atomic level.
- Develop a fundamental understanding of the relationships between materials, structure, and properties.

Anticipated Outcome

A fundamental understanding of how layering, constituent materials, and interfacial characteristics of a structure determines its properties and performance. Predictive tools that can be used to design materials for specific applications.

Insights into basic mechanical properties of nanolayered materials can be applied to reliability issues in the magnetic recording and microelectronics industry, to address such problems as thermomechanical fatigue, adhesion, and abrasive wear resistance in fine layered structures.

Calculation

Young's modulus, E, can be calculated by dividing the tensile stress by the tensile strain :

Where

$$E \equiv \frac{\text{tensile stress}}{\text{tensile strain}} = \frac{\sigma}{\varepsilon} = \frac{F/A_0}{\Delta L/L_0} = \frac{FL_0}{A_0 \Delta L}$$

E is the Young's modulus (modulus of elasticity) F is the force applied to the object;

A_o is the original cross-sectional area through which the force is applied;

L is the amount by which the length of the object changes;

L_o is the original length of the object.

Force exerted by stretched or compressed material:

$F = \frac{EA_0 \Delta L}{L_0}$ The Young's modulus of a material can be used to calculate the force it exerts under a specific strain.

$$F = \frac{EA_0 \Delta L}{L_0}$$

Where F is the force exerted by the material when compressed or stretched by L.

Hooke's law can be derived from this formula, which describes the stiffness of an ideal spring:

$$F = \left(\frac{EA_0}{L_0}\right) \Delta L = kx$$

Where

$$k = \frac{EA_0}{L_0}$$

Elastic potential energy

The elastic potential energy stored is given by the integral of this expression with respect to L :

$$U_e = \int \frac{EA_0\Delta L}{L_0}\, dL = \frac{EA_0}{L_0}\int \Delta L\, dL = \frac{EA_0\Delta L^2}{2L_0}$$

Where Ue is the elastic potential energy.

The elastic potential energy per unit volume is given by :

$$\frac{U_e}{A_0L_0} = \frac{E\Delta L^2}{2L_0^2} = \frac{1}{2}E\varepsilon^2 \ ,$$

$$\varepsilon = \frac{\Delta L}{L_0}$$

where is the strain in the material

This formula can also be expressed as the integral of Hooke's law:

$$U_e = \int kx\, dx = \frac{1}{2}kx^2.$$

Relation among elastic constants

For homogeneous isotropic materials simple relations exist between elastic constants (Young's modulus E, shear modulus G, bulk Modulus K, and Poisson's ratio) that allow calculating them all as long as two are known :

$$E = 2G(1+\nu) = 3K(1-2\nu).$$

Photoluminescence Properties

Introduction

It is one of the more often used units for very small lengths i.e. nanometer (nm). One nanometer equals to one thousandth of a micrometer or one millionth of a millimeter or ten angstrom. It is often associated with the field of nanotechnology and the wavelength of light. Nanotechnology can offer potential solution to many problems using emerging nano techniques. Depending upon the interdisciplinary character of nanotechnology, there are many research fields and several potential applications. But irrespective of the field, materials are getting importance. Nanomaterials means, material with one (or) two (or) zero dimensions and with an internal structure which could exhibit novel characteristics compared to the same material without nanoscale features. Any physical substance with structural dimensions between 1-100 nm can be defined as nanomaterials.

Luminescence

Luminescence is the term applied to the re- emission of previously absorbed radiation. In molecular photoluminescence, photons of electromagnetic radiation are absorbed by molecules raising them to some existed state and then, on returning to the ground state, the molecules re-emit the radiation back and this is known as luminescence. Some important luminescences are photoluminescence, thermo luminescence, bioluminescence and cathodoluminescence. In this presentation details of photoluminescence will be discussed.

Photoluminescence from bulk matter

If a photon has energy greater than that of the band gap energy of a material, then it can be absorbed by that material (mostly semiconducting type). The absorbed energy raise an electron from the valence band (leaving behind the holes) up to the conduction band across the forbidden energy gap. In this process of photoexcitation, the electron generally has excess energy and that will be mostly lost before coming to rest at the lowest energy level in the conduction band. From this point, the electron eventually falls back down to the valence band and recombine with the holes. As it falls down, the energy it loses is converted back into a luminescent photon. This energy of the emitted photon is a direct measure of the band gap energy.

Photoluminescence includes fluorescence and phosphorescence. In fluorescence the energy transitions do not involve a change in electron spin, on the other hand phosphorescence involves a change in electron spin and is therefore much slower than the fluorescence process. Fluorescence emission involves a lifetime of 10^{-8} sec (electron life time in the excited state or conduction state) and therefore happens quickly after initial photon absorption. [Some fluorescent nanomaterials have been successfully used as fluorescent labels for a variety of bio-analytical purposes such as detection of DNA, proteins and other biomolecules, cellular labeling etc]. In phosphorescence type emission the life of time of excited electrons is about 10^{-3} sec and therefore happens slowly after initial photon absorption.

Photoluminescence excitation and emission

In physics, excitation means elevation in energy level above an arbitrary baseline energy state. The below Fig. 3 schematically represent the excitation process. An incoming photon of energy greater than that of band gap supplies energy to the valence band electron and therefore this valence electron is removed from the valence band [Valence band: A representation of bond state, here electrons are bound to individual atoms and therefore they can't make a free movement inside the material (it means no current is possible)]. The as removed electrons are moving to the conduction band [Conduction band: A representation of a free state, here the electrons are bound to the individual atoms of crystal network but belongs to the whole network, so that they can move freely around the material but definitely not out of the material] and spends its time there for a while. After spending an initial time of 10^{-8} or 10^{-3} Secs, the electrons return back to the valence band. Since energy is conserved; this returning process gives back the photon absorbed and is known as "emission" process.

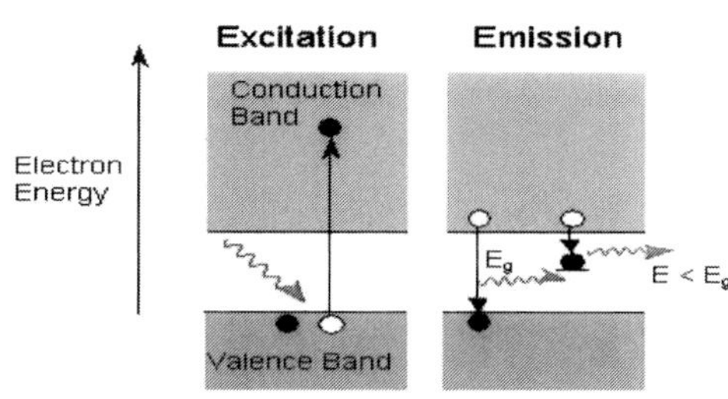

Fig. 3: Continuous energy levels (bands)

Effect of Size reduction

When the materials (mostly semi-conductors) are in the bulk form then band theory is applicable as discussed above. But when the materials size is reduced to nanoscale level then it is impossible to see such a bands (that is valence and conduction bands). Instead of bands there will be discrete energy levels[1] as shown below in Fig. 4. These discrete energy levels have the nomenclature as valence and conduction energy levels (well separated lines in the conduction and valence

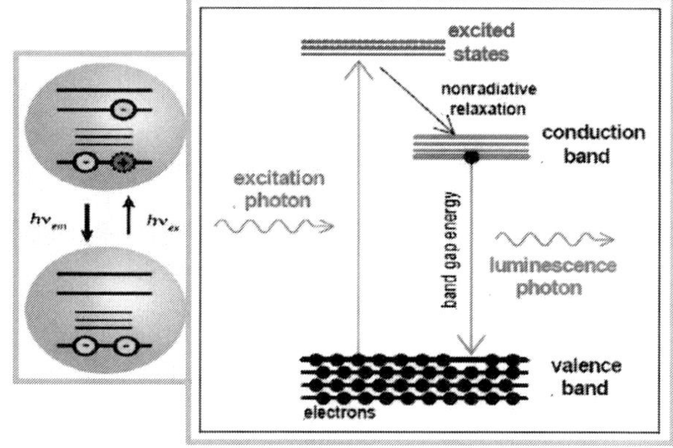

Fig. 4: Discrete Energy level upon size reduction

band). Here also there exists a forbidden gap and therefore energy absorption is required for an electron to get excited into the conduction energy levels (take care not conduction bands).

What makes discreteness?

A simple answer for the conversion of a "band" into "discrete energy level" is a cut in the number of atoms (upon size reduction). We all know that matter is formed by the joining of atomic orbital's together and this joining, also known as over lapping of energy levels, occurs at the outer most levels known as valence levels of atoms. The over lapping of energy levels may not be so perfect due to the thermal vibration of atoms and therefore there will be a minimum energy level separation between the overlapping energy levels. When the size of the matter is big then it means more number of atomic orbital's or energy levels are overlapping. This in turn results a collection of closely packed energy levels also known as bands. The bound states are known as valence band and the free states are known as conduction band. If one can closely observe the bands there also exists discreteness but with a very minimum energy separation. Because of the minimum energy separation, just room temperature energy of about 23 to 25 meV (milli electron volt) is sufficient for the electrons to get promoted from one level to the other and therefore, within in the band, either conduction or valence band, the electrons can make free movement from bottom to top. That is getting energy from room temperature the electrons can move from one level to the other within the band. Once reached top of the band the electrons loses their energy by collision and therefore comes down to the bottom of the band. As this process is random the net current will be zero. However for the electrons to move from valence band to conduction band a minimum energy, equivalent to that of forbidden gap is required. Getting this energy the electrons from valence band can make a movement towards conduction band, leaving behind holes in valence bands. [Within the conduction band the electrons can move freely from bottom to top levels getting energy from room temperature and therefore these motion with mostly will be random and therefore the net current will be zero. However by applying a biasing voltage, we can give directionality to the electrons and therefore a non-zero net current will be resulted. Remember, here also there exists randomness and therefore the electron movement towards the positive end of the potential is slowed down. This slowing down procedure gives raise to heat and that's why most of the bulk devices and their products produce lot of heat.] These excited electrons in the conduction band, in the absence of external biasing voltage, will recombine with the holes in the valence band and thereby re-emits the radiation.

When the size of the matter is reduced, obviously the number of atomic levels overlapping to form a band of states is reduced. If this size reduction process continues up to nanoscale level then there will be fewer atoms and hence fewer overlapping of energy levels. In such a situation, the band (of states) appears with discrete states. Remember, now also we have valence states, conduction states (not bands) and forbidden gap. So we need to supply the energy externally to excite the electrons from the valence states into conduction states. One point, one has to remember is that upon size reduction the forbidden gap width is increased and therefore the resulting recombination (in the absence applied biasing voltage) gives a high energy photon. Therefore by closely studying the emission wavelength one can indirectly obtain the size of the nanoparticle. However, the relation between the size and the emission wavelength is unique to given material of particular composition and crystal structure and therefore one has to take care in co-relating the emission wavelength to the size. For example, factors such electron and hole effective masses etc has to be taken into account for the calculation size of the nanoparticle from the wavelength.

Conclusion

In conclusion, one can measure the emission wavelength using spectrofluorimeter and therefore can obtain the size of the nanoparticle indirectly. Many scientists are using this luminescence technique to find the nanoparticle size. In particular this emission mechanism will be very much useful to semiconducting nanoparticles only. However, in the case of metals, since there is no forbidden gap, the emission process will be different.

7

Properties of Nano-materials – II

R.T. Rajendra Kumar

Nano-materials are to be extensively studied and characterized prior to their application in material science, biological science, electrical and electronics. In addition to the basic physical properties of the nano-materials, other characteristics such as paramagnetism, ferromagnetism, supermagnetism, optical properties, quantum size effects and electrical properties have to be studied. With a view to provide the fundamental information on these aspects this chapter is dedicated.

Diamagnetism, paramagnetism and ferromagnetism

There are three categories of magnetism that we need to consider: diamagnetism, paramagnetism and ferromagnetism. Diamagnetism is a fundamental property of all atoms (molecules), and the magnetization is very small and opposed to the applied magnetic field direction. Many materials exhibit paramagnetism, where a magnetization develops parallel to the applied magnetic field as the field is increased from zero, but again the strength of the magnetization is small. In the language of the physicist, ferromagnetism is the property of those materials which are intrinsically magnetically ordered and which develop spontaneous magnetization without the need to apply a field. The ordering mechanism is the quantum mechanical exchange interaction. It is this final category of magnetic material that will concern us in this chapter. A variation on ferromagnetism is ferrimagnetism, where different atoms possess different moment strengths but there is still an ordered state below a certain critical temperature. We will now define some key terms. The magnetic induction B has the units of tesla (T). The magnetic field strength H can be defined by

$$B = \mu_o H \qquad (1)$$

where m0 = 4p x 10^{-7} Hm^{-1} is the permeability of free space. This gives the horizontal component of the earth's magnetic field strength as approximately 16 Am^{-1} in London. The flux j = BA can be defined, where A is a cross-sectional area. Flux has the units of weber (W).

If a ferromagnetic material is now placed in a field H, Equation (1) becomes

Table 1: Classification of magnetic materials by susceptibility

Diamagnet ($\chi < 0$)	Paramagnet ($\chi > 0$)	Ferromagnet ($\chi >> 0$)
Cu: - 0.11 x 10^{-5} Au: - 0.19 x 10^{-5} Pb: - 0.18 x 10^{-5}	Al: 0.82 x 10^{-5} Ca: 1.40 x 10^{-5} Ta: 1.10 x 10^{-5}	Fe:>10^2

$$B = \mu_o(H+M) = \mu_o(H + xH) = \mu_o\mu_r H \qquad (2)$$

where M is the magnetization of the sample (the magnetic dipole moment per unit volume). We define c as the susceptibility of the magnetic material, and m_r = 1 + c as the relative permeability of the material (both m and c are dimensionless). Table 1 summarizes this classification scheme, and gives values for typical susceptibilities in each category.

Hysteresis Loops

Most of the magnetic properties of a material can be derived from its hysteresis loop. Figure 1 schematically illustrates a magnetization vs field (M– H) hysteresis loop. When the external magnetic field is sufficiently large, all the spins within a magnetic material align with the applied magnetic field. In this state, the magnetization of the material achieves its maximum value, and t his value is called the saturation magnetization, Ms. When the external magnetic field becomes weaker, the spins in the material cease to be aligned with the external

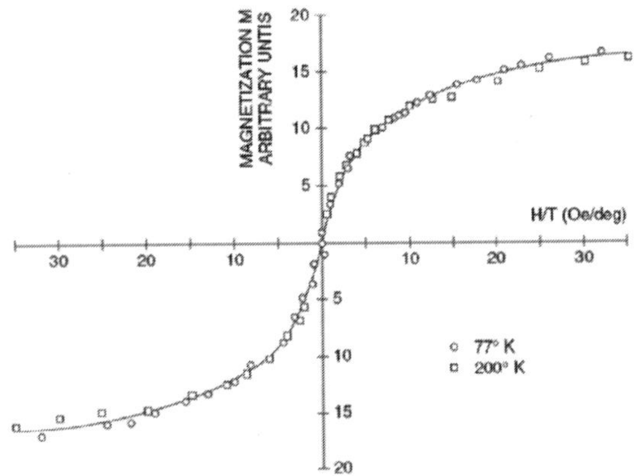

Fig. 1: A Schematic diagram of magnetization loop

magnetic field, so the total magnetization of the material decreases. For a ferromagnetic material, when the external magnetic field.

Remanence is the magnetization remaining when the field is reduced to zero from that required to saturate the sample. Coercivity is the field required to bring the magnetization to zero from remanence. Decreases to zero, the material still has a residual magnetic moment, and the value of the magnetization at zero field is called the remanent magnetization.

Figure 2 A hypothetical situations in which ferromagnetic particles with a range of sizes from nanometer up to micron scale are injected into a blood vessel. The magnetic responses associated with different classes of magnetic material are illustrated by their corresponding M–H curves. The biomaterials in the blood vessel are diamagnetic (DM) or paramagnetic (PM). Depending on their sizes, the injected particles can be ferromagnetic (FM) or superparamagnetic (SPM). Ferromagnetic materials can be multi-domain (- - - - in FM diagram) or single-domain (______ in FM) is defined as the ratio of the remanent magnetization to the saturation magnetization, M_r/M_S, which varies fromparamagnets is usually in the range of 10"6 to 10-1, while the ÷ value for diamagnets is usually in the range "10"60 to 1. To bring the material back to zero magnetization, a magnetic field in the negative direction should be applied, and the magnitude of the field is called the coercive field, H_C. The re-orientation and growth of spontaneously magnetized domains within a magnetic material depends on both microstructural features such as vacancies, impurities or grain boundaries, and intrinsic features such as magnetocrystalline anisotropy as well as the shape and size of the particle. In most cases, the hysteresis loop of a magnetic material should be experimentally measured using, a vibrating sample magnetometer (VSM) or superconducting quantum interference device (SQUID) magnetometer, and it is not possible to predict a priori what the hysteresis loop will look like.

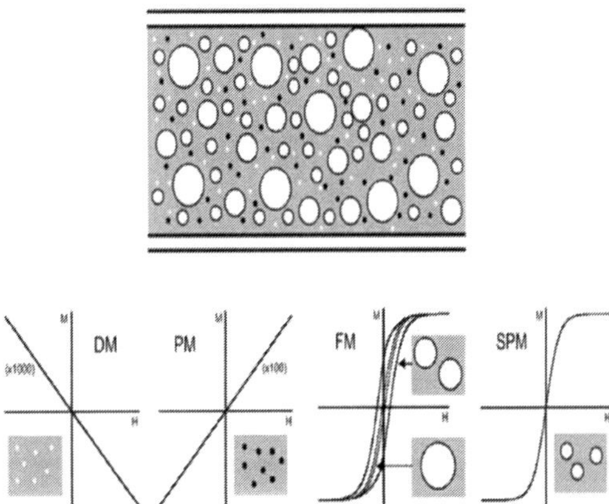

Fig. 2: Hypothetical situation with ferromagnetic particles

Materials with different magnetic properties have different shapes of hysteresis loops. Figure 2 shows a schematic diagram of a blood vessel into which some magnetic nanoparticles have been injected, and the magnetic properties of both the injected particles and the ambient biomolecules in the blood stream. Generally speaking, the blood vessel and the biomaterials in the blood vessel are either diamagnetic or paramagnetic, while the injected magnetic particles are either ferromagnetic or superparamagnetic, depending on their sizes.

Broadly, all materials can be regarded as magnetic materials because all the materials respond to magnetic fields to some extent. However, they are usually classified based on their volumetric magnetic susceptibility, ÷, describing the relationship between the magnetic field H and the magnetization M induced in a material by the magnetic field: to “10^{-3}. Negative ÷ value of diamagnets indicates that in such materials, the magnetization M and the magnetic field H are in opposite directions. However, some materials exhibit ordered magnetic states, and they are usually classified as ferromagnets, ferrimagnets and antiferromagnets. The prefixes of these names refer to the nature of the coupling interactions between the electrons within the material. Such couplings may lead to large spontaneous magnetizations, and this is the reason why ordered magnetic materials usually have much larger ÷ values than paramagnetic or diamagnetic materials.

It should be noted that the susceptibility in ordered materials also depends on applied magnetic field H. This magnetic field gives rise to the characteristic sigmoidal shape of the M–H curve, with M approaching a saturation value at high magnetic field. In ferromagnetic and ferrimagnetic materials, hysteresis loops can be observed, as shown in Figure 1. The shape of a hysteresis loop is partly determined by the particle size. A particle in the order of micron size or more usually has a multi-domain structure. As it is easy to make the domain walls move, the hysteresis loop of such particles is narrow. In a smaller particle, the single-domain structure leads to a broad hysteresis loop. When particle size becomes even smaller, in the order of tens of nanometers or less, superparamagnetism can be found. The magnetic moment of a super paramagnetic particle as a whole is free to fluctuate in response to thermal energy, while the individual atomic moments maintain their ordered state relative to each other. As shown in Figure 2, the M–H curve of a superparamagnetic particle is anhysteretic, but still sigmoidal.

Magnetic anisotropy

$$M = \chi H \tag{3}$$

In the SI unit system, ÷ is dimensionless, while both M and H are expressed in Am`[1]. Most materials display little magnetism, and these are classified either as paramagnets or diamagnets. The ÷ value for dMost materials contain some type of anisotropy affecting their magnetization behaviors. The magnetic anisotropy of a material can be modeled as uniaxial in character and represented by:

$$E = KV\, sin^2\theta \tag{4}$$

where K is the effective uniaxial anisotropy energy per unit volume, è is the angle between the moment and the easy axis, and V is the particle volume.

Single-domain Particles

A domain is a group of spins whose magnetic moments are in the same direction, and in the magnetization procedure, they act cooperatively. In a bulk material, domains are separated by domain walls, which have a characteristic width and energy associated with their formation and existence. The movement of domain walls is a primary means of reversing magnetization and a major source of energy dissipation.

Figure 3 schematically shows the relationship between the coercivity in particle systems and particle sizes. In a large particle, energetic considerations favor the formation of domain walls, forming a multi-domain structure. The magnetization of such a particle is realized through the nucleation and motion of these walls. As the particle size decreases toward a critical particle diameter, D_C, the formation of domain walls becomes energetically unfavorable. So there is no domain wall in such a particle and this particle is called a single-domain particle. For a single-domain particle, the magnetization procedure is realized through the coherent rotation of spins. The particles with size close to D_C usually have large coercivities. As the particle size is much smaller than D_C, the spins in this particle

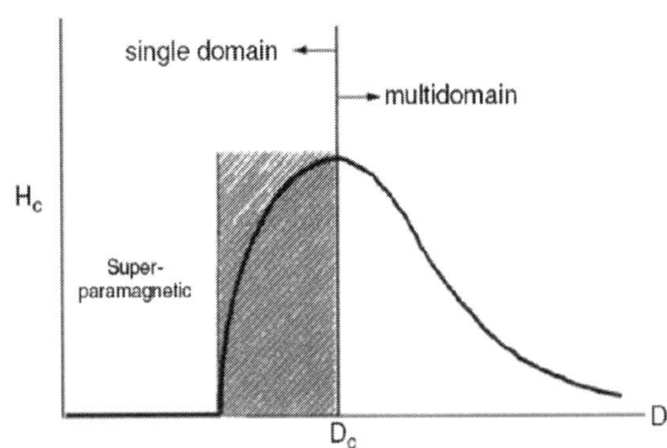

Fig. 3: The relationship between the coercivity in ultrafine particle systems and particle sizes.

are affected by thermal fluctuations, and such a single-domain particle is usually called a superparamagnetic particle, which will be discussed in later section.

Frenkel and Dorfman (1930) theoretically predicted the existence of single-domain particles. The Dc values for some typical magnetic materials of spherical shape are listed in Table 2. It should be noted that particles with significant shape anisotropy usually have a larger effective critical single-domain diameter than corresponding spherical particles

Table 2. Critical single-domain sizes, Dc, for spherical particles with no shape anisotropy

Material	Dc (nm)
Co	70
Fe	14
Ni	55
Fe_3O_4	128
g-Fe_2O_3	166

Time Dependence of Magnetization

The time dependence of magnetization of a material is important for its engineering applications and for investigating the fundamental mechanisms of magnetism. The variation of magnetization with time of a magnetic material can be described by:

$$\frac{dM(t)}{dt} = -\frac{M(t) - M(t=\infty)}{\tau} M(t=\infty)\tau \quad (5)$$

Where M (t = ") is the magnetization at the equilibrium state, and is a characteristic relaxation time given by:

$$\tau = \frac{1}{f_0} exp\left(\frac{\Delta E}{kT}\right) \quad (6)$$

For a uniaxial anisotropy, the energy barrier, ΔE, is equal to the product of the anisotropy constant and the volume. In most cases, f_0 is often taken as a constant of value 10^9 s" 1. As the behavior of is dominated by the exponential argument, the accurate value of is usually not necessary. However, the particle size greatly affects the relaxation time. If we choose typical values = 10^9 s"1, K= 10^6 erg/cm^3, and T = 300 K, the relaxation time of a particle with diameter of 11.4 nm is 0.1 s, while the relaxation time of a particle with diameter of 14.6 nm is 10^8s.

If all components of a system have the same relaxation time, Equation (5) offers the simplest solution. However this assumption is not applicable to real systems because of the distribution of energy barriers in real systems. The energy barrier distribution may be related to the variations of a lot of parameters, such as part icle siz es, anisotropies or compositional inhomogeneity, and the distribution of energy barriers causes a distribution of relaxation times. If the distribution of energy barriers is constant, the magnetization decays logarithmically:

$$M(t) = M(t=0) - S\ln(t) \tag{7}$$

where the magnetic viscosity, S, is related to the energy barrier distribution. If the distribution of energy barriers is constant, deviations from the behavior can be observed. To keep Equation (7) applicable, the magnetic viscosity, should be accordingly modified.

Superparamagnetism

Néel theoretically demonstrated that H_C approaches zero when particles become very small because the thermal fluctuations of very small particles prevent the existence of a stable magnetization. This is a typical phenomenon of superparamagnetism. There are two experimental criteria for super paramagnetism. First, the magnetization curve exhibits no hysteresis, and seconds the magnetization curves at different temperatures must superpose in a plot of. Figure 4 shows the magnetization curves of iron amalgam on bases. Measurements were made at 77K and 200K respectively, and the magnetization curves at 77K and 200K superpose each other. The imperfect superposition may be due to a broad distribution of particle sizes, changes in the spontaneous magnetization of the particle as a function of temperature or anisotropy effects.

$$\tau = \tau_0 \exp\left(\frac{\Delta E}{k_B T}\right) \qquad \Delta E = KV$$

V

ΔE

$k_B T$

The basic mechanism of superparamagnetism is based on the relaxation time of the net magnetization of a magnetic particle:

$$M \ vs \ H/T \tag{8}$$

where is the energy barrier to moment reversal, and is the thermal energy. For non-interacting particles the pre-exponential factor is in the order of 10^{-10} 10^{-12} s and only weakly dependent on temperature. The energy barrier has several origins, including both intrinsic and extrinsic effects such as the magnetocrystalline and shape anisotropies, respectively. However, in the simplest cases, it is given by, where is the anisotropy energy density and is the particle volume. For small particles, is comparable to at room temperature, so superparamagnetism is important for small particles.

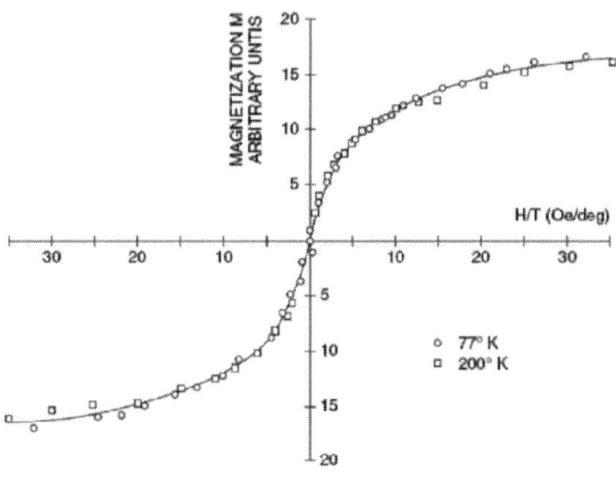

Fig. 4: Magnetization curves on H/T bases, as a demonstration of the superparamegnetism of iron amalgam.

It should be noted that, for a given material, the observation of superparamagnetism is dependent not only on temperature, but also on the measurement time of the experimental technique used. As shown in Figure 5, if, the flipping is fast relative to the experimental time window and the part icles appear t o be paramagnetic; while if , the flipping is slow, and such a state is called a blocked state. In a block state, the quasi- static properties of the material can be observed. The blocking temperature can be obtained by assuming . In typical experiments, the measurement time can range from the slow timescale of 10^2s for DC magnetization, and medium timescale of $10^{-1} – 10^{-5}$ s for AC susceptibility, through to the fast timescale of $10^{-7}–10^{-9}$ s for ^{57}Fe Mössbauer spectroscopy.

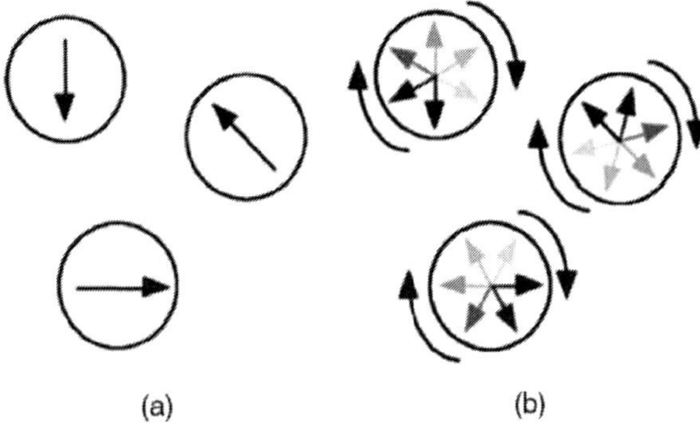

Fig. 5: Observation of superparamagnetism The circles depict three magnetic nanoparticles, and the arrows represent the net magnetization direction in these particles. In case (a), at temperatures well below the measurement-technique dependent blocking temperature of the particles, or for relaxation times (the time between moment reversals) much longer than the characteristic measurement time , the net moments are quasi-static. In case (b), at temperatures well above, or for much shorter than , the moment reversals are so rapid that in zero external field the time-averaged net moment on the particles is zero.

Optical Properties

The reduction of materials' dimension has pronounced effects on the optical properties. The size dependence can be generally classified into two groups. One is due to the increased energy level spacing as the system becomes more confined, and the other is related to surface plasmon resonance.

Surface plasmon resonance

Surface plasmon resonance is the coherent excitation of all the " free" electrons within the conduction band, leading to an in-phase oscillation. When the size of a metal nanocrystals is smaller than the wavelength of incident radiation, a surface plasmon resonance is generated and Fig. 6 shows schematically how a surface plasmon oscillation of a metallic particle is created in a simple manner. The electric field of an incoming light induces a polarization of the free electrons relative to the cationic lattice. The net charge difference occurs at the nanoparticles boundaries (the surface), which in turn acts as a restoring force. In this manner a dipolar oscillation of electrons is created with a certain frequency. The surface plasmon resonance is a dipolar excitation of the entire particle between the negatively charged free electrons and its positively charged lattice. The energy of the surface plasmon resonance depends on both the free electron density and the dielectric medium surrounding the nanoparticle.

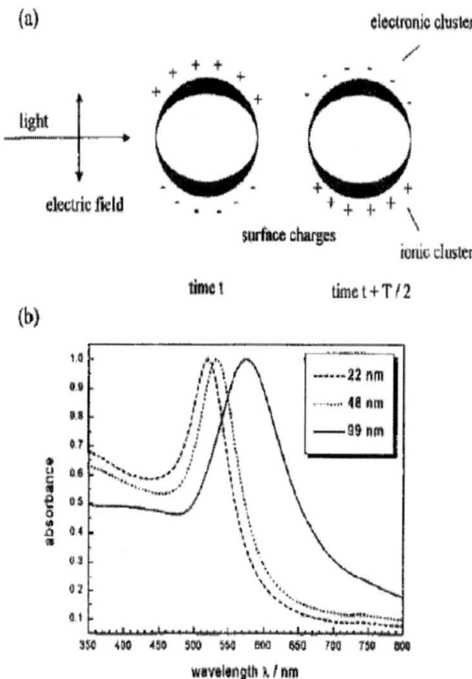

Fig. 6: Surface plasmon absorption of spherical nanoparticles and its size dependence. **(a)** A schematic illustrating the excitation of the dipole surface plasmon oscillation. The electric field of an incoming light wave induces a polarization of the (free) conduction electrons with respect to the much heavier ionic core of a spherical metal nanoparticle. A net charge difference is only felt at the nanoparticle surfaces, which in turn acts as a restoring force. In this way a dipolar oscillation of the electrons is created with period T. **(b)** Optical absorption spectra of 22, 48 and 99nm spherical gold nanoparticles. The broad absorption band corresponds to the surface plasmon resonance.

The width of the resonance varies with the characteristic time before electron scattering. For larger nanoparticle, the resonance sharpens as the scattering length increases. Noble metals have the resonance frequency in the visible light range.

Quantum size effects

Unique optical property of nanomaterials may also arise from another quantum size effect. When the size of a nanocrystal (i.e. a single crystal nanoparticle) is smaller than the de Broglie wavelength, electrons and holes are spatially confined and electric dipoles are formed, and discrete electronic energy level would be formed in all materials. Similar to a particle in a box, the energy separation between adjacent levels increases with decreasing dimensions. Figure 7 schematically illustrates such discrete electronic configurations in nanocrystals, nanowires and thin films; the electronic configurations of nanomaterials are significantly different from that of their bulk counterpart. These changes arise through systematic transformations in the density of electronic energy levels as a function of the size, and these changes result in strong variations in the optical and electrical properties with size. Nanocrystals lie in between the atomic and molecular limit of discrete density of electronic states and the extended crystalline limit of continuous band. In any material, there will be a size below which there is substantial variation of fundamental electrical and optical properties with size, when energy level spacing exceeds the temperature. For a given temperature, this occurs at a very large size (in nanometers) in semiconductors as compared with metals and insulators. In the case of metals, where the Fermi level lies in the center of a band and the relevant energy level spacing is very small, the electronic and optical properties more closely resemble those of continuum, even in relatively small sizes (tens or hundreds of atoms). In semiconductors, the Fermi level lies between two bands, so that the edges of the bands are dominating the low-energy optical and electrical behavior. Optical excitations across the gap depend strongly on the size, even for crystallites as large as 10,000 atoms. For insulators, the band gap between two bands is already too big in the bulk form.

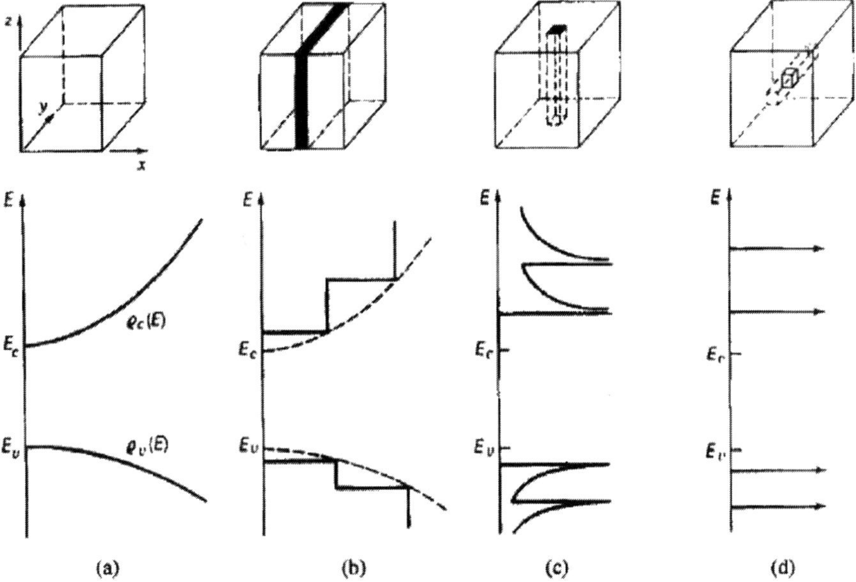

Fig. 7: Schematic illustrating discrete electronic configurations in nanocrystals, nanowires and thin films and enlarged band gap between valence band and conduction band.

The quantum size effect is most pronounced for semiconductor nanoparticles, where the band gap increases with a decreasing size, resulting in the interband transition shifting to higher frequencies. In a semiconductor, the energy separation, i.e. the energy difference between the completely filled valence band and the empty conduction band is of the order of a few electrovolts and increases rapidly with a decreasing size. It is known that both the absorption edge and the luminescence peak position shift to a higher energy as the particle size reduces in the optical absorption and luminescence spectra of the InP nanocrystals. This size dependence of absorption peak has been widely used in determining the size of nanocrystals.

When the diameter of nanowires or nanorods reduces below the de Broglie wavelength, size confinement would also play an important role in determining the energy level just as for nanocrystals. For example, the absorption edge of Si nanowires has a significant blue shift with sharp, discrete features and silicon nanowires also have shown relatively strong "band-edge" photoluminescence.

Measurement method of optical properties of nanoparticles

PL spectroscopy concerns monitoring the light emitted from atoms or molecules after they have absorbed photons. It is suitable for materials that exhibit photoluminescence, PL spectroscopy is suitable for the characterization of both organic and inorganic materials

Electromagnetic radiation in the UV and visible ranges is utilized in PL spectroscopy. The sample's PL emission properties are characterized by four parameters: intensity, emission wavelength, bandwidth of the emission peak, and the emission stability. The PL properties of a material can change in different ambient environments, or in the presence of other molecules. Many nanotechnology-enabled sensors are based on monitoring such changes. Furthermore, as dimensions are reduced to the nanoscale, PL emission properties can change, in particular a size dependent shift in the emission wavelength can be observed. Additionally, because the released photon corresponds to the energy difference between the state, PL spectroscopy can be utilized to study material properties such as band gap, recombination mechanisms, and impurity levels.

In a typical PL spectroscopy setup for liquid samples is shown in Figure 8, a solution containing the sample is placed in a quartz cuvette with a known path length. Double beam optics are generally employed. The first beam passes through an excitation filter or monochromator, then through the sample and onto a detector. This impinging light causes photoluminescence, which is emitted in all directions. A small portion of the emitted light arrives at the detector after passing through an optional emission filter or monochromator. A second reference beam is attenuated and compared with the beam from the sample. Solid samples can also be analyzed, with the incident beam impinging on the material (thin film, powder etc.). Generally an emission spectrum is recorded, where the sample is irradiated with a single wavelength and the intensity of the luminescence emission is recorded as a function of wavelength. The fluorescence of a sample can also be monitored as a function of time, after excitation by a flash of light. This technique is called time resolved fluorescence spectroscopy.

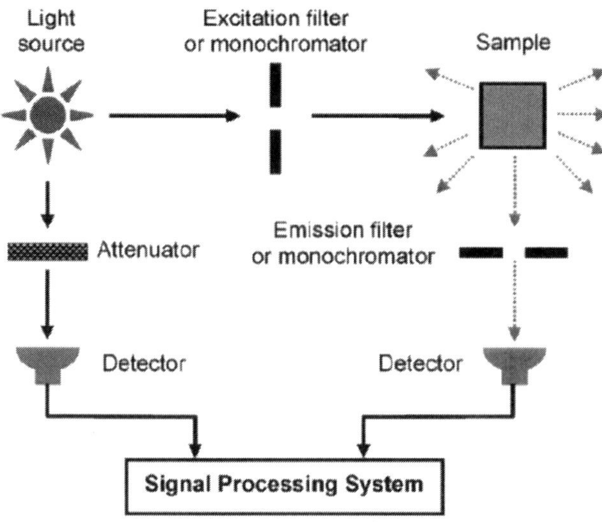

Fig. 8: A typical fluorescence spectrophotometer instrument setup. of virtually any size, and the samples can be in solid, liquid, or gaseous forms.

Nanomaterials with PL effects, in particular nanocrystals, can reveal many interesting and improved optical properties. These include brighter emission, narrower emission band, and broad UV absorption. For example, semiconductor nanocrystals produce narrower emission peaks than luminescent organic molecules, with bandwidth of around 30-40 nm. Having a smaller bandwidth, it is much easier to discriminate individual wavelengths emanating from multiple sources, such as in an array of nanocrystals. The PL emission intensities and wavelengths are dependent on particle size. Hence, PL spectroscopy directly enables particle size effects, in particular those in the nanoscale, to be observed and quantified.

A. Electrical Properties

The electrical conductivity is a material- dependent property which for metallic conductors is independent of the applied voltage or the flowing electrical current. In contrast, for semiconductor or insulators the conductivity usually increases with increasing applied voltage. When reducing the geometric dimensions of a wire to nanometer or molecular dimensions. Ohms law is no longer valid in any case. Rather, the strictly linear relationship between current and voltage is replaced by a nonlinear, non-ohmic characteristic. In order to understand these phenomena, it is necessary first to consider the mechanism of electrical conductivity, the conventional macroscopic case.

$V = I R = I (1 / G); \quad G = I / V \alpha 1$

Where V= Applied voltage; I= electric current, R= Resistance; G= Electrical conductance

Note : G depends on geometric parameters (length & cross section). But ó (electrical conductivity) is material dependent property and it is independent of the applied voltage or the flowing electric current (for metal). For semiconductor ó usually increases with increasing the applied voltage.

Conduction Mechanism of Bulk

Let us consider an electrical conductor such as metallic wire is concerned to an electrical circuit and electrons start to move, driven by the electrical field. Within the wire there are a huge number of electrons moves slowly from one end of the wire to the other end. In this way, the electrons experience scattering processes that lead to a change in the momentum by interactions with electrons, phonons, impurities or other imperfections of the lattice, which are responsible for the electrical losses. In metallic wires, electrical conductivity is characterized by mean free path of the electrons. In an electrical field the electrons exhibit a type of "drift movement", and such a process of electrical conductivity is termed " diffusive conductance". Reducing the size of the conducting wire changes the mechanism of electrical conductivity. Hence when the geometric dimensions reach the mean free path length of the electrons, the mechanism of conduction changes from a diffusive to a "ballistic".

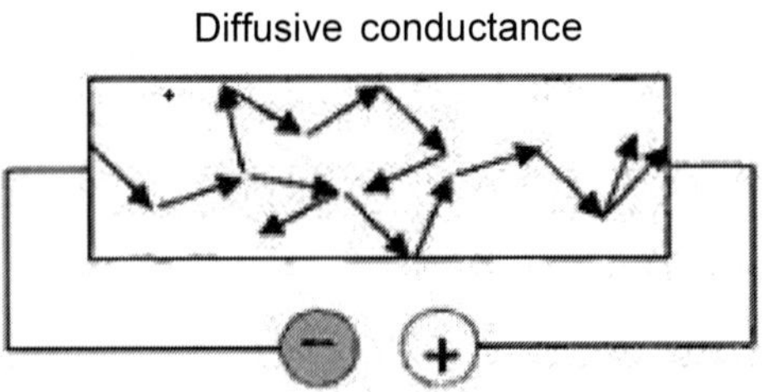

Fig 9: Diffusive electric conductance, as observed in conventional metallic conductors. Diffusive conductance is characterised by a scattering of free electrons in the conductor. The electrical current is transported by a slow drift movement of the electrons.

Conduction Mechanism of Nanoparticles

In the ballistic conduction the scattering phenomena are no longer observed, classically zero resistivity is expected, but this is not observed because now quantum mechanical phenomena are occurring. In order to understand this ballistic conductivity Eq: 1 must be rewritten in a form which takes into account the transport of electricity by electrons. That is the electrical current I transports within a time interval "t the charge Q.

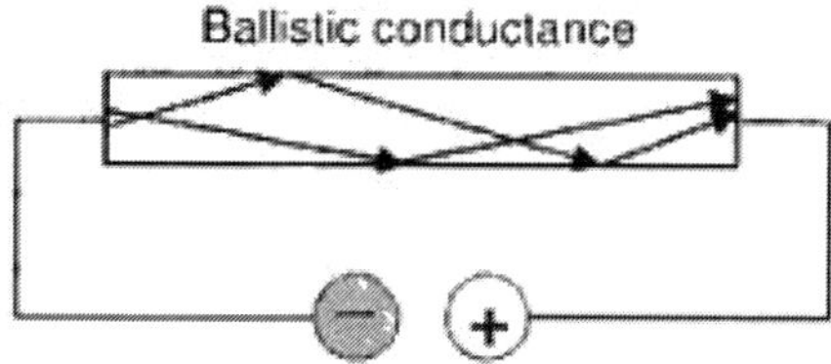

Fig 10: Ballistic conductivity of an electrical current in a small electrical conductor. Ballistic conductivity is not characterised by scattering of the free electrons in the lattice, as the geometric dimensions of the conductor are smaller than the mean free path length of the electrons.

Finally we get $G = (Ne^2/h) \times (\lambda/L)$

Here $(L/\lambda) = n$ is the electron wave mode number. Each electron wave mode can have two modes (spin up and spin down) leading to $N = 2n$; therefore, one finally obtains for the conductance of a short, thin wire with one moder. $G = 2e^2 /h$. Assuming m active modes in a wire, the conductance is

$$G = 2e^2 /h.$$

Pesticide Optimum Parameters

Regression equation results From above discussion, there are no longer any variables depending on the material or the geometry of the wire. It is clear that the electrical conductance of a small, thin wire increases with the increment $G_O = 2e / h = 7.72 \times 10^{-5}$ S. Hence the conductance decreases with increasing voltage.

Electrical Parameters

Nanoparticles array shows environm ent– dependent electrical properties (conductivity). These properties are modified by the chemical species present in its vicinity. The Conductivity of nanoparticles is believed to occur due to:

1. Tunneling of electrons through the metal core.
2. Hopping of the electrons along the atoms constituting the chain of the legend molecule encapsulating the nanoparticle.

By changing the parameters of the nanoparticle such as its particle diameter, space between the particles and the number of layers, the conductivity of the system can be altered. The analyte can be made to interfere with any one of the processes and hence can help vary the conductivity. This could lead to a sensing of the analyte.

Application

In recent years, several experimental groups have reported measurements of the current-voltage (I-V) characteristics of individual or small numbers of molecules. Even three-terminal measurements showing evidence of transistor action has been reported using carbon nanotubes as well as self-assembled monolayer of conjugated polymers. These developments have attracted much attention from the semiconductor industry that are actively looking for ways to progress from gigabit to terabit integration by complementing or even replacing present day CMOS circuitry. There is great interest therefore from an applied point of view to model and understand the capabilities of molecular conductors.

Let us consider for an example CNT. The unusual properties of carbon nanotubes make many applications ranging from battery electrodes, to electronic devices, to reinforcing fibres, which make stronger composites.

8

Characterization of Nano-materials using PSA and Raman Spectroscopy

K. Gunasekaran and S. Thirunavukkarasu

Raman Spectroscopy

It is a spectroscopic technique based on inelastic scattering of monochromatic light, usually from a laser source. Inelastic scattering means that the frequency of photons in monochromatic light changes upon interaction with a sample. Photons of the laser light are absorbed by the sample and then reemitted. Frequency of the reemitted photons is shifted up or down in comparison with original monochromatic frequency, which is called the Raman effect. This shift provides information about vibrational, rotational and other low frequency transitions in molecules. Raman spectroscopy can be used to study solid, liquid and gaseous samples.

A Raman system typically consists of four major components:

1. Excitation source (Laser).
2. Sample illumination system and light collection optics.
3. Wavelength selector (Filter or Spectrophotometer).
4. Detector (Photodiode array, CCD or PMT).

A sample is normally illuminated with a laser beam in the ultraviolet (UV), visible (Vis) or near infrared (NIR) range. Scattered light is collected with a lens and is sent through interference filter or spectrophotometer to obtain Raman spectrum of a sample.

Since spontaneous Raman scattering is very weak the main difficulty of Raman spectroscopy is separating it from the intense Rayleigh scattering. More precisely, the major problem here is not the Rayleigh scattering itself, but the fact that the intensity of stray light from the Rayleigh scattering may greatly exceed the intensity of the useful Raman signal in the close proximity to the laser wavelength. In many cases the problem is resolved by simply cutting off the spectral range close to the laser line where the stray light has the most prominent effect. People use commercially available interference (notch) filters which cut-off spectral range of ± 80-120 cm^{-1} from the laser line. This method is efficient in stray light elimination but it does not allow detection of low-frequency Raman modes in the range below 100 cm^{-1}.

Stray light is generated in the spectrometer mainly upon light dispersion on gratings and strongly depends on grating quality. Raman spectrometers typically use holographic gratings which normally have much less manufacturing defects in their structure then the ruled once. Stray light produced by holographic gratings is about an order of magnitude less intense then from ruled gratings of the same groove density. Using multiple dispersion stages is another way of stray light reduction. Double and triple spectrometers allow taking Raman spectra without use of notch filters. In such systems Raman-active modes with frequencies as low as 3-5 cm^{-1} can be efficiently detected.

In earlier times people primarily used single-point detectors such as photon-counting Photomultiplier Tubes (PMT). However, a single Raman spectrum obtained with a PMT detector in wavenumber scanning mode was taking substantial period of time, slowing down any research or industrial activity based on Raman analytical technique. Nowadays, more and more often researchers use multi-channel detectors like Photodiode Arrays (PDA) or, more commonly, a Charge-Coupled Devices (CCD) to detect the Raman cattered light. Sensitivity and performance of modern CCD detectors are rapidly improving. In many cases CCD is becoming the detector of choice for Raman spectroscopy.

Particle Size Analyzer (PSA)

Type of analysis can be done using particle size analyzer

Particle size distribution, Zeta potential and Molecular weight measurement

Type of sample

Liquid sample only used

Particle Size Analyzer

Principle

In a sample, particles are moving in a Brownian motion, when laser light comes into the sample, the light is scattered by the particles when the phases of scattered light by each particles are the same, the intensity of detected scattered light become strong, when the phases are opposite they cancel each other then the scattered light intensity fluctuates. Particle size analyzer is used to characterize the size distribution of particles by laser diffraction method in our sample of interest.

Particle Size Analyzer study

The particle size distribution of the sample was obtained by laser diffraction method using particle size analyzer (Horiba, SZ-100). In the particle size analyzer, particle size measured by photon correlation method in the range of 0.30 nm to 8.00 μm and a data acquisition time is less than two minutes. The cell holder temperature can be controlled from 1 to 90 °C whereas for electrode and plastic cell it was 1 to 70 °C. The laser light goes through ND-filter and lenses to be at optical strength when input to the sample particles in the cell. For particle size measurement, the light scattered at 90 or 173 degree angle by the sample particle is collected as pulses is sent from CPU board to personal computer by USB connection and calculated.

Components

Laser light, Photomultiplier tube, Detector, sample holder and Temperature controller

Zeta Potential

Principle

For surface charged particles, the surface potential from the real surface of the particles to the bulk is divided in different areas called the electric double layer. The surface potential at the boundary of two of these areas, the so- called shear plane, is the zeta potential. Zeta potential depends on interaction between particle and colloidal dispersion system. So, zeta potential depends on sample's dispersion stability. To get the zeta potential the most direct way is an electrophoretic light scattering (ELS) measurement. By applying an alternative potential to the solution an oscillating movement is induced to the particles. The frequency of the scattered light by the particles is then shifted compared to the fundamental beam. This is well known Doppler Effect. This technique is known among the laser - Doppler velocimetry techniques. It detects this signal and reference light interference, from measured frequency variation, electric voltage charge of particles can be get.

Zeta potential study

Zeta potential of the sample measured using particle size analyzer (Horiba, SZ-100). In the particle size analyzer, zeta potential measurement varied from -200 mV to 200 mV.

As a data acquisition time, usually less than one minutes for zeta potential measurement. In the zeta analyzer, laser light is divided into two beams as input light and reference light. Scattered light by sample particles and reference light modulated by the modulator interfere in the prism and are detected. The detected signals are changed into digital signal to be calculated.

Dispersing agent

Sodium hexametaphosphate used as dispersing agent to break the agglomeration in our sample. Before going to characterize the sample in particle size analyzer, sample should be dispersed in dispersing agents.

Refractive index

Before characterizing the sample in particle size analyzer, we should know the refractive index of our sample of interest.

Type of cuvette used

i. Deionized water – plastic cuvette or use and through cuvette

ii. Solvents – use quartz glass cuvette

Volume of sample in cuvette

One centimeter of sample should be filled in cuvette

References

Hare, C.R. 1972. Visible and Ultraviolet Specroscopy In: Guide to modern methods of Instrumental analysis. (Gouw, T.H. ed). Wiley Interscience, New York. p.161-194.

Musakin, A, Khrapovsky, A. Shaikvid, S. and Efros. S. 1975. Problems in qualitative analysis. Mir Publications, Moscow. p.170-176.

Meits, L. and Thomas, H.C. 1958. Advanced analytical chemistry. McGraw-Hill Book Company Inc., New York. p. 231-303.

9

X-ray Diffraction Spectroscopy

K. Srinivasan

Wilhelm Conrad Roentgen, German physicist, born on November 8, 1895 discovered X-rays and was awarded the first ever Nobel Prize in Physics in 1901 in recognition of the extraordinary services he has rendered by the discovery of the remarkable rays subsequently named after him. These short-wave electromagnetic radiations have the wavelengths that are matching with both the atomic sizes and shortest interatomic distances. But, the index of refraction of X-rays is near unity for all materials and hence they could not be focused by a lens in order to observe small objects such as atoms, as it is done by glass lenses in a visible light microscope or by magnetic lenses in an electron microscope. Thus, in general, X-rays cannot be used to image individual atoms directly. As an alternative to this difficulty, Max von Laue in 1912 proved that the periodicity of the crystal lattice allows atoms in a crystal to be observed with exceptionally high resolution and precision by means of X-ray diffraction for the first time by using a single crystal of hydrated copper sulfate ($CuSO_4$ $5H_2O$). Particles in motion, such as neutrons and electrons, may also be used as an alternative to X-rays however these direct imaging methods require sophisticated equipments and their accuracy in determining atomic positions is substantially lower than that possible by means of diffraction techniques. Hence, direct visualization of a structure with atomic resolution is invaluable in certain applications, but the three-dimensional crystal structures are determined exclusively from diffraction data. Hence immediately after their discovery, X-rays were put to use to study the internal structure of objects that are opaque to visible light but transparent to X-rays, for example, parts of a human body using radiography, which takes advantage of varying absorption: bones absorb X-rays stronger than surrounding tissues.

Nature of X-rays

Electromagnetic radiation is generated every time when electric charge accelerates or decelerates. It consists of transverse waves where electric (E) and magnetic (H) vectors are perpendicular to one another and to the propagation vector of the wave (k). The X-rays have wavelengths from ~0.1 to ~100 Å, which are located between λ -radiation and ultraviolet rays. The wavelengths, most commonly used in crystallography, range between ~0.5 and ~2.5 Å since they are of the same order of magnitude as the shortest interatomic distances observed in both organic and inorganic materials.

Production of X-rays

The X-rays are usually generated using two different methods or sources. The first is a device, which is called an X-ray tube, where electromagnetic waves are generated from impacts of high-energy electrons with a metal target. These conventional X-ray sources usually have a low efficiency, and their brightness is fundamentally limited by the thermal properties of the target material. The latter must be continuously cooled because nearly all kinetic energy of the accelerated electrons is converted into heat when they decelerate rapidly (and sometimes instantly) during the impacts with a metal target. The second is a much more advanced source of X-ray radiation – the synchrotron, where high energy electrons are confined in a storage ring. When they move in a circular orbit, electrons accelerate towards the center of the ring, thus emitting electromagnetic radiation. The synchrotron sources are extremely bright since thermal losses are minimized, and there is no target to cool. Their brightness is only limited by the flux of electrons in the high energy beam.

Detection of X-rays

The detector is an integral part of any diffraction analysis system, and its major role is to measure the intensity and, sometimes, the direction of the scattered beam. The detection is based on the ability of X-rays to interact with matter and to produce certain effects or signals, for example, to generate particles, waves, electrical current, etc., which can be easily registered. In other words, each photon entering the detector generates a specific event, better yet, a series of events that can be recognized, and from which the total photon count (intensity) can be determined. Obviously, the detector must be sensitive to X-rays and should have an extended dynamic range and low background noise. An important characteristic of any detector is how efficiently it collects X-ray photons and then converts them into a measurable signal. Detector efficiency is determined by first, a fraction of X-ray photons that pass through the detector window (the higher, the better) and second, a fraction of photons that are absorbed by the detector and thus result in a series of detectable events (again, the higher, the better). The product of the two fractions, which is known as the absorption or quantum efficiency, should usually be between 0.5 and 1. The linearity of the detector is critical in obtaining correct intensity measurements (photon count). The detector is considered linear when there is a linear dependence between the photon flux (the number of photons entering through the detector window in one second) and the rate of signals generated by the detector (usually the number of voltage pulses) per second. In any detector, it takes some time to absorb a photon, convert it into a voltage pulse, register the pulse, and reset the detector to the initial state, that is, make it ready for the next operation. This time is usually known as the dead time of the detector – the time during which the detector remains inactive after it has just registered a photon. The presence of the dead time always decreases the registered intensity. This effect, however, becomes substantial only at high photon fluxes. When the detector is incapable of counting every photon due to the dead time, some of them could be absorbed by the detector but remain unaccounted, that is, become lost photons. It is said that the detector becomes nonlinear under these conditions. Thus the linearity of the detector can be expressed as: (i) the maximum flux in photons per second that can be reliably counted (the higher the better); (ii) the dead time (the shorter, the better), or (iii) the percentage of the loss of linearity at certain high photon flux (the lower percentage, the better). The three most commonly utilized types of X-ray detectors today are gas proportional, scintillation, and solid state detectors, all of which are true counters. Conventional gas proportional, scintillation, and solid-state detectors do not support spatial resolution and therefore, they are also known as point detectors. A point detector registers only the intensity of the diffracted beam, one point at a time.

Fundamentals of diffraction

When X-rays propagate through a substance, the following processes occur in the phenomenon of diffraction: Coherent scattering produces beams with the same wavelength as the incident (primary) beam. In other words, the energy of the photons in a coherently scattered beam remains unchanged when compared to that in the primary beam. Incoherent (or Compton) scattering in which the wavelength of the scattered beam increases due to partial loss of photon energy in collisions with core electrons. Incoherent scattering is not essential when the interaction of X-rays with crystal lattices is of concern, and it is generally neglected. When absorption becomes significant, it is usually taken into account as a separate effect. Absorption of the X-rays, in which some photons are dissipated in random directions due to scattering, and some photons lose their energy by ejecting electron(s) from an atom (i.e., ionization) and/or due to the photoelectric effect (i.e., X-ray fluorescence).

Diffraction can be observed only when the wavelength is of the same order of magnitude as the repetitive distance between the scattering objects. Thus, for crystals, the wavelength should be in the same range as the shortest interatomic distances, that is, somewhere between ~0.5 and ~2.5 Å. This condition is fulfilled when using electromagnetic radiation, which within the mentioned range of wavelengths, are X-rays. It is important to note that X-rays scatter from electrons, so that the active scattering centers are not the nuclei, but the electrons, or more precisely the electron density, periodically distributed in the crystal lattice.

Bragg law

There are two geometrical facts that are worth remembering during diffraction; the first one is that the incident beam, the normal to the reflecting plane, and the diffracted beam are always coplanar and the second one is that the angle between the diffracted beam and the transmitted beam is always 2q. This is known as the diffraction angle, and it is this angle, rather than q, which is usually measured experimentally. Diffraction in general occurs only when the wavelength of the wave motion is of the same order of magnitude as the repeat distance between scattering centers. This requirement follows from the Bragg law. Since sinq cannot exceed unity, we may write

$$[n\lambda/2\theta] = \text{sinq} < 1$$

Therefore, në must be less than 2d'. For diffraction, the smallest value of n is 1 (n corresponds to the beam diffracted in the same direction as the transmitted beam). Therefore the condition for diffraction at any observable angle 2θ is

$$\lambda < 2d'$$

If λ is very small, then the diffraction angles are too small to be conveniently measured. Hence the Bragg law may be written in the form

$$\lambda = [2(d'/n)\sin\theta]$$

The coefficient of λ is unity and d = d'/n and hence the Bragg law attains its final form which is

$$\lambda = 2d\sin\theta$$

Diffraction methods

Diffraction can occur whenever the Bragg law, $\lambda = 2d \sin \theta$, is satisfied. This equation puts very stringent conditions on λ and θ for any given crystal. With monochromatic radiation, an arbitrary setting of a single crystal in a beam of x-rays will not in general produce any diffracted beams. Some way of satisfying the Bragg law must be devised, and this can be done by continuously varying either λ or q during the experiment. The ways in which these quantities could be varied led to three main diffraction methods. The first one is the Laue method in which λ is varied and θ is fixed. The second one is the Rotating crystal method and in this λ is fixed whereas, θ is varied in parts. Finally the third one is the Powder method and in this ë is kept fixed and q is varied.

The Laue method was the first diffraction method ever used, and it reproduces von Laue's original experiment. A beam of white radiation, the continuous spectrum from an x-ray tube, is allowed to fall on a fixed single crystal. The Bragg angle θ is therefore fixed for every set of planes in the crystal, and each set picks out and diffracts that particular wavelength which satisfies the Bragg law for the particular values of δ and θ involved. Each diffracted beam thus has a different wavelength. There are two variations of the Laue method, depending on the relative positions of source, crystal, and film. In each, the film is flat and placed perpendicular to the incident beam. The film in the transmission Laue method (the original Laue method) is placed behind the crystal so as to record the beams diffracted in the forward direction. This method is so called because the diffracted beams are partially transmitted through the crystal. In the back-reflection Laue method the film is placed between the crystal and the x-ray source, the incident beam passes through a hole in the film, and the beams diffracted in a backward direction are recorded. In either method, the diffracted beams form an array of spots on the film. The positions of the spots on the film, for both the transmission and the back-reflection method, depend on the orientation of the crystal relative to the incident beam, and the spots themselves become distorted and smeared out if the crystal has been bent or twisted in any way. These facts account for the two main uses of the Laue methods: the determination of crystal orientation and the assessment of crystal perfection.

In the rotating-crystal method, a single crystal is mounted with one of its axes, or some important crystallographic direction, normal to a monochromatic x-ray beam. A cylindrical film is placed around it and the crystal is rotated about the chosen direction, the axis of the film coinciding with the axis of rotation of the crystal. As the crystal rotates, a particular set of lattice planes will, for an instant, make the correct Bragg angle for reflection of the monochromatic incident beam, and at that instant a reflected beam will be formed. The reflected beams are again located on imaginary cones but now the cone axes coincide with the rotation axis. Since the crystal is rotated about only one axis, the Bragg angle does not take on all possible values between and 90 for every set of planes. The chief use of the rotating-crystal method and its variations is in the determination of unknown crystal structures.

In the powder method, the crystal to be examined is reduced to a very fine powder and placed in a beam of monochromatic X-rays. Each particle of the powder is a tiny crystal oriented at random with respect to the incident beam. Just by chance, some of the particles will be correctly oriented so that their (100) planes, for example, can reflect the incident beam. Other particles will be correctly oriented for (110) reflections, and so on. The result is that every set of lattice planes will be capable of reflection. The mass of powder is equivalent, in fact, to a single crystal rotated, not about one axis, but about all possible axes. The method is especially suited for determining lattice parameters with high precision and for the identification of phases, whether they occur alone or in mixtures such as poly phase alloys.

Directions of diffracted beams

Scattering by Electrons

The origin of the electromagnetic wave elastically scattered by the electron is due to the fact that the electrons are charged particles. Thus, an oscillating electric field from the incident wave exerts a force on the electric-charge (electron) forcing the electron to oscillate with the same frequency as the electric-field component of the electromagnetic wave. The oscillating electron accelerates and decelerates in concert with the varying amplitude of the electric field vector, and emits electromagnetic radiation, which spreads in all directions. In this respect, the elastically scattered X-ray beam is simply radiated by the oscillating electron; it has the same frequency and wavelength as the incident wave, and this type of scattering is also known as coherent scattering. The scattering of X-rays by a single electron yields an identical scattered intensity in every direction and when more than one point is affected by the same incident wave, the overall scattered amplitude is a result of interference among multiple spherical waves.

Scattering by atoms and atomic scattering factor

When an atom is considered instead of an electron, it is easy to see that no path difference is introduced between the waves for the forward scattered X-rays. Thus, intensity scattered in the direction of the propagation vector of the incident wave front is proportional to the total number of core electrons, Z, in the atom. For other angles, the presence of core electrons results in the introduction of a certain path difference, δ, between the individual waves in the resultant wave front. The amplitude of the scattered beam is therefore, a gradually decaying function of the scattered angle and this intrinsic angular dependence of the X-ray amplitude scattered by an atom is called the atomic scattering function (f).

Applications of X-ray diffraction

- Determination of grain size
- Determination of particle size
- Determination of crystal perfection
- Determination of depth of X-ray penetration
- Determination of crystal orientation
- Determination of degree of crystallinity and crystallite size of amorphous materials
- Determination of crystal structure
- Determination of lattice parameters
- Determination of phases in mixed samples
- Determination of residual stress

References

Advanced X-ray Techniques in Research and Industry, Ashok Kumar Singh, IOS Press, Amsterdam, The Netherlands, 2005.

Diffraction Analysis of the Microstructure of the Materials, E. J. Mittemeijer and P. Scardi, Springer-Verlag Berlin Heidelberg, 2004.

Elements of X-ray Diffraction, B. D. Cuilty, Addison - Wesley Publishing Company, Inc., 1956.

Industrial Applications of X-ray Diffraction, Frank H. Chung and Deane K. Smith, Marcel Dekker, Inc. 2000.

Powder Diffraction-Theory and Practice, R. E. Dinnebier and S. J. L. Billinge, The Royal Society of Chemistry, 2008.

Theory in X-ray Diffraction in Crystals-William H. Zachariasen, Dover Publications, Inc., 31 East 2nd Street, Mineola, N.Y. 11501, 1945.

X-ray Diffraction by Macromolecules, M. Kasai and M. Kakudo, Kodansha Ltd. and Springer-Verlag Berlin Heidelberg, 2005.

X-ray Diffraction in crystals, Imperfect crystals and Amorphous Bodies, A. Guinier, General Publishing Company, Ltd., 30 Lesmill Road, Don Mills, Toronto, Ontario, 1994.

X-ray Diffraction, B.E. Warren, Dover Publications, Inc., 31 East 2nd Street, Mineola, N.Y. 11501, 1994.

X-ray Diffraction-A Practical Approach, C. Suryanarayana and M. Grant Norton, Pleenum Press, New York, 1998.

10

Basics of Electron Microscopy

M.L. Sharma

Electron microscopy is a specialized branch of science, which uses the electron microscope as a tool. The electron microscope is an instrument which magnify smallest object thousands times. The magnification is not only requirement to see the smallest otherwise invisible object but the details of the object are most important. The details of the object are depending upon the resolving power of the microscope, which is known as the resolution of the microscope. This resolution or power of the microscope is depends upon the wavelength of the illuminating source. The smallest the wavelength of the illuminating sources is the best resolution of the microscope. Electron microscopes use electrons as an illumination source to take advantage of the very short wavelengths attainable in electron beam, which permit a considerable improvement in resolution over the light microscope.

The Electron Microscopy has changed our view of universe. It has extended our vision to the sub cellular, molecular and atomic level. It is continuously changing our concept of material and cellular structure. The modern concept is inconceivable without this in the field of chemistry, material sciences, industries, medicine, diagnostic techniques, agriculture, food technology, forensic sciences, and life sciences and so on. Hence, it become a indispensable tool of modern science but itself also a fascinating subject of study. The desire to resolve finer and finer detail led to the development of modern electron microscope. The scientists are able to get high- resolution images that are not available with any other techniques. The modern application of this field in Fisheries, Food

Technology, Packaging Technology, Agriculture Sciences, Veterinary Sciences, Forensic Sciences and Quality Analysis of industrial products of Metal Industry, Textile, Paper and Pulp, Ceramic and Semiconductor and so on opens the new channels of specialization.

Electron microscopes

Electron Microscopes are scientific instruments that use a beam of highly energetic electrons to examine objects on a very fine scale. The electron microscopic examination can yield the following information.

Topography : The surface features of an object or "how it looks", its texture; direct relation between these features and materials properties (hardness, reflectivity...etc.)

Morphology : The shape and size of the particles making up the object; direct relation between these structures and materials properties (ductility, strength, reactivity...etc.)

Composition : The elements and compounds that the object is composed of and the relative amounts of them; direct relationship between composition and materials properties (melting point, reactivity, hardness...etc.)

Crystallographic Information: How the atoms are arranged in the object; direct relation between these arrangements and materials properties (conductivity, electrical properties, strength...etc.)

An Electron Microscope is a complex instrument consists of different scientific systems integrated into one functional unit. The special requirements of electron beam behavior are taken into consideration to incorporate these systems. These systems are electron generating system, lens system, specimen manipulation system, vacuum system and recording system.

Where did Electron Microscopes come from? Electron Microscopes were developed due to the limitations of Light Microscopes, which are limited by the physics of light to 500x or 1000x magnification and a resolution of 0.2 micrometers. In the early 1930's this theoretical limit had been reached and there was a scientific desire to see the fine details of the interior structures of organic cells (nucleus, mitochondria...etc.). This required 10,000x plus magnification, which were just not possible using Light Microscopes.

The Transmission Electron Microscope (TEM) was the first type of Electron Microscope to be developed and is patterned exactly on the Light Transmission Microscope except that a focused beam of electrons is used instead of light to "see through" the specimen. It was developed by Max Knoll and Ernst Ruska in Germany in 1931.

The first Scanning Electron Microscope (SEM) debuted in 1942 with the first commercial instruments around 1965. Its late development was due to the electronics involved in "scanning" the beam of electrons across the sample.

How do electron microscopes work?

Electron Microscopes (EMs) function exactly as their optical counterparts except that they use a focused beam of electrons instead of light to "image" the specimen and gain information as to its structure and composition.

The basic steps involved in all EMs

A stream of electrons is formed (by the Electron Source) and accelerated toward the specimen using a positive electrical potential This stream is confined and focused using metal apertures and magnetic lenses into a thin, focused, monochromatic beam. This beam is focused onto the sample using a magnetic lens. Interactions occur inside the irradiated sample, affecting the electron beam. These interactions and effects are detected and transformed into an image. The above steps are carried out in all EMs regardless of type.

There are different types of microscopes for various applications. The microscopes are divided into two families; Transmission Electron Microscopes (TEM) and Scanning Electron Microscopes (SEM). Each family has various models depending upon the applications. Electron Microscopy is fast expanding field, which contribute the new models of the equipment based upon latest finding.

Transmission electron microscope (TEM)

A TEM works much like a slide projector. A projector shines a beam of light through (transmits) the slide, as the structures and objects on the slide affect the light passes through it. These effects result in only certain parts of the light beam being transmitted through certain parts of the slide. This transmitted beam is then projected onto the viewing screen, forming an enlarged image of the slide. TEMs work the same way except that they shine a beam of electrons (like the light) through the specimen (like the slide). Whatever part is transmitted is projected onto a phosphor screen for the user to see.

It is analogues to the optical microscope. It can achieve a resolution of 0.2 nanometer, thousand times better resolution, cannot be reached by the light microscope. The beam of electrons passes through the specimen and analyzes the internal structure of the specimen in the form of images. The electron has the poor penetrating capability and get absorbed in the thick specimen. Therefore the thickness of the specimen should not be more than few hundred Angstroms [one angtron = 10^{-10} m] However sometimes, slightlyA stream of electrons is formed (by the Electron Source) and accelerated toward the specimen using a positive electrical potential This stream is confined and focused using metal apertures and magnetic lenses into a thin, focused, monochromatic beam. This beam is thickens samples are used in High Voltage Electron Microscope.

The transmission electron microscope (TEM) is similar in function to an optical microscope. The structural differences between the two lies in the following aspects.; the fundamental function of both the microscope is to magnify minutes objects invisible to the naked eye. Basically, component terminology of an electron microscope is similar to that of an optical microscope. The light source in the OM is visible light from an incandescent lamp, in EM a beam of electrons emitted from a tungsten filament (cathode) by electrical heating. In the light microscope, the beam passes through different lens of glass, and the specimen. The lenses are condenser lens, objective lens and eyepiece. In the electron microscope, the electrons are drawn from a filament, because of a voltage difference between cathode and anode. The shaft of electrons beam drawn toward an anode and pass through an aperture. The beam traverses the aperture and next moves through an electromagnetic condenser, and objective lenses. The focused electron beam passed through a thin specimen loaded on a grid inserted in the path and manipulated by goneometer. The part of the beam absorbed, scattered and passes through it and projected by projector lens after corrected by intermediate lenses on the fluorescence screen. The image is observed with the help of optical binocular attached to its viewing window. The image is recorded on plate camera or 35mm camera or on computer. Telectron are not visible with naked eye. The differentially scattered electron image is projected into a fluorescent screen.

Although the TEM resembles a bright field light microscope in construction and operation, the interaction between specimen points and the illuminating beam in the two instruments is significantly different. In the bright field light microscope development of contrast in the image depends upon primarily on differential absorption of light by structure within the cells. In the TEM electron scattering rather than differences in absorbance produces contrast in the image. Scattering results from an interaction between specimen atoms and electrons of the illuminating beam. Atomic nuclei in the specimen, which are positively charged, scattered electron by attracting them from their paths in the illuminating beam. The negatively charged electron clouds around atomic nuclei scatter electrons by repelling them. These effects increase as electrons pass closer to specimen atoms. Nuclei of high atomic number, as in atom of heavy metals such as lead and uranium, cause a widest scattering.

The TEM can easily resolve structure such as ribosome, microtubules, microfilaments and large molecules such as proteins. Even images of individual heavy metal atom have been produced under the special operating conditions.

Resolution is ability to distinguish two closely placed points into two separate entities. The resolution is depends on the wavelength of the illuminating source. The resolving power of the optical microscope is limited by the wavelength of the illuminating beam; light beam. With the illuminating source and lenses available these days is somewhat below the 0.2u. However with electron microscope, the wavelength of the greatly accelerated electrons beam is a negligible factor in the ability to distinguish two points. Thus, the electron microscope in use today have a theoretical resolving power in the order of 0.05A. However, because of limitation of specimen thickness and electromagnetic lenses aberrations, most electron microscopes have a workable resolving power around 2 to 5A.

For a TEM operated at 50,000V, the wavelength of the electron beam is 0.005nm and gives an expected resolution of about 0.7 to 0.8 nm. These days commercially available TEMs regularly achieve resolution in this range. which is about far far better than the best resolution of the light microscope.

Resolving power and magnification are inter- related. Only magnified image of an object has no meaning until unless the every small particle is clear. The distinguishable magnification is useful otherwise a hazy image has no use. Hence useful magnification of a microscope is only, up to which it can resolve its minute structure.

Scanning electron microscope (SEM)

It is similar to the optical stereo-binocular microscope to observe the morphology and shape of the specimen. It differs extensively in its construction and operation from the TEM. Only the illumination source and condenser lenses are similar. An electron gun produces an electron beam in the SEM, which is focused into an intense spot on Interaction of Electron Beam with Specimen, the specimen surface by a magnetic lens system similar to the condenser lens of a TEM. The electron beam sweeps like television raster across the surface of the sample and the final image is built point by point. The beam deflectors, charged plates between the condenser lenses and the specimen, accomplish the scanning movement. The best SEM now has a resolution close to the standard TEM. The SEMs have proved extremely useful in studying the surface of the specimen. The electrons beam interacts with the surface of the sample and number of different signals are generated at the point of interaction. These signals are used for different types of analysis. These signals are trapped by detectors and transformed into data based upon the information generated from the specimen. This information provides sample's elemental composition, structural variation and morphology.

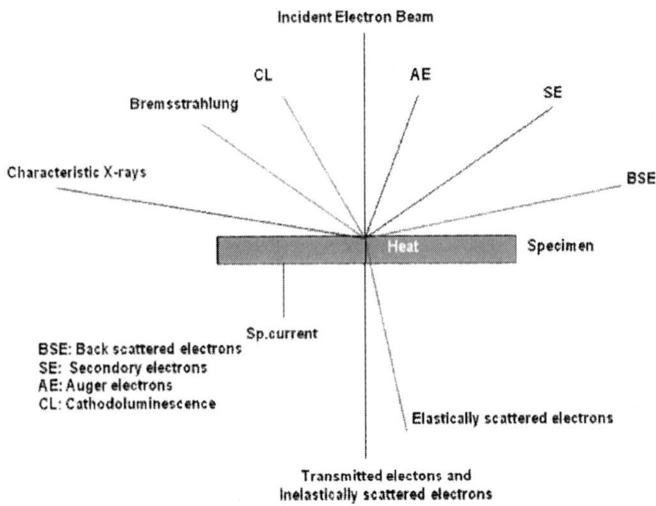

Fig. 1: Interaction of Electron Beam with Specimen

Analytical Electron Microscope [AEM] and Scanning Transmission Electron Microscope [STEM], these microscopes use bright sources and thin samples to permits the better resolution in microanalysis. The two signals most frequently used to provide analytical information are the x-rays generated in the sample and the electrons that loss energy in traversing the sample. These signals provide not only identification of the elements present but also information of the chemical state in which they are found.Although the information obtained by electron microscopes in generally is static, dynamic studies are possible through the use of special stages for cooling, heating, straining or electrical testing. It is even practical to place specimen in aqueous or gaseous environments, especially in the higher village microscopes.

The following phenomena take place on the interaction of electrons and specimen.

Transmitted electrons: are used in the conventional transmission electron microscope to reveal the ultra structural information about the morphology of a thin specimen. These are also used to determine the elemental composition with the help of electron energy loss spectrometer (EELS) because some electrons pass through the specimen with the loss of some energy. Other electrons pass through without any loss or no energy. Such elastically scattered electron may also pass through the specimen as transmitted electrons.

Backscattered electron: Those electrons, which are deflected back in the direction of the beam. The special detector in scanning and transmission electron microscope traps these signals. These are used to discriminate areas of different atomic numbered elements. Higher atomic numbered elements gives off more backscattered electrons and appear brighter than lower numbered elements. It has the resolution to the level of 1000 nm. These electrons has high energy.

Secondary Electrons: These electrons are also collected with a special type of detector used in SEM and TEM. They are used primarily to reveal topographical feature of a specimen. It has the resolving power <10 nm. These electrons have low energy.

Auger Electrons: These are special types of low energy electrons that carry the information about the chemical nature (atomic composition) of the specimen. These are generated from the upper layer of specimen. It is a powerful tool in the material sciences for studying the distribution of the lighter numbered atomic elements on the surface of the specimen. It has limited application in biological sciences. It is a specialized equipment known as scanning auger electron spectrometer.

Cathodoluminescent: This effect results when the energy of the impinging electrons in converted into visible light. Certain types of compounds are capable of cathode luminescence and detected by a special types of detector. The resolution is the similar to the light microscope.

Bremsstrahlung: Two important types of x-ray may be generated when the beam electron encounters the atoms of the specimen, continuous or bremssstrahlung x-ray and characteristic x-ray are generated when incoming, beam passing close to the atomic nucleus is slowed by the coulomb field of the nucleus with the release of x- ray energy. The intensity of x-ray energy released depends on how close the electron comes to the nucleus closer. The closer passes decelerate the electron more and yield higher energy x-rays. These are used to measure specimen mass thickness when quantitative analysis performed on thin sections. These are continuous x-rays also known as background or white radiation.

Characteristic X-rays: When high energy beam electrons interact with the shell electrons of the specimen atoms so that an inner shell electron is ejected. The removal of this electron temporarily lionizes the atom until an outer shell electron drops into the vacancy to stabilize the atom. Since this electron comes from a higher energy level, a certain amount of energy must be given off before it will be accommodated in the inner shell. The energy is released as an x-ray, the energy which equals the difference in energy between the two shells. Since this x-ray is of a discrete energy level, rather than a continuous, this event may be plotted as discrete peaks. Different elements will fill the vacancies in shells in unique ways. This means that since each element will generate a unique series of peaks, the spectrum may be used to identify the elements, such discrete x-rays are termed characteristic x-rays. The equipment for detection x-rays are energy disporsive x-ray (EDX) detector and Wavelength Disporsive X-ray (WDX) Detector.

The traditional electron microscopy is closely associated with the functions of observation and analytical techniques of S canning Tunnelling microscopy, Scanning lon Microscopy, Secondary Ion Microscopy, X-ray Microscopy and Photoelectron Microscopy. The recent development of video-enhanced light microscopy makes use of many electronic imaging techniques that are commonly used in electron microscopy. Computers are widely used for the operation of instruments and for the acquisition and analysis of images and spectroscopic data. The many branches of microscopy are merely branches of microscopy and merging into a common ground of microscopy.

Types of analysis

In the electron microscopy two types of analysis can be performed one is qualitative and another is quantitative. Qualitative-analysis incl udes imaging (Photography) of ultrastructure, enzyme cytochemistry, Autoradiography, Tracers and labeling technique and other miscellaneous localization- techniques. The comparison or identification with standard is also includes in qualit ative analysis besides the Morphological and structural studies.

Quantitative Analysis: It is performed by the electron microscope by using different detectors, and spectroscopes like x- ray probes, diffractions, stereo images form volumes size, EELS etc. Mostly qualitative analysis is done in material sciences specimens. as compared to life science specimen.

Specimen preparation

The various techniques are available for specimen preparation that improves the visibility and resolution of structures inside the cell. Although some of these methods do allow living material to be observed in freezing conditions, most are designed to kill and preserve cells in forms as close to the living state as possible.

Specimen preparation (conventional protocol)

The procedure starts with the excision of a sample from the whole organism. The time between removing tissue and fixing it should be as short as possible. Animal tissues is often more susceptible to observable artifacts induced by isolation trauma than is plant material. The post mortem changes occur in different organs at the different intervals of time. The most tedious tissue is brain, which is very soft and starts deteriorating as soon as the blood supply stopped. To avoid this damage there are certain techniques are available to avoid the damage at the ultra structure level. The perfusion technique is performed under ideal conditions. The incision should be clean and in one stroke with a sharp edged blades without any tearing of the tissue. The immediate transfer to the fixing fluid is also most important without drying in air. The handling of tissue is also critical due to its unpreserved nature. Due to some specific reason very small portion of the material is separated from the organ for processing. The size of the specimen should not be more than 1 mm cube.

The good preservation of ultra structures of cells is depends upon different factors like method of collection of material, types of Fixation and Fixatives, Concentration, Size of specimen, Buffer and pH, Duration, Temperature, Tonicity, Embedding and types of embedding and Handling of specimen. The most important part is the chemicals used and solution preparation for processing of the specimens for life sciences.

The samples of life sciences are in the form of tissue or partials, which are processed chemically in different way. The particulate material is processed either with centrifugation method or gel method. Tissue preparation for TEM can be divided into eleven major steps: primary fixation, washing, secondary fixation, dehydration, infiltration with transitional solvents, infiltration with resin, embedding, curing, sectioning, staining and observation. The process begins with living hydrated tissue and ends with tissue that is virtually water-free and preserved in a static state within a plastic resin matrix. The plastic resin mixture permeates the tissue, replacing all water within the cell and making the cell firm enough for sectioning to be cut. The following steps are involved in sample preparation.

Chemical preservation

Fixation : In preserving the sample to make it more realistic. Glutaraldehyde: for hardening and Osmium Tetraoxide : which stains lipids black.

Dehydration : Replacing water with organic solvents such as ethanol or acetone.

Embedding : Infiltration of the tissue with a resin such as araldite or epoxy for sectioning.

Sectioning : Produces thin slices of specimen, semitransparent to electrons. These can be cut on an ultramicrotome with a diamond knife to produce very thin slices. Glass knives are also used because they can be made in the lab and are much cheaper.

Staining : Uses heavy metals such as lead, uranium or tungsten to block electrons to give contrast between different structures, since many (especially biological)

Mechanical preservation

Cryofixation : freezing a specimen so rapidly, to liquid nitrogen or even liquid helium temperatures, that the water forms vitreous (non-crystalline) ice. This preserves the specimen in a snapshot of its solution state. An entire field called cryo-electron microscopy has branched from this technique. With the development of cryo-electron microscopy. it is now possible to observe virtually any biological specimen close to its native state.

Freeze-fracture or freeze-etch : The fresh tissue or cell suspension is frozen rapidly (cryofixed), then fractured by simply breaking The cold fractured surface (sometimes "etched" by increasing the temperature to about -100° C for several minutes to let some ice sublime) is then shadowed with platinum or gold at an average angle of 45° in a high vacuum evaporator. A second coat of carbon, evaporated normal to the average surface plane is often performed to improve stability of the replica coating. The specimen is returned to room temperature and pressure, then the extremely fragile "pre-shadowed" metal replica of the fracture surface is released from the underlying biological material by careful chemical digestion with acids, hypochlorite solution or SDS detergent. The still- floating replica is thoroughly washed from residual chemicals, carefully fished up on EM grids, dried then viewed in the

Ion Beam Milling: A Focused Ion Beam (IFB) milling, where gallium ions are used to produce an electron transparent membrane in a specific region of the sample, for example through a device within a microprocessor.

Conductive Coating: Thin film deposition

Evaporation: Thin-film deposition, or sputtering of carbon, gold, gold/ palladium, platinum or other conductive material to avoid charging of non conductive specimens in a scanning electron microscope.

General Animal Tissue Preparation Scheme for Electron Microscopy

Activity	Chemicals	Time involved
Primary Fixation	1m m 3 tissue fixed with 2 -4%	1-2 hr for soft tissue,m ore time to hard tissue
	Glutaraldehyde in 0.2M Sod, Cacodylate Buffer, pH 7.4 at Room Tempt or 4°C.	
Washing	Rinsing buffer 0.1m Sod. Cacodylate, p H 7.4 Add 7% sucrose.	1-12 hr
Secondary or Post Fixation	Osmium tetraoxide (1 -2%: buffered with Sod. Cacodylate	1-2h r
Washing	With rinsing buffer	1-3h rs several change
	30% ethanol or Acetone	5 m in ts
	50% ethanol or acetone	5-15 m in
	70% ethanol or Acetone	5-15 m in
	95% ethanol or Acetone 2 changes	5-15 m in
	Absolute ethanol 2 changes	20 m in each
Transitional Solvent	Propylene Oxide 3 changes	10 m in each
In filtration of resin	P O and resin mixture gradually increasing concentration of resin	Overnight
Embedding	Pure resin mixture (Spur's)	2-4 hr
Curing	At 60°-70°C	1-3 days

Resin: Spurr

	Component	Firm Mixture	Soft Mixture	Rapid Cure
Resin	VCD	10 .0g	1 0.0g	10.0 g
Flexibilizer	DE R 736	6.0 g	7 .0g	6.0g
Hardener	NS A	26 .0g	2 6.0g	26.0 g
Accelerator	DM AE	0.4 g	0 .4g	1.0g
Curing time at 70°C		8h r	8 hr	3hr

General classification of specimen for SEM examinations and processing steps involved

A wide variety of objects can be possible to examine under transmission electron microscope. The objects are roughly categories into non-biological and biological objects. The formers objects are come under material sciences and later come under life sciences. Ineach categories there is further classification on the basis of low atomic number specimen and high atomic number specimen. One can say the soft material and hard material. These objects are examined directly or indirectly. Directly observation is possible where the objects itself or part of it put directly in the path of electron beam The protocol of specimen preparation depends upon its softness or hardness.

Specimens	Nature			Steps Involved
Non -Biological	Wet			Selection Surface preparation Dehydration Mounting Coating
	Dry	Conductive		Selection Surface preparation Mounting
Biological dry (air dried)		Non conductive	Selection	
Biological hydrated				Surface preparation Mounting Coating
	Fresh			Selection Mounting
	Fixed	Mechanically fixed (frozen)	Uncoated	Selection Surface preparation Mounting Freezing
			Coated	Selection Surface preparation Mounting Freezing Dehydration Coating

1. **Non-Biological, Wet Specimen:** The samples like Asbestos in beverage, Clay type moisture containing samples, Wet clay and soil samples and Cement etc. are processed differently. The basic step is removing the water from the sample via a dehydration step. Hence the dehydration of any type can be applied. Some time the cryo-techniques employed to observe these specimen.

2. **Non-Biological, Dry, Conductive Specimen:** A clean, dry specimen of non-magnetic, non- radioactive metal may be viewed in the SEM with little preparation provided size requirement of the specimen are met. The specimen may be affixed to a metal stub with conductive adhesive, or may be clamped onto some specimen holders for viewing. The standard metallographic techniques of grinding, polishing, and etching a particular alloy are usually suitable preparation for SEM examination of fine microstructure. Non- conductive plastic or resinous mounts are often use to facilitate grinding and polishing procedures. With these samples coating may be necessary to maintain the flow of current from metal part to the ground.

3. **Non-Biological, Dry, Non-conductive Specimen:** Sample like dry leather, organic gels, particles of various sizes and machine components etc. are processed by exposing the surface, dry it and make it conductive for examination.

4. **Biological, air dried fresh specimen:** These specimens are observes without elaborate preparatory procedures. These samples require only collection, mounting and coating prior to their viewing in the SEM. Samples like pollens, insect and wood.

5. **Biological, Hydrated, Fresh Specimen:** for this sample the equipment should be in a position to have the low vacuum facility and lower accelerating voltage and fast pumping system. These type of equipment now commercially named as Environmental Scanning Electron Microscopes (ESEM).

6. **Biological, hydrated, frozen, uncoated specimen:** That specimen which has the low melting point can be viewed in SEM by freezing and mounting on cooled stage. This type of the facility is an additional attachment with the basic equipment.

7. **Biological, hydrated, frozen and coated specimen:** The biological specimen that has not been chemically fixed may be viewed without dehydration in a frozen state. These samples are generally viewed at low magnification, low accelerating voltage, large working distance and less probe current.

8. **Biological, hydrated, chemically fixed specimen:** The sample of life sciences material are usually chemically processed where the steps like selection, surface preparation, fixation, dehydration, mounting and coating take place. These samples are viewed in SEM as a routine.

The generalized protocol includes the following steps.

- Collection of specimen.
- Specimen selection.
- Surface cleaning.
- Fixation.
- Dehydration or drying.
- Mounting the specimen.
- Coating the specimen with a thin, electrically conductive layer.

Some specific preparatory techniques

Sem protocols (Selected)

Diatomaceous earth

Put the diatomaceous earth in amortar and slowly grind in into pieces. Put the fine pieces of the diatomaceous earth into a test tube and suspend it in distilled water or alcohol. Clean it by ultrasonic cleaner for 20 minutes. Take a drop of it with a pipette from the skin of the suspension onto the hydrophilic treated silver tape, which is sticked to the stub. Dry and coat the specimen with carbon and metal.

Free cells

Put free cells into 1% glutaraldehyde (0.1M Phosphate buffer, pH (7.4); Shake well and carry out fixation for 20-30 minutes. Separate the free cells from the fixative by using a centrifugal separator (1500 rpm) and discard the supernatant liquid. Put 0.1 M phosphate buffer (pH 7.4) into the test tube containing the precipitated free cells' shake and rinse the mixture to remove the fixative sticking to the free cells. Then perform centrifugation and discard the supernatant liquid. Dehydrate the free cells with 60,70,80,90,100% acetone for 10 minutes each. Suspend the free cells in 100 acetone. Then drop the suspension on silver tape; rapidly dry the drop in a stream of hot air for at about 45 °C. The dried free cells are used a specimen for examination.

Note: *Free cells (especially erythrocytes, spermatozoa etc.) are sensitive to osmotic pressure of the fixative. In a fixative with high osmotic pressure some erythrocytes show one or several wart like elevations on their cell surfaces. And in a fixative high low osmotic pressure, so we erythrocytes become spherical. Thus it is necessary to pay special attention to the concentration of the fixative.*

Insects

Since the integument of insect is covered with a hard cuticular layer which can be observed the specimen is hard enough to hold its water contents while in the vacuum or is capable of withstanding a loss of water contents in the body. It is also interesting to note that the fixative may not permeate the integument due to considerable secretion of the lipid, thus leaving the problem of fixing the cuticular layer unsolved. The fixation and the dehydration methods using chemicals however are often for comparatively for soft specimens, such as insect bellies, viscera, larvae, parasites and the like as in the case of soft tissues of animals. In the field of parasites, the integument of paragonius, cercaria have been observed.

Bacteria (*Staphylococcus*)

- Collect cells from aeriated nutrient broth by centrifugation.
- Apply a loopful of the above suspension via a platinum loop on a thin layer of agar formed on a small piece (5mm square) of glass fiber filter. (Whatman glass fiber filter; FG/F),
- Incubate the pieces of filter for several hours at 37 °C on a nutrient agar plate.
- Fix specimens initially with 1% glutaraldehyde in 0.15M cacodylate buffer in a refrigerator overnight.
- Dehydrate specimen with a grades series of ethanol amyl acetate, critical point dry, coat and view.

Bacteria (*Myxococcus disciformis*), Hook, 1978

- Select individual specimen or cut out agar blocks with specimens attached to them.
- Immense in Formalin-acetic acid-alcohol fixative (FAA) for 24 hours at room temperatutre.

Composition of FAA

Ethyl alcohol	356 ml
Glacial acetic acid	72 ml
Commercial formaldehyde (40%)	72 %
Glycerol	36 ml
Water	500 ml

- Rinse by immersion is distilled water.
- Dehydrate for at least 24 hours, depending on whether or not the specimen is attached to agar, in 2,2 dimethoxypropane acidified with 1 drop of concentrated HCl/50 ml DMP.
- Transfer the specimen to acetone. If specimen is attached to agar, leave in acetone for 48 hours. Keep in a sealed container.
- Critical point dry, mount and coat.

Solution preparation

General Preparation of Fixatives: Please note that all fixatives are dangerous and should only be used under controlled conditions.

Aldehyde fixatives: Formaldehyde-Formaldehyde forms the basis for many coagulative and non- coagulative fixatives. Commercially available solutions of formalin or formaldehyde have been routinely used in the preparation of most coagulative fixing solutions, but the percentage of methanol (11-16%) present in this chemical has made it unsuitable for most ultrastructural studies. Formaldehyde to be used in a non-coagulative fixative for ultrastructural studies is prepared in the laboratory from powdered paraformaldehyde by the following procedure:

1. In a fume hood, add 20 g of paraformaldehyde powder to 100 ml of distilled water.
2. Heat the solution to 65°C with continuous stirring.

   The paraformaldehyde power will continually settle out and does not, at this stage, seem to go into solution.
3. Add a few drops of 1N NaOH slowly until the solution becomes clear.
4. Allow the solution to cool.

The paraformaldehyde prepared in this manner may now be added to a prepared buffer.

Formalin-Acetic Acid-Alcohol (FAA)

Ethyl alcohol (50%)	90 ml
Clacial acetic acid	5 ml
Commercial 40% Formaldehyde (or made up as above)	5 ml

Glutaraldehyde

Glutaraldehyde can generally be purchased in a highly purified liquid form from any of a number of commercial sources. It is highly recommended that the purified form of this chemical be used as it is very unstable and subject to oxidation and polymerization. Such chemical degradation will result in poor fixation. To prepare a solution of glutaraldehyde it is only necessary to dilute the fixative with a suitable amount of buffer prepared at the proper pH.

Osmium tetroxide

(OsO_4)-Osmium is a highly dangerous fixative which must be prepared and used under the most careful of conditions in a fume hood. This fixative has been used in both a vapor and liquid form to prepare tissue for scanning electron microscopy. Osmium tetroxide can be obtained from commercial suppliers in both the liquid and crystalline forms. The liquid form is generally prepared in 4% aqueous solutions and is added directly to the appropriate amount of prepared buffer. The crystalline form is prepared as follows:

2% OsO_4 stock solution

1. Under a fume hood, break open a vial containing 1 g of OsO_4 and add to 50 ml of glass-distilled water in a stoppered flask.
2. Tightly seal the flask and place into a refrigerator for a few days until the osmium goes into solution.
3. Store at 4°C.

1% OsO_4 in buffer : Add a vial of 1 g of osmium to a glass bottle. Under the hood, break the vial with a glass stirring rod (often it is helpful to score the vial first). After it is broken, add 100 ml of the buffer at the desired pH. Store in a tightly sealed flask in a refrigerator until the osmium goes into solution.

Parducz Fixative : Make up a 2% OsO_4 stock solution in distilled water as above. In a fume hood, add a saturated solution of $HgCl_2$ (mercuric chloride) in the ratio of 6 parts of 2% OSO_4 to 1 part of the $HgC1_2$ solution. As with all fixation solutions, extreme precautions should be taken to avoid contact with these chemicals on the skin, in the eyes or nose, etc.

General preparation of buffers

0.1M Phosphate Buffer

Solution A

- Prepare stock solutions of 0.1 M KH_2PO_4 (13.6 g/liter) and 0.1 M Na_2HPO_4 (14.19 g/liter) in distilled water.
- Adjust phosphate buffer to the proper pH by adding the following stock solution in the prescribed ratio for the desired pH.

pH	KH_2PO_4	Na_2HPO_4
6.33	72	28
6.41	68	32
6.53	62	38
6.61	56	44
6.70	52	48
6.81	48	52
6.91	40	60
7.00	34	66
7.10	28	72
7.24	22	78
7.30	20	80
7.42	16	84
7.57	12	88

- Check pH with a meter and adjust with NaOH or HCl if necessary.

0.1 M Phosphate buffer

Solution B

- Prepare three stock solutions as follows:

  a. 0.2 M NaH_2PO_4

  b. 0.2 M Na_2HPO_4

  c. 1% $CaC1_2$

- Combine : 23 ml of A

  77 ml of B

  100 ml H_2O

- Add 1.0 ml of C slowly while stirring vigorously.
- Adjust pH to 7.3-7.4.

0.1 M Cacodylate-HCI Buffer

- Prepare a 0.4 M stock solution by adding 21.4 g of sodium cacodylate to 250 ml of distilled water.
- Add 50 ml of the 0.4 M stock solution to 8 ml of M HC1 and distilled water to make 100 ml.
- Adjust the pH with HC1 to reach the required value.

Note: Cacodylate buffer contains arsenic and is poisonous.

0.1 M s-Collidine Buffer (2,4,6-Trimethylpyridine)

- Dissolve 5.34 ml of pure s-collidine in 100 ml of distilled water.
- Add the correct amount of 1.0 M hydrochloric acid from the chart below in order to obtain the desired pH.

pH	ml of HC1
7.74	10.0
7.67	12.0
7.59	14.0
7.50	16.0
7.41	18.0
7.33	20.0

- Dilute the mixture to 200 ml with distilled water.
- Determine the pH and adjust it if necessary before use.

Note: *If upon addition to water the s-collidine forms yellowish crystalline or oily substances, the chemical probably was contaminated and will not be effective as a buffer with osmium tetroxide. Highly purified s-collidine is readily obtainable through most electron microscopy suppliers.*

Veronal acetate buffer

Prepare Solution A as follows

- Sodium veronal (barbital) 2.94 g
- Sodium acetate 1.94 g
- Sodium chloride 3.40 g

Add water to bring the volume to 100 ml. Store in a foil wrapped container in the cold.

- To make the buffer, combine :

Solution A	50 ml
Water	130 ml
1 M $CaC1_2$	2.5 ml
0.1 N HCl	70 ml

- Adjust to pH 6 with 1 N HC1 if necessary. Some General Protocols

Tissues

Fixation

1. Fix tissue quickly into 2.5% gluteraldehyde and 2% paraformaldehyde in 0.1M sodium cacodylate buffer (pH 7.4).
2. Chop into small 1 mm3 cubes in the hood and leave to fix for 1 hour @ RT.
3. 2.Wash with 0.1M sodium cacodylate buffer 3 x 20 minutes.

Postfixation

1. Postfix in 1% osmium tetroxide in 0.1M cacodylate buffer for 1 hour @ RT in the hood.

2. 2 wash with 0.1M Sodium cacodylate buffer 3 X 20 minutes

Enbloc staining

1. Wash in 50mM sodium maleate buffer (pH 5.2) 3 x 15 minutes.
2. Stain in 2% uranyl acetate in maleate buffer for 1 hour @ RT in the dark.

Dehydration

1. Wash in water 3 x 5 minutes.
2. Dehydrate in the following order: 50% ethanol 2 x 5 minutes

   70% ethanol 2 x 5 minutes

   90% ethanol 2 x 5 minutes

   100% ethanol 3 x 10 minutes
3. Replace ethanol with propylene oxide
4. Leave in fresh propylene oxide for 10 minutes with the lid closed.

Infiltration & embedding

1. 1 Replace with 50% propylene oxide/50% Epon mix. Leave on the wheel for 2 hours with the lid closed.
2. 2 Replace with pure Epon and leave on the wheel for 2 hours with lid open. Repeat once.
3. Transfer samples to fresh Epon in moulds or Beem capsules (remove air bubbles). Add computer- printed labels. Cure in the oven overnight @ 60 °C.

Embedding of cell monolayers

Fixation

1. Wash cells with PBS buffer 3 times at room temperature (optional).
2. Replace PBS buffer (medium) with 2.5% gluteraldehyde in 0.1M sodium cacodylate buffer (pH 7.4) for 1 hour @ RT.
3. Wash cells with 0.1M sodium cacodylate buffer 3 x 5 minutes.

Postfixation

Postfix cells in 1% osmium tetroxide in 0.1M cacodylate buffer for 1 hour @ RT in the hood.

En bloc staining

1. Wash in 50mM sodium maleate buffer (pH 5.2) 3 x 5 minutes.
2. Stain in 2% uranyl acetate in maleate buffer for 1 hour @ RT in the dark.

Dehydration and infiltration

1. Wash in water 3 x 5 minutes.
2. Dehydrate cells in the following order: 50% ethanol 2 x 5 minutes
   - 70% ethanol 2 x 5 minutes
   - 90% ethanol 2 x 5 minutes
   - 100% ethanol 3 x 10 minutes
3. Replace ethanol with propylene oxide.

Cells that have not been scraped from culture dishes will be removed from the plastic surface of the dish at this stage. The propylene oxide will dissolve the plastic and the cell layer will float off. Remove the cells quickly because the propylene oxide continues to dissolve the plastic. Transfer the cell layer, in propylene oxide, to Eppendorf tubes, making sure the cells do not dry at any stage.

1. Wash several times in propylene oxide (4-5 times) to remove plastic residues.
2. Replace with 50% propylene oxide / 50% Epon. Leave on the wheel for 2 hours with the lid closed.
3. Replace with pure Epon and leave on the wheel for 2 hours with lid open. Repeat once.
4. Transfer the cell pellets to fresh Epon in moulds or Eppendorf tubes, add computerprinted labels. Cure in the oven overnight @ 60 °C.

Embedding of cells in suspension FIXATION

1. Spin the cells at 2,000g for 5 min at 4 °C and remove medium.
2. Add 2.5% gluteraldehyde in 0.1M sodium cacodylate buffer (pH 7.4) and fix for 1 hour@ RT.
3. Wash cells with 0.1M sodium cacodylate buffer 3 x 5 minutes. Each time spin the cells at 2,000 g for 5 min, RT.
4. After final wash, spin the cells at 5,000 – 10,000 RPM in an Eppendorf centrifuge, 5 min at RT.

Postfixation

1. Postfix cells in 1% osmium tetroxide in 0.1M cacodylate buffer for 1 hour @ RT in the hood (dislodge the pellet from de wall of the tube with needle to optimize penetration of osmium).

En bloc staining

1. Wash in 50mM sodium maleate buffer (pH 5.2) 3 x 5 minutes.
2. Stain in 2% uranyl acetate in maleate buffer for 1 hour @ RT in the dark.

Dehydration

1. Wash in water 3 x 5 minutes.
2. Transfer pellets into glass vials.
3. Dehydrate pellets in the following order: 50% ethanol 2 x 5 minutes
    - 70% ethanol 2 x 5 minutes
    - 90% ethanol 2 x 5 minutes
    - 100% ethanol 3 x 10 minutes
4. Replace ethanol with propylene oxide
5. Leave in fresh propylene oxide for 10 minutes with the lid closed.

Infiltration

1. Replace with 50% propylene oxide / 50% Epon mix. Leave on the wheel for 2 hours with the lid closed.
2. Replace with pure Epon and leave on the wheel for 2 hours with lid open. Repeat once.
3. Transfer the cell pellets to fresh Epon in moulds or Beem capsules (remove air bubbles). Add computer-printed labels. Cure in the oven overnight @ 60°C.

Embedding of scraped cells (Tissue culture) FIXATION

1. Wash cells with PBS buffer 3 times at room temperature (optional).
2. Replace PBS buffer (medium) with 2.5% gluteraldehyde in 0.1M sodium cacodylate buffer (pH 7.4) for 1 hour @ RT.
3. Wash cells with 0.1M sodium cacodylate buffer 3 x 5 minutes.

4. Rinse the plate once with a solution of 1% gelatin (bovine) in cacodylate buffer, then remove the buffer and add back 1 ml of the gelatin/cacodylate solution.
5. Scrape cells with piece of Teflon and collect into an Eppendorf tube.
6. Spin cells for 5 minutes @ 5,000 RPM @ RT in an Eppendorf centrifuge. Wash 3 times with 0.1 M cacodylate buffer to remove the gelatin. Last spin can be done at higher speed to obtain a tight pellet for postfixation (10,000 RPM – maximum speed).

Postfixation

Postfix cells in 1% osmium tetroxide in 0.1M cacodylate buffer for 1 hour @ RT in the hood (dislodge the pellet from de wall of the tube with needle to optimize penetration of osmium).

En bloc staining

1. Wash in 50mM sodium maleate buffer (pH 5.2) 3 x 5 minutes.
2. Stain in 2% uranyl acetate in maleate buffer for 1 hour @ RT in the dark.

Dehydration

1. Wash in water 3 x 5 minutes.
2. Transfer pellets into glass vials.
3. Dehydrate pellets in the following order: 50% ethanol 2 x 5 minutes.
    - 70% ethanol 2 x 5 minutes
    - Ethanol 2 x 5 minutes
    - 100% ethanol 3 x 10 minutes
4. Replace ethanol with propylene oxide.
5. Leave in fresh propylene oxide for 10 minutes with the lid closed.

Infiltration

1. Replace with 50% propylene oxide / 50% Epon. Leave on the wheel for 2 hours with the lid closed.
2. Replace with pure Epon and leave on the wheel for 2 hours with lid open. Repeat once.

3. Transfer the cell pellets to fresh Epon in moulds or Beem capsules (remove air bubbles). Add computer-printed labels. Cure in the oven overnight @ 60°C.

Embedding of Yeast in Spurr resin

Fixation

1. Filter 10 OD units of cells (~1 x 10^8 cells) on a 0.45 mm Millipore filter unit without letting the cells dry.
2. Wash on the filter with 10ml of 0.1M sodium cacodylate buffer (pH 6.8).
3. Add 10ml of 3% gluteraldehyde in 0.1M sodium cacodylate buffer pH 6.8 and immediately disconnect filter unit from vacuum line. Transfer to a 15ml centrifuge tube. Continue fixation for 1 hour @ RT, then overnight @ 4 °C.

Zymolase digestion

1. Wash twice in 10ml of 50mM KPO_4 pH 7.5, each time centrifuging the cells @ 2,000 RPM @ 4 °C.
2. Make a 0.25mg/ml solution of zymolase in 50mM KPO_4 pH 7.5. Incubate mixture @ 37 °C for 15 min while shaking. Remove insoluble particles by centrifuging @ 2,000 rpm for 5 min @ RT (20 °C). Resuspend cells in 2 ml of zymolase supernatant. Incubate tubes for 25 minutes @37 °C while shaking.
3. Stop zymolase digestion by centrifuging cells @ 2,000 RPM for 5 min @ 4°C.

Postfixation

1. Wash 2 x in 5ml of ice-cold 0.1M sodium cacodylate pH 6.8, each time centrifuging them @ 2,000 RPM @ 4 °C. Resuspend them in 1 ml of buffer and transfer to Eppendorf tube.
2. Pellet the cells @ 14,000 RPM for 2 min @ RT.
3. Add 0.5ml of cold 2% osmium tetroxide in 0.1M sodium cacodylate pH 6.8. Dislodge the pellet from the bottom of the tube using a needle to ensure good penetration of the osmium. Incubate for 1 hour on ice in the hood.
4. Wash 3 x 5 minutes with water.

En bloc staining

1. Add 1ml of 2% aqueous uranyl acetate and incubate for 1 hour @ RT in the dark.
2. Wash 2 x 5 min with water.

Dehydration

1. Dehydrate in the following order: 50% ethanol 2 x 5 minutes
   - 70% ethanol 2 x 5 minutes
   - 90% ethanol 2 x 5 minutes
   - 100% ethanol 3 x 5 minutes
2. Wash 4 more times with 100% ethanol from a fresh bottle, 5 min each.
3. Wash with 100% acetone for 5 min.

Infiltration

1. Add 50% acetone / 50% Spurr and incubate for 2 hours on wheel. Change to 100% Spurr and incubate overnight.
2. Change Spurr 2 x over the next day. Bake @ 80 °C for at least 24 hours.

Embedding and Polymarisation should be done as described and demonstrated in the laboratory. The ultrathin sections cut from resin blocks with Ultramicrotome using glass knives in the rage of 50 to 80nm thickness. The grids prepared and loaded with sections and stored in grid boxes or gelatin capsule for subsequent staining.

Staining

In light microscopy, various parts of a tissue can be identified by wide variety of stains based on colour and intensity. In electron microscopy, the cell structures can only be identified by their morphology and ability to transmit or scatter electrons (electron opacity). thus, the contrast enhancement of a ultrathin section of biological material is based on electron scattering of the cell constituents. The need for contrast in TEM studies of biological specimen is affected by two inherent problems. First, cells and tissues have a low intrinsic electron scattering power since the common elements of biological materials are carbon, hydrogen, oxygen and nitrogen; therefore unstained sections yield little details. Second, epoxy plastics cause a large amount of random electron scattering, reducing specimen contrast.

To work effectively the stain must be able to penetrate the resin surface. If more than one stain is to be used, the stain must be compatible or synergistic with each other and differentially stain the cellular components. Unstained biological tissues have little density difference compared to the surrounding environment and contrast is usually quite low.

It is important to increase the contrast of most biological specimens by reacting various cellular components with heavy metals. The use of salts of heavy metals such as lead citrate and uranyl acetate increases contrast of tissue when metal ions of salt react with various cellular components. These heavy metal salts on termed as 'stains'.

The Stains should :

- Have adequate contrast,
- Allow deposition of heavy metals by selective reaction
- Not cause artifacts by precipitation or distortion of fine structure.
- Not extract cellular constituents
- Not react with the fixatives
- Allow uniform staining.

There are two types of staining technique like positive staining technique used for section and negative staining technique which helps in delineeate the details of particulate samples. Positive staining techniques enhance the density of constrast of biolotgical structure as compared to the back ground. The molecules of heavy metal salts attch to various organeles or macromolecules within the specimen to increase their density and thereby increase contrast differentially. This differ from the negative staining contrast situation where the background are surrounding the specimen is made dense by a heavy metal salt so that the specimen appears lighter in contrast to the darkly stained background.

Basis of Contrast: The heavy metal salts used as stains in the electron microscope consist of ions of a high atomic number with a large number of protons and electrons that scatter the beam electrons. When beam electrons encounter the atomic nucleus of a heavy metal ion they will be deflected with minimal energy loss (elastic scattering) at such a wide angle that they will not enter the imaging lenses. This subtractive action gives rise to the various tonalities or shades evident on the screen. This contrast, which is dependent on the elimination of a number of electrons. It is termed as amplitude contrast. The electrons of the heavy metal stain may also deflect the beam electrons to give rise to amplitude constrast.

The two most commonly used positive stains are uranyl acetate (WM-422) and lead citrate (WM=1054). It is known that uranyl ions reacts strongly with phosphate and amino groups so that nucleic acid and certain proteins are highly stained. With lead stains, it is through that lead ions bind to negatively charged components such as hydroxyl groups and osmium reacted areas. Phosphate groups may also be involved in this phenomenon, since thenuse of phosphate buffers often enhances overall staining with lead ions.

Uranyl and lead stains are known as general or non specific stains when used in the routine manner since they will stain many different cellular components.

On two occasions when the treatment of heavy metal salts are given one at pre-embedding stage and secondly post embedding stage including sectioning. In the pre- embedding stage, which is also termed as enblock staining, the reaction of the tissue with uranyl stains may take place during the dehydration. It has an advantage of uranyl acetate which serve as fixative also and preserve the membranes structure. The adjustment of the urnyl stain to a pH of 5.2 also leads to that fine structural preservation of DNA filaments as well as junctions, mitochandria, myofibrils, nucleoproteins, and phospholipids. Depending on the specimen, ultrastructural preservation is so much better when uranyl ions are used following aldehyde/osmium fixation, that it should be considered for use peior to embedding.

Post embedding staining: Most uranyl staining is carried out on sectioned material that has been embedeled in plastic. However, it is possible to stain a trimmed plastic specimen block before sectioning. Only lead staining will be needed after the sections are cut.

Procedure

1. Trim specimen block so that specimen is exposed on as many surface as possible.
2. Immerse specimen block in a small container of 2% uranyl acetate dissolved in 95% ethanol. Tightly sealed the container and place in a 60 °C oven for 12 to 24 hours.

   Note: *Phosphotungestic acid may be used in place uranyl acetate.*
3. Rinse the specimen block quickly in several changes of ethanol, allow the block to dry, and proceed to cut ultrathin sections that may then be stained with lead citrate.

Depending on the type of plastic used, the staining extends to a depth of 10 to 15 um so that many sections can be cut that will be need subsequent uranyl staining. This method is particularly useful for thick sections to be viewed in a high voltage transmission electron microscope.

Staining ultrathin sections

Staining Solutions : The most frequently used staining solutions are uranyl acetate and lead citrate. The former is made up in either water or alcohol at a concentration ranging from 1% (w/v) to saturation. The stain is usually taken up more quickly and more effectively from alcoholic solutions. It reacts principally with nucleic acids, but proteins also become stained. Lead citrate increases the general contrast of membranes, proteins, nucleic acids and glycogen, responsible for this are phosphate, carboxyl, and sulphydryl groups which bid lead cations. Both lead and uranium are poisonous heavy metals. In addition, uranium is weakly radioactive.

Uranyl acetate solutin

Saturated aqueous solution: Solid uranyl acetate is added to 100 ml double distileld H_2O in a glass- stoppered, brown bottle (Or a polyethylent bottle, wrapped in alumin foil) and the bottle is frequently sheken until the uranyl acetate dissolves. This does several time over 2-3 h until saturation conditions are reached. The solution is allowed to settled overnight. The supernatant is then carefully pipetted out and either centrifuged (bench centrifuge, 15 min) or passed through a Millipore filter. The final solution has a pH of 4.0 and should be stored in the dark. If the solution becomes turbid it should be discarded.

Saturated Methanolic Solution: A small amount of uranyl acetate is added to 2-4ml 70% (w/v) methanol in a centrifuge tube, shaken vigorously (vortex mixer) for a few minutes and then centrifuged to remove undissolved salt. The supernant is carefully pipetted into a brown glass bottle. The solution is stable for about a week at -4 °C and has a pH of 3.5-4.0.

Lead Citrate Solutin : 1.33g lead hitrate and 1.76g sodium citrate are dissolved in 30-ml doubel distilled H_2O in 50-ml ground-glass stoppered bottle or a polyethylene bottle. The solution is shaken virorously for 1 min and then at intervals over the next 30 min. 8 ml 1M NaOH (free of sodium carbonate is then added together with water to make a volume of 50 ml. The solution is clear and has pH of 12.

Procedure for double staining sections

Double staining of sections from conventionally embedded material with uranyl acetate and lead citrate is a routine procedure in TEM. Sections are stained on the grids which were used to pick them up from the trough liquid.

The staining of sections adhering to grids is best carried out in a glass petri-dish fitted with moist filter paper and a piece of dental wax or parafilm. When staining with lead citrate, it may be necessary to place one or two moistened pellets of KOH or NaOH on the wax in order to maintain a CO_2 free atomsphere.

Usually grids are stains first with uranyl acetate. Using a clean pasteur pipette, the stain is taken from roughly the middle of the stock bottle. The first few drops are discarded and a row of three or five drops is made on the wax. the grids are then placed on the drops of uranyl acetate with section side downwards. The Petri-dish is covered for the duration of the staining.

The grid are then washed of excess stain by dipping in a series (usually four) of 50 -ml breaker filled with double distilled water. Excess water is removed from the surface of the grids with filter paper before placing them on to lead citrate. Lead citrate staining is carried out in a petri-dish. After the lead staining is completed the grids are washed by holding them in a steam of 0.02 M NaOH from a wash bottle for a second or two, followed by double-distilled water. The grids are finally dried with filter paper and stored with section side uppermost on filter paper in a covered petri-dish.

Optimal staining periods have to be worked out for each individual object; they are usually in the range of 5 min for uranyl acetate and lie between 2 and 10 min for lead citrate.

The stain should

- have adequate contrast.
- allow deposition of heavy metals by selective reactive.
- Not cause artifacts by preparation or distortion of five structure.
- Not extract cellular.
- Not react with the fixatives.
- Allow uniform staining.
- Be simple in application and preparations.
- Be stable under normal conditions.

The interpretation of stain reactions must be considered in the light of a number of factors that include the fixative used, pH, temperature, concentration of the stain and inhibitors.

Factors such as pH, temperature, and concentration will influence the reactive sites and level of stain reaction. A higher pH level, for example, will cause most heavy metals to precipitate as hydroxides or hydrous oxides. A number of staining procedures call for higher temperatures to insure penetration and reactivity. Lower concentrations of the staining solution, usually results in a higher level of differential staining at "strong" and "weak" sites. In the same context, duration of staining will allow differentiation between sites of strong stain reactivity and secondary weakly reactive sites.

Thick section staining

Toluidine blue Staining

Solution: 0.5-3% (w/v) toludine blue in 0.1 NaK- phosphate buffer. The solution should be heated ot 90 °C for 15 min and then fltered.

Staining Procedure: The sections on the microscope slide are covered with the stain and then warmed to 50 °C for 5 min. During this period it is important that an evaporation of water from the solution be prevented in order to avoid an unspecific precipitation of stain. The sections are then washed with distilled water to remove unreacted stain. If the specimens are too deeply stained, excess stain can be removed by extracting wiht 70 (v/v) ethanol. The extent, depth and intensity of the staining can be varied by adjusting the stain concentration or extending the length of the staining period.

Methylene blue staining

Solutions: 1% (w/v) periodic acid in distilled water (100ml)

1. 1% (w/v) sodium tetraborate plus 1% (w/v) methylene blue in 50ml distilled water.
2. 1% (w/v) azure-II in 50 ml distilled water.

Solutions 2 and 3 are mixed 1:1 (v/v) and 50g sucrose is added.

Staining Procedure: Add several drops of the periodic acid solution to the sections and allow to stand for 5 min. Wash off and dry the borders of the glass slide. Add several drops of the stain to the sections and warm to 60°-80°C heating plate for 5-15 min. Quickly wash the stain off the sections and dry on the heating plate. Cover the sections with Caedax or paraffin oil. Examine immediately because the staining tends to decrease in intesity with time.

Processing particulate samples

The biological particulate sample like Unicellular algae, bacteria and all fragments are examined under the transmission Electron Microscope with the technique of whole mount and sectioning. The particulate material loaded on nitrocellulose coated grids is treated with heavy molecules chemicals or through the shadow casting to delineate its morphology. Sections are made out of pallet formal of the particulate sample through centrifugation and complete processing like the tissue up to embedment.

Whole Mount Method: Whole mount technique is very old technique but still in use for bacteria, viruses, isolated organelles and macromolecules. Negative staining and shadow casting are the commonly used methods of whole mount.

Negative Staining: The whole specimen is deposited on a grid. Several conditions must be satisfied in order to achieve optimal results. A firm structure less substrate is essential to support the specimen. Collodion or Formvar substrates stablized with carbon are satisfactory. Secondly this techniques involve the following steps:

- Fragmentation or separation.
- Coating of substrate on the grid.
- Mounting.
- Negative staining.
- Shadow casting.

Fragmentation or Separation: There are some lengthy technique are evolved in the process of fragmentation and isolation of particulate material from the whole unit of biological specimen. This process is limited to material which represent a large proportion of tissue volumes. (mitochandria, Chloroplast, Collagen fibrils, cellulose Microfibriles) and which can be fragmented without loss of their identity.

The process of fragmentation by chemical disruption: The dissolution of cementing substances to isolate different cell constituent of animal as well as plant cell.

Animal Cell: Majority of organelles are isolated with the help of a density gradient after homogenizing in the buffer. The organelles pelleted and resuspended in the buffer in concentration form. And a drop of this concentrate contain number of particles which are loaded on the coated grids for further treatment.

The plant cells are digested with polysaccharides enzymes. Since the digestion of the cell wall occurs under plasmolyzing condition (presence of 0.4-0.8M mannital or sorbital). The plasmolyzing agent can be rebound by centrifugation. The suspension is made in a buffer for further steps.

Mechanical disruption: The following methods are used to fragment the organelles.

- Ultrasonic vibration.
- Chopping action of blender.
- Smashing of thick frozen material.
- Grinding between two plates of ground glass.
- Quick freezing and slamming.

The cellular ingredients should be kept in a proper buffer with relation of toxicity and pH as well as temperature.

Mounting: Loading of sample upon grid is essential as the small biological organelles very small to the size of openings of the grid mesh. The grid must have a continuous coating. The types of coated grids are commonly used as specimen support.

The preparation of specimen grid support films

Specimens which are to be observed in the electron microscopy must be mounted on a rigid support. This cannot be a glass slide as in microscopy since this would be opaque to electrons. Instead, a fine copper grid or mesh (300-400 mesh) is used, onto which a thin film of plastic or carbon is placed. Such support films are thin enough to eb transparent to electrons but strong enough to withstand irradiation. In order to obtain satisfactory images of viruses or other biological specimens in the electron microscope, it is necessary to treat them in some way to enhance their electron scattering property. The staining method most used in virus detection is negative staining in which electron dense materials usually heavy metal salts, surround but don't impregnate the virus particles. Thus, the virus particles standout as lighter objects against a dark background, hence the expression negative stain.

The supporting films used in electron microscopy today are (1) Collodion, prepared 0.5% solution in amyl acetate (2) Formavar, prepared either from a 0.25 or 0.50% solution of formvar (polyvinayl formal) resin, in ethylene dichloride (3) Carbon, which is evaporated onto mica and floated free on a water surface to be picked up on grid (4) Carbon, evaporated onto a collodion film, which is then extracted. Formvar films are simple to prepare and are slightly more stable then Parlodian films. However, larger quantities of grids can be prepared more easily with Parlodian.

Where ultrahigh resolution micrographs are required, a film or evaporated carbon alone is superior, as it gives a stable support with minimum of background grain.

Method of preparation of support film : In order to observe the whole mount the support film is required on which the sample is to be loaded. The support film is made of either carbon or formvar and collodion. Formvar is dissolved in the chloroform and collodion in isiamyle acetate and the solution is spreaded on the surface of the water. It is picked up on the copper grid. There are three methods of preparing support film on the grid.

- Drop method
- Slide method
- Funnel method

Thickness of the support film should not be more than 200 Å and should be stable in the column of the microscope. The thickness varies with the concentration of the substrate and speed of releasing the substrate solution on water surface. The substrate film further supported with the carbon to give contrast and firmness to it.

Collodion support films

Collodion solution: Preparation of collodion (nitro cellulose) solution is as follow:

The gauze (with one corner bent up for ease of banding) is lowered to the bottom of the water-filled vessel and moved to one side. Using fine forceps, copper grids are placed onto the gauze until it is covered. Grids should not drop and should always be placed with either their dull or shiny side uppermost: adopt a standard convention.

Floating off support film: Raise gauze loaded with grids slowly upwards underneath support film catching it correctly oriented to cover the grids. or

Allow the water to drain slowly from the specially manufactured vessel, steering the floating film so that it is lowered onto the grids on the gauze. This method is most convenient.

Formvar film

- This is prepared using the commercially available material in a solution of -0.2% w/v in chloroform of 1.2-dichloroethane (this should be stored in a refrigerator).
- A glass microscope slide (new unscratched) is cleaned using a chamois leather (not too clean or film or film may stick to well).
- Using forceps dip the slide in the solution, withdraw it vertically and allow to dry in a staining position.
- Scrape the edges of the slide to give a cut to the film, using the back of a scalpel.
- Carefully and slowly, lower the slide at an angle of about 30° into glass distilled water which should have a dust-free surface. The film should separate from the slide and float on the surface of the water.
- Mount the film as described above, on the grid by putting the dull side downward on the floating film. Take cut the film along with grids on slide or filter paper.

Collodian film

- Prepare a solution of 2% Collodion in amyl acetate.
- Allow a single drop of solution to fall onto the surface of clean glass-distilled water in a suitable container. The solution spread leaving a film as the solvent evaporates.
- Skim off the first film and discard it, this serves to clean the water surface.
- Cast a second film as above.
- Mount as described above.

Carbon films

Preparation of carbon films requires the use of a vacuum coating unit. Within the evacauted bell jar of the coating unit passing the electrical current between two electrodes of carbon produces an arc. Carbon vapor is formed which condenses on a suitable flat surface below.

Negative staining

The goal of the negative staining is to stain the background around the specimen with a compound that strongly scatters electrons such as heavy metal salt solution. As the solution dries, it forms a glassy film on the support film coated on the grid interrupted by "cast" of the electron lucent biological material. At the edge of the specimen the area of the stained film that differs in thickness, corresponding to the surface irregularities of the specimen. These area of the surface irregularities of the specimen. These area of different thickness scattered electrons differentially and provide contrast necessary for viewing.

The material used for negative staining should contain a heavy metal atom. This material should be

- Highly soluble
- Have a high melting point (beam stability)
- Have a small molecular size for good penetration
- Dry into a smooth (glassy) film
- Not react with (positive stain) the specimen. Specimen for negative Stain:
- Virus and bacterophages
- Bacteria
- Cell fragments.

Procedure

The particulate specimen in suspension is applied on the surface of the subsequent negative staining. There are three common procedure for negative staining.

- Drop method
- Double drop method
- Mixture method

Drop method : Put a drop of the stain on the clean surface of the dental wax sheet. The grid loaded wiht the specimen on its support folm is put on a drop f stain facing sample side toward the drop of the stain on it. It is allowed to incubate for 30 seconds or more. The grid then dry with filter paper to examine under the microscope for its examination.

Double drop method : The coated grids placed on the first drop of specimen suspended solution. After some time dry the grid and put on the surface of second drop of the stains, specimen side towards the stain drop. Then again dry the grid with help of filter paper and view under the microscope.

Mixture method : Stain and specimen are mixed in equal volume. Use the one drop of this mixture for destining as well as loading of the specimen on the grid. Dry the grid for viewing under the microscope.

Protocols

Virus and bacteriophage

Method 1

- Purify the specimen thoroughly unto a concentration of approximately 10^{-7} particles/m;.
- Put a drop of the suspension on a carbon-coated grid and allow it to remain for 5-10 minutes (or place a drop of this suspension on a clean dental wax sheet and expose the coated face of grid on it for 5 minutes.)
- Sap off the drop slowly by touching the tip of cutting of a filter paper allowing a thin film to remain on the grid.

With the aid of the pipette, place a drop of the stain on the grid and after 10-15 seconds drain off the stain with a filter paper. (Or place a drop of the stain on the clean surface of the dental wax and put the specimen coated side of the grid toward the stain drop for 1-3 minutes

Note: *If the stain does not spread evenly over the carbon film, add 0.005-0.5% bovine serum albumin, glycerin, or propylene glycol to the particle suspension.*

Method 2

- Purify the preparation to a concentration of 10/6 particles /ml.
- Blend the preparation with an equal volume of the stain.
- With the aid of a glass nubulizer (fine atomizer), spray the mixture directly onto a carbon-coated grid. (If the drops are too dense, rinse the grids with carbontetrachloride solution prior to spraying or add a wetting such as serum albumin to the mixture).

Bacteria

- Purify the preparation of a concentration of 10/10 particles/ml.
- Suspend in a 2% solution of potassium phosphotungstate PTA or any other suitable negative stain so that the final concentration of bacteria in the mixture ranges between 10/5 and 10/6/ml.
- With the aid of pipette, place a drop of mixture onto a carbon-coated grid.
- After a few seconds, drain off the mixture with a pointed piece of filter paper and air dry (some workers prefer to dry the grid in the electron

microscope). After drying, the concentration of bacteria can be determined wiht the aid of a phase contrast light microscope.

Cell Fragments

- Suspend the maerial either in 0.32M sucrose or in 1-2% ammonium acetate solution.
- Mix the suspension wiht an equal volume of ice- cold 2% potassium phosphotungstate.
- With the aid of a nebulizer, spray the mixture onto a carbon-coated grid and allow to air dry.

Negative Stains

Phosphotungstic Acid (PTA) : PTA dissolved in glass distilled water to give 2% w/v solution, pH 5-8 adjusted with 1 N KOH or NaOH.

For optimum spreading of purified preparations, add approx. 0.01% Bovine serum albumin (BSA) to the negative stain solution. Store in 1ml syringe at 4°C.

Uranyl Acetate : Prepare shortly before use (requires 15-30 minutes to dissolve salt); pH is not adjusted and has a volume of approx. 4.0. Concentration should be 1-2% w/v in glass distilled water. Stable for a few days in the dark.

Caution : This solution is slightly radioactive and therefore should not be discarded in drains.

Ammonium Molybodate : pH range 3-9 adjusted with HC1 or NaOH as desired. Concentration 1-4%.

Uranyl Formate : pH 3.0; solid dissolved in glass distilled water to 1-2% w/v solution. Store in cool and dark place. Prepare fresh regularly.

Sodium Silico-Tungstate : Concentration 1-4% in glass distilled water, pH range 6-8 (adjusted with NaOH).

Methyl Amine-tungstate : pH range 5-8; Concentration 2% w/v solution in glass distileld water (adjust pH by drop wise addition of 1 N NaOH>

Bovne serum base, phosphotungsti c acid stain: Preparation

- 5N potassium hydroxide (5N KOH)
- KOH 28g
- Distilled water to make 100 ml
- Buffer 2% phosphotungstate (PTA)

2% phosphotungstic acid is adjusted to the desired pH (usually 4.5-7.2) with 5N KOH. The solution is then filtered and can be kept for about a month.

Buffered phosphotungstate with albumin (PTA with BSA)

2 parts of buffered PTA 1 part of 0.1% bovine serum albumin. this solution should be kept in the refrigerator to retard growth of fungi which might be confused in the final preparation with the object being examined.

Staining

Taking appropriate precautions with infectious microorganisms, one drop of suspension of the specimen is mixed in a depression slide with two to four drops of PTA with BSA. A drop of the mixture is picked up by touching the support film side of the grid to the surface of the mixture. After 30 sec, excess fluid is drained off from the grid by touching its edge to filter paper, leaving a very thin film of liquid. It is allowed to dry, and the preparation is ready for examination.

Choice of pH for the suspension medium and for the PTA is a matter of trial and error. The final choice will be a compromise between the pH that seems physiological for the specimen, that which may give some positive staining of internal structures, and that pH which gives the best spreading properties.

Plant virus

Electron microscopy is frequently used in plant virology to locate the symptoms on the parts of plants showing the presence of infection of any sort of virus e.g., leaves showing mosaics and mottle, using breaking of colour of petals contains higher concentration of viruses, petals and leaf veins and roots etc.

Sample preparation: There are different methods for preparation of extracts for viruses. Some of the methods are described below:

Leaf Squash method: Leaf material is best prepared by using this method but other types of material (petals) may also be tried.

Procedure

- Select a small piece of tissue showing symptoms like mosaics or mottles or lesion margin of about 4-10 mm^2. Squash the tissue on a glass slide in 2 drops of chosen stains (PTA or water when uranyl acetate is used) using a rounded glass rod until sap is extracted.

- Draw up a small amount with Pasteur pipette of the stain/sap mixture and place onto the carbon coated side or a grid.
- Carefully move excess liquid from the grid using a small piece of filter paper touched to the edge of the grid.
- After 1-2 minutes of drying, the grids may be examined in the electron microscope.

Disadvantages of the above method are only the debris accompanies the virus particles on the grid. To eliminate the above debris following modification can be made :

Squeezing of the tissue in stain or buffer between two microscopic glass slides gives more efficient maceration where the amount of tissue is limited. After rotating them accurate between the two slides they can be pressed a apart at one and end the sap gathered at the other end. This sap can be further diluted if necessary, so that it is only pale green in colour. Load on the grid as described above.

In certain cases better preparation has been achieved when the support grid is held coated side downward and touched onto prepared stain-sap mixture before being drained and dried as usual. This avoids the loading of larger pieces of plant materials on the grid.

Leaf-Dip method : This technique is used principally but not exclusively for the preparation of leaf samples.

Procedure

- Clamp support grid in the forceps, coated slide up.
- Put one drop of stain (or water,if uranyl acetate stain is to be used)
- Cut a small (1-3 mm wedge shaped piece of the leaf and hold in a second pair of the forceps.
- Run the freshly cut edge of the leaf through the drop on the grid over a period of 30 seconds (take care not to damage the support films).
- Remove leaf wedge and drain the grid and allow drying before examination.
- Modification can be made to use the epidermal strip in place of the leaf wedge. These are using fine forceps from the underside of the leaf, and allowed to float in the stain drop on the grid. After 30 seconds the epidermal strips are removed and the grid drained as before. Dry the grid and examine under the microscope.

Good results may also be obtained by placing the drops of the stain onto the part of the leaf from which the epidermal strip has been removed, then transferring them to the grid in the usual way.

Other methods

Grinding small amounts of the tissue, frozen in the liquid nitrogen is an efficient method of extraction of virus from the fibrous tissue. Tougher tissue may be grounded in small quantity in a carborandoum/buffer paste. Further addition of the buffer allows removal of the cerborandum by centrifugation leaving the extracted virus.

Artifacts can develop during specimen preparation

Step	Artifacts	Possible remedy
During Fixation		
Chemical fixation	• Due to selective nature of a fixative substance can be washed out dring fixation	• Select proper fixative as per experiment
	• Chemical reaction can occure which may lead to changes in conformation and reactivity.	• Complete washing in between primary and secondary fixatives and after secondary fixation
	• Since the fixations are not instantaneous there is always the possibility of cellular changes through cytolysis after cell death.	• Perfussion fixation is always prefered otherwise try to fox tissue as quickly as possible.
	• Since fixation are not always isotonic with the cell fluid and membrane permeability can be changed during fixation, there is a distinct possibility that a swelling or shrinkage may occur	• Careful attention has to be paid to the concentration, composition and osmolarity of the fixation solution
Physical Fixation	• Deformation may occur through ice crystal formation	• Soak tissue in Cryo Protectant before freezing, keep size of tissue to minimum, plunging should be instantaneous
	• Phase separation may occur	• Not always, repeat
Dehydration		
Air drying	• Deformation of tissue due to surface tension	• Avoid air contact during whole process
	• Substances leach out	• Dehydration should not be unnecessarily prolonged less period in low concentration
Critical point drying	• Contamination on the surface of the specimen	• Gas is not pure contains debris
	• Drying is not complete	• Pressure released very quickly
	• Smell of amyl acetate	• Remove amyl acetate completely by flushing

Step	Artifacts	Possible remedy
	• Only patches are properly dried	• Generally it happened use those patches for examination
Freeze drying	• Deformation may occur through ice crystal formation • Drying isn't complete • Use cryo protactant • Drying rate is not properly set	
Embedding		
Capsule embedding	• Deformation due to volume changes during polymerization • Bubble formation • Blocks softened after some time	• Gradual embedding should be done Gradual polymerization in ascending order • Due to humidity and improper mixing of different ingredients of resin
Flat embedding	• Tissue comes at the surface • Tissue is not in the matrix	• Do complete infiltration • Put a thin layer of resin in each mould so that resin become slightly viscous then put tissue and resin to cover it

General Precautions

- Try to work, whenever possible, in a fume cupboard.
- The various containersfor stocks and waste solution etc. should only opened for short period.
- All glassware should be rinsed immediately after use.
- Hand gloves should be worn.
- Distilled water rinse should be given to glass ware before and after use.
- Fixatives and other solutions should be kept in a separate place ear marked for this.
- Care should be taken not to contaminate objects with soiled gloves. A common mistake is to touch books, papers, reagent bottles, telephones, or doorknobs with soiled gloves, thereby contaminating them.
- Gloves should be removed carefully.
- Eye protection is necessarywhen working with cryogenic agents.
- Suitable clothing should be worn , opened toe shoes or chapple are unsafe.
- Eating, drinking, smoking are forbidden.
- Work area must be kept clean, uncluttered and free of physical obstruction.

- Be familiar with the location and operation of all safety equipment or reagents
- Label all reagent
- Post material data sheets must be consulted before using any reagent
- Wash hands and arms before leaving the laboratory
- Verify the security of the laboratory before leaving:
- Extinguish all flames or source of ignition
- Turn off all unnecessary gase, water, vacuum, and electricity
- Lower the sash to the fume hood
- Turn off all non essenrial lights
- Lock all door.
- Pathogen and radioisotopes: These should not be processed in EM premises. These should be processed in specified, controlled and safe separate laboratory then after fixing bring it to EM laboratory.

11

Scanning Electron Microscope

N. Natarajan, C. Sharmila Rahale, R. Sunitha

Electron Microscopes are scientific instruments that use a beam of highly energetic electrons to examine objects on a very fine scale. This examination can yield information about the topography (surface features of an object), morphology (shape and size of the particles making up the object), composition (the elements and compounds that the object is composed of and the relative amounts of them) and crystallographic information (how the atoms are arranged in the object).

Electron Microscopes were developed due to the limitations of Light Microscopes which are limited by the physics of light to 500x or 1000x magnification and a resolution of 0.2 micrometers. In the early 1930's this theoretical limit had been reached and there was a scientific desire to see the fine details of the interior structures of organic cells (nucleus, mitochondria...etc.). This required 10,000x plus magnification which was just not possible using Light Microscopes.

The first Scanning Electron Microscope (SEM) debuted in 1942 with the first commercial instruments around 1965. Its late development was due to the electronics involved in "scanning" the beam of electrons across the sample.

The scanning electron microscope (SEM) is an essential tool for a large part of the scientific community with applications ranging from archeology and geology, to biology and material science. The SEM's breadth of utility comes from its versatility in imaging a large variety of samples over a wide range of magnifications from about 10 times (about equivalent to that of a powerful hand-lens) to more than 500,000 times, about 250 times the magnification limit of the best light microscopes. It can obtain three-dimensional images as well as elemental analysis and crystal structure.

The versatility of the SEM stems from its ability to detect a wide array of signals with secondary electrons, back scattered electrons, and characteristic X-rays being some of the most common. The SEM also benefits from the ability to image and characterize a diverse field of specimens ranging from the organic world of plants, animals and insects, to inorganic materials such as crystals and polymers.

The SEM instrument is made up of two main components, the electronic console and the electron column. The electronic console provides control knobs and switches that allow for instrument adjustments such as filament current, accelerating voltage, focus, magnification, brightness and contrast. The recent model electron microscopes use a computer system in conjunction with the electronic console making it unnecessary to have bulky console that houses control knobs, CRTs and an image capture device. All of the primary controls are accessed through the computer system using the mouse and keyboard. The image that is produced by the SEM is usually viewed on the computer monitor. Images that are captured can be saved in digital format or printed directly.

Electron column

The electron column is where the electron beam is generated under vacuum, focused to a small diameter, and scanned across the surface of a specimen by electromagnetic deflection coils. The lower portion of the column is called the specimen chamber. The secondary electron detector is located above the sample stage inside the specimen chamber. Specimens are mounted and secured onto the stage which is controlled by a goniometer. The manual stage controls are found on the front side of the specimen chamber and allow for x-y-z movement, 360 rotation and 90.

Electron gun : In a typical SEM, an electron beam is thermionically emitted from an electron gun fitted with a tungsten filament cathode. Tungsten is normally used in thermionic electron guns because it has the highest melting point and lowest vapour pressure of all metals, thereby allowing it to be heated for electron emission, and because of its low cost. Other types of electron emitters include lanthanum hexaboride (LaB_6) cathodes, which can be used in a standard tungsten filament SEM if the vacuum system is upgraded and field emission guns (FEG), which may be of the cold-cathode type using tungsten single crystal emitters or the thermally assisted Schottky type, using emitters of zirconium oxide.

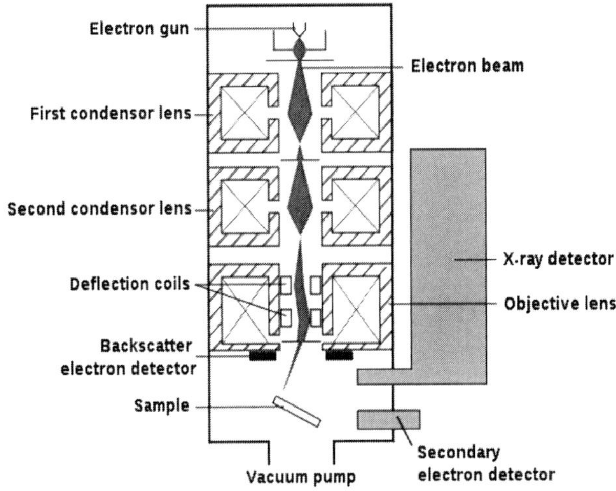

Fig. 1: Scanning electron microscope column

Condenser Lenses : After the beam passes the anode it is influenced by two condenser lenses that cause the beam to converge and pass through a focal point. What occurs is that the electron beam is essentially focused down to 1000 times its original size. In conjunction with the selected accelerating voltage the condenser lenses are primarily responsible for determining the intensity of the electron beam when it strikes the specimen (Postek et al., 1980).

Apertures : Depending on the microscope one or more apertures may be found in the electron column. The function of these apertures is to reduce and exclude extraneous electrons in the lenses. The final lens aperture located below the scanning coils determines the diameter or spot size of the beam at the specimen. The spot size on the specimen will in part determine the resolution and depth of field. Decreasing the spot size will allow for an increase in resolution and depth of field with a loss of brightness.

Scanning System : Images are formed by rastering the electron beam across the specimen using deflection coils inside the objective lens. The stigmator or astigmatism corrector is located in the objective lens and uses a magnetic field in order to reduce aberrations of the electron beam. The electron beam should have a circular cross section when it strikes the specimen, however, it is usually elliptical thus the stigmator acts to control this problem (Postek et al., 1980 and Watt, 1985).

Specimen Chamber : At the lower portion of the column the specimen stage and controls are located. The secondary electrons from the specimen are attracted to the detector by a positive charge.

Vacuum system

The ability for a SEM to provide a controlled electron beam requires that the electronic column be Electron Beam-Specimen Interactions : In SEM the electrons interact with the atoms that make up the sample producing signals that contain information about the sample's surface topography, composition, and other properties such as electrical conductivity. The types of signals produced by an SEM include secondary electrons, back-scattered electrons (BSE), characteristic X-rays, light (cathodoluminescence), specimen current and transmitted electrons. In scanning electron microscopy visual inspection of the surface of a material utilizes signals of two types, under vacuum at a pressure of at least 5×10^{-5} Torr. A secondary and backscattered electrons. Secondary and high vacuum pressure is required for a variety of reasons. First, the current that passes through the filament causes the filament to reach temperatures around 2700K (Lyman et al., 1990). A hot tungsten filament will oxidize and burn out in the presence of air at atmospheric pressure. Secondly, the ability of the column optics to operate properly requires a fairly clean, dust-free environment. Third, air particles and dust inside the column can interfere and block the electrons before the ever reach the specimen in the sample chamber (Postek, 1980). In order to provide adequate vacuum pressure inside the column, a vacuum system consisting of two or more pumps is typically present.

Most SEMs are designed to operate in three backscattered electrons are constantly being produced from the surface of the specimen while under the electron beam however they are a result of two separate types of interaction. Secondary electrons are a result of the inelastic collision and scattering of incident electrons with specimen electrons. They are generally characterized by possessing energies of less than 50 eV (Postek, 1980). They are used to reveal the surface structure of a material with a resolution of ~10 nm or better.

Backscattered electrons are a result of an elastic collision and scattering event between incident electrons and specimen nuclei or electrons. Backscattered electrons can be generated further from the surface of the material and help to resolve topographical contrast separate vacuum modes, high vacuum (10^{-6} to 10^{-7} Torr), low vacuum (0.1 to 20 Torr) and ESEM (0.1 to 20 Torr). In the low vacuum and ESEM mode of operation the pressure inside the specimen chamber operates at a lower pressure than that inside the actual column which is always at high vacuum. The high vacuum mode is typically where most SEMs operate. In this mode the highest resolutions and magnifications can be achieved although, it is not suitable for all specimens. Generally, anything that is conductive and has a high density works well in high vacuum mode. Specimens that are non- conductive or have a low density are more suitable for the low vacuum or ESEM mode.

The ESEM mode of operation is necessary for wet, non-conductive samples. and atomic number contrast with a resolution of >1 micron.

While there are several types of signals that are generated from a specimen under an electron beam the x-ray signal is typically the only other signal that is used for scanning electron microscopy. The x-ray signal is a result of recombination interactions between free electrons and positive electron holes that are generated within the material. The x-ray signal can originate from further down into the surface of the specimen surface and allows for determination of elemental composition through EDS (energy dispersive x-ray spectroscopy) analysis of characteristic x-ray signals. Figure 2 is a diagram which displays a cross section of the volume of primary excitation illustrating zones from which signals may be detected.

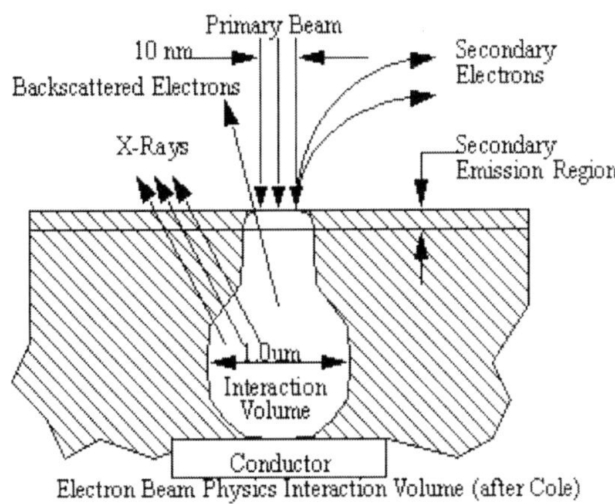

Fig. 2: Electron Beam Physics Interaction Volume (after code)

Operation of SEM

In SEM, a source of electrons is focused in vacuum into a fine probe that is rastered over the surface of the specimen. The electron beam passes through scan coils and objective lens that deflect horizontally and vertically so that the beam scans the surface of the sample.

As the electrons penetrate the surface, a number of interactions occur that can result in the emission of electrons or photons from or through the surface. A reasonable fraction of the electrons emitted can be collected by appropriate detectors, and the output can be used to modulate the brightness of a cathode ray tube (CRT) whose x-and y-inputs are driven in synchronism with the x-y voltages rastering the electron beam. In this way an image is produced on the CRT; every point that the beam strikes on the sample is mapped directly onto a corresponding point on the screen. As a result, the magnification system is simple and linear magnification is calculated by the equation:

M=L/l

where L is the raster's length of the CRT monitor and l the raster's length on the surface of the sample.

SEM works on a voltage between 2 to 50kV and its beam diameter that scans the specimen is 5nm-2μm. The principle images produced in SEM are of three types : secondary electron images, backscattered electron images and elemental X-ray maps. Secondary and backscattered electrons are conventionally separated according to their energies. When the energy of the emitted electron is less than about 50eV, it is referred as a secondary electron and backscattered electrons are considered to be the electrons that exit the specimen with an energy greater than 50eV (Watt, 1985). Detectors of each type of electrons are placed in the microscope in proper positions to collect them.

Advantages and disadvantages

Electrons in scanning electron microscopy penetrate into the sample within a small depth, so that it is suitable for surface topology, for every kind of samples (metals, ceramics, glass, dust, hair, teeth, bones, minerals, wood, paper, plastics, polymers, etc). It can also be used for chemical composition of the ample's surface since the brightness of the image formed by backscattered electrons is increasing with the atomic number of the elements. This means that regions of the sample consisting of light elements (low atomic numbers) appear dark on the screen and heavy elements appear bright. Backscattered are used to form diffraction images, called EBSD, that describe the crystallographic structure of the sample. In SEM, X- rays are collected to contribute in Energy Dispersive X- Ray Analysis (EDX or EDS), which is used to the topography of the chemical composition of the sample.

Consequently, SEM is only used for surface images and both resolution and crystallographic information are limited (because they're only referred to the surface). Other constraints are that the samples must be conductive, so non-conductive materials are carbon- coated and secondly, that materials with atomic number smaller than the carbon are not detected with SEM.

Environmental SEM (ESEM)

The major growth of SEMs is in the development of specialized instruments. Environmental SEM uses differential pumping to permit the observation of specimens at low-pressure gaseous environments (e.g. 1-50 Torr), at high relative humidity (up to 100%) and at higher pressures. In this type of SEM, there's no need for conductive coating, the secondary electron detector operates in the presence of water vapour, and in the microscope's column there are pressure-limiting apertures. The ESEM is ideal for non-metallic surfaces, such as biological materials, plastics and elastomers.

References

FEI. The Quanta 200 User's Operation Manual 2nd ed. (2004).

Goldstein, J.I., H. Yakowitz, D.E. Newbury, E. Lifshin, J.W. Colby, J.W. Colby and J.R. Coleman, Pratical Scanning Electron Microscopy: Electron and Ion Microprobe Analysis, edited by J.I. Goldstein and H. Yakowitz (Pelnum Press. New York, N.Y., 1975).

JEOL. Guide to scanning electron Microscopy [Online]. Available at http://www.jeol.com/sem/docs/sem_guide/tbcontd.html

Lyman, C.E., D.E. Newbury, J.I. Goldstein, D.B. Williams, A.D. Romig, J.T. Armstrong, P. Echlin, C.E. Fiori, D.C. Joy, E. Lifshin and Klaus-Ruediger Peters, Scanning Electron Microscopy X-Ray Microanalysis and Analytical Electron Microscopy: A Laboratory Workbook, Press. New York, N.Y., 1990).

S. Howard, A.H. Johnson and K.L. McMichael, Scanning Electron Microscopy: A Student's Handbook, (Ladd Research Ind., Inc. Williston, VT., 1980).

Watt, I.M. The Principles and Practice of Electron Microscopy. Cambridge Univ. Press. Cambridge, England, 1985.

12

Transmission Electron Microscope

C.R. Chinnamuthu and K. Brindha

Transmission electron microscope (TEM)

The transmission electron microscope (TEM) operates on the same basic principles as the light microscope but uses electrons instead of light. What you can see with a light microscope is limited by the wavelength of light. TEMs use electrons as "light source" and their much lower wavelength makes it possible to get a resolution a thousand times better than with a light microscope.

We can see objects to the order of a few angstrom (10^{-10} m). For example, we can study small details in the cell or different materials down to near atomic levels. The possibility for high magnifications has made the TEM a valuable tool in both medical, biological and materials research.

Magnetic lenses guide the electrons

A "light source" at the top of the microscope emits the electrons that travel through vacuum in the column of the microscope. Instead of glass lenses focusing the light in the light microscope, the TEM uses electromagnetic lenses to focus the electrons into a very thin beam. The electron beam then travels through the specimen you want to study. Depending on the density of the material present, some of the electrons are scattered and disappear from the beam. At the bottom of the microscope the unscattered electrons hit a fluorescent screen, which gives rise to a "shadow image" of the specimen with its different parts displayed in varied darkness according to their density. The image can be studied directly by the operator or photographed with a camera (Fig. 1).

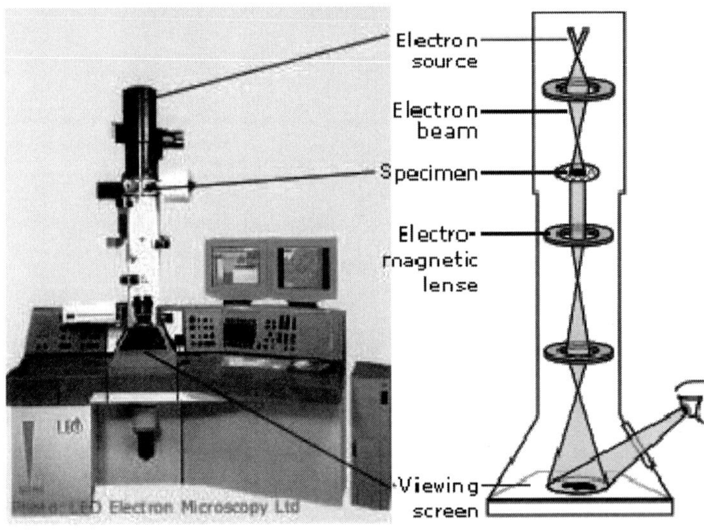

Fig. 1: Schematic diagram of TEM

Preparation of specimen for TEM

1. Biological samples

In a TEM, the specimen you want to look at must be of such a low density that it allows electrons to travel through the tissue. There are different ways to prepare your material for that purpose. You can cut very thin slices of your specimen from a piece of tissue either by fixing it in plastic or working with it as frozen material. Another way to prepare your specimen is to isolate it and study a solution of for example viruses or molecules in the TEM.

We can also stain the specimen in different ways and use markers to locate specific things in the tissue. It can for example, be stained with heavy metals like uranium and lead, which scatters electrons well and improves the contrast in the microscope.

Sections of embedded material

Biological material contains large quantities of water. Since the TEM works in vacuum, the water must be removed. To avoid disruption as a result of the loss of water, you preserve the tissue with different fixatives. These cross-link molecules with each other and trap them together as stable structures. The tissue is then dehydrated in alcohol or acetone.

After that, your specimen can be embedded in plastic that polymerize into a solid hard plastic block. The block is cut into thin sections by a diamond knife in an instrument called ultramicrotome. Each section is only 50-100 nm thick.

The thin sections of your sample is placed on a copper grid and stained with heavy metals. The slice of tissue can now be studied under the electron beam.

Whole mounts

Small or very thin objects can be examined directly by mounting them onto a support film and introducing them directly into the electron beam. Contrast is provided by heavy metal precipitation in one of three ways.

1. **Positive staining:** The object is chemically stained with a solution of the metal salt and appears dark on a bright background.
2. **Negative staining:** The object remains unstained but is embedded in a dried film of the heavy metal salt. The specimen appears light on a dark background. This method of visualization has been used extensively in the study of virus particles but is also useful for cell fractions (e.g. coated vesicles).
3. **Shadowing:** A thin layer of heavy metal atoms is deposited on the specimen by evaporation in a vacuum chamber. Shadowing from one direction only produces a pseudo-three-dimensional image. Rotary shadowing, where the specimen is uniformly coated with heavy metal, is used to visualize nucleic acids and proteins.

Ultrathin sectioning

The most popular technique for examining biological materials is to embed the material under study in plastic and cut ultrathin sections that can be examined in a TEM. The material is stabilized by chemical fixation (usually with aldehydes such as formaldehyde or gluteraldehyde), contrasted with solutions of heavy metal salts (osmium tetroxide and uranyl acetate), dehydrated in ethanol or acetone, and embedded in plastic (epoxy resin). Ultrathin sections (60 nm) cut with glass or diamond knives using an ultramicrotome are floated on water, transferred to specimen support grids and examined in the TEM. Often the sections are further contrasted with uranyl acetate and lead citrate prior to examination in the microscope.

In some cases, macromolecules can be specifically labelled prior to embedding and sectioning. For example, the location of some enzymes can be visualized by incubating the tissue with a substrate whose reaction with the enzyme leads to the local deposition of electron opaque material. Alternately, antibodies can be coupled to such enzymes, and the electron opaque reaction product is used to localize the antigens recognized by the antibodies. Some embedding resins (e.g. Lowicryl resins and LR White resin) have been designed to enable antibodies and electron opaque markers (such as colloidal gold particles) to be applied to the ultrathin sections. In this way, subcellular antigens recognized by the antibodies can be localized with the TEM.

Another sectioning technique that is increasing in popularity is cryosectioning (the sectioning of vitrified, frozen material). After chemical fixation, the tissue is immersed in cryo-protectant (usually sucrose) and then quickly frozen in liquid nitrogen. The cryo-protectant allows the biological material to be frozen without the formation of ice crystals, which would damage ultrastructure. This type of freezing, or vitrification, is possible in the absence of cryo-protectants but is technically demanding. Sections cut from the vitrified block can be thawed and incubated with antibodies specific to subcellular antigens. Electron opaque markers allow the antibodies to be seen in the TEM.

Colloidal gold coupled to protein A (a protein from bacterial cell walls which binds to the Fc portion of some immunoglobulins) has been used extensively in recent years to localize antibodies on resin and frozen sections of biological materials. The ability to produce homogeneous populations of colloidal gold with different particle sizes has enabled researchers to use these probes to colocalize different structures on the same section.

Cryofixation

It is possible to freeze biological material fast enough to vitrify the water present inside the cells. Vitrification of water occurs when the freezing has occurred so fast that ice crystals have no time to form. Vitrified biological material can be sectioned at low temperatures. Thin films of vitrified water and sections of vitrified material can be examined in transmission electron microscopes that are equipped with specimen stages that can be kept cold.

Rapid Freezing Methods

There are seven main rapid freezing methods presently available. They are

1. Immersion freezing - the specimen is plunged into the cryogen.
2. Slam (or metal mirror) freezing - the specimen is impacted onto a polished metal surface cooled with liquid nitrogen or helium.
3. Cold block freezing - two cold, polished metal blocks attached to the jaws of a pair of pliers squeeze-freeze the specimen.
4. Spray freezing - a fine spray of sample in liquid suspension is shot into the cryogen (usually liquid propane).
5. Jet freezing - a jet of liquid cryogen is sprayed onto the specimen.

6. High pressure freezing - freezing the specimen at high pressure to subcool the water.

7. Excision freezing - a cold needle is plunged into the specimen, simultaneously freezing and dissecting the sample.

Freeze-fracture followed by freeze etch and replication

If, for some reason, the object to be studied cannot be examined in the TEM, then a thin replica can be made. This is usually made by evaporating a thin layer of a heavy metal (usually platinum) onto the specimen and then coating this with a thin layer of carbon. The object and the replica are separated either by floating off the replica or by digesting away the object. There are four basic steps to follow:

1. The specimen is frozen (often without regard to vitrification).

2. The specimen is fractured, while still frozen, under vacuum.

3. The fractured specimen can then be etched by leaving it frozen and under vacuum. Depending on the time of exposure, more or less water sublimes from the specimen (freeze drying).

4. A replica of the fractured surface is made which is then examined in the electron microscope.

A recent modification of this method employs rapid freezing achieved by slamming cells against a copper block cooled to -269°C with liquid helium. If these frozen cells are then exposed to extensive freeze drying (deep etching), very impressive images of the internal structures of cells are uncovered.

General Schedule of Sample Preparation for Plant Tissue

Steps	Chemical	Temperature	Time	Repetitions
Primary fixation	2.5% glutaraldehyde in buffer	room or 0-4°C	2-4 hours or microwave	1
Wash	buffer	room or 0-4°C	30 minutes	3-5
Secondary fixation	1-4% osmium tetroxide in buffer	room or 0-4°C	2-4 hours	1
Wash	buffer or distilled water**	room or 0-4°C	30 minutes	3-5
en bloc staining (optional)***	0.5% uranyl acetate	0-4°C	overnight	1
Wash after *en bloc* staining	distilled water	room or 0-4°C	10-15 minutes	2
Dehydration	25% ethano	room or 0-4°C	20 minutes	1
	150% ethano		20 minutes	1
	170-75% ethano		20 minutes	1
	190-95% ethano		20 minutes	1
	1100% ethanol		30 minutes	2
	Transition solvent if embedding resin is not miscible with ethanol			
Infiltration	1 part resin/2 parts solvent	room	1 hour-overnight	1
	1 part resin/1 part solvent (optional)	room	1 hour-overnight	1
	2 parts resin/1 part solvent100% resin	room room	1 hour-overnight 1 hour	1 1
Embedding	Place in 100% resin in suitable container	1	1	1
Degassing (optional)	Place in vacuum desiccator or vacuum oven	room-60°C	3-30 minutes	
Polymerization	1	60-70°C	> 8 hours	

**Wash in water if *en bloc* staining is used.

*** If omitting *en bloc* staining, proceed to dehydration at end of washes.

2. Material samples

The properties of materials are determined by their structure which depends on composition and growth conditions. The formation and development of defects of the crystal structure and their influence on the materials properties is also essential to be studied. Thus, it is necessary to characterize both composition and microstructure at the highest levels of resolution possible in order to understand materials behavior and to facilitate the design of new improved materials. Such a characterization requires advanced methods of analysis using microscopic, diffraction, and spectroscopic techniques. The short wavelength of the electrons enables atomic scale resolution. In this regard the electron microscope is an instrument providing all the capabilities necessary to obtain both qualitative information on the crystal structure and chemical concentration profiles (quantitative analysis) of different materials systems.

There are two basic modes of TEM operation: diffraction patterns and imaging modes (Fig. 2). As the beam of electrons passes through a crystalline specimen, it is scattered according to the Bragg's law. The beams that are scattered at small angles to the transmitted beam are focused by the objective lens to form a diffraction pattern at its back focal plane. The scattered beams are recombined to form an image in the image plane. In order to see the diffraction pattern you have to adjust the imaging system lenses so that the back focal plane of the objective lens acts as the object plane for the intermediate lens. Then the diffraction pattern is projected onto the viewing screen .While for the imaging mode, you adjust the intermediate lens so that its object plane is the image plane of the objective lens. The image is then projected onto the viewing screen.

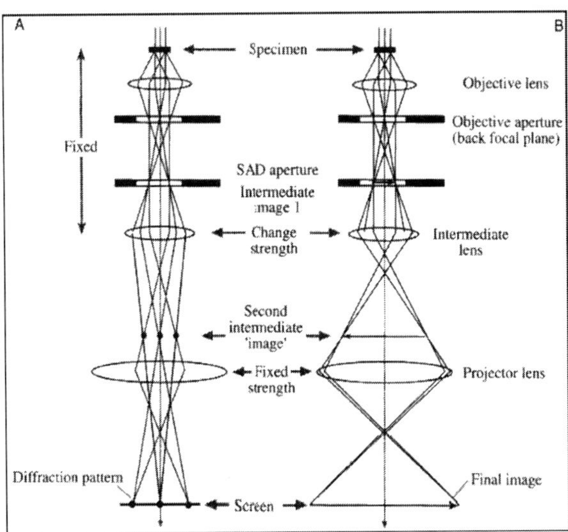

Fig. 2: Diffraction patterns and imaging modes in TEM

The diffraction pattern contains electrons from the whole area of the specimen that we illuminate with the beam. Such a pattern is not very useful because the sample and the viewing screen can be damaged. So there is the need to select a specific area of the specimen to form the diffraction pattern. There are two way to reduce the illumination area. Either to make the beam smaller or to insert an aperture which could permit only the electrons that passes through it to form the diffraction pattern. In the first case, one can make the lenses (C2 and/or C3) in order to converge the beam at the specimen forming convergent-beam diffraction (CBED) patterns. Converging the beam destroys any coherence and the spots in the pattern are not sharply defined but spread into disks. On the other hand, if we wish to obtain a diffraction pattern with a parallel beam of electrons, then the standard way is to use a selecting aperture. This aperture is placed in the image plane of the objective lens, so a virtual aperture at the plane of the specimen is created. The so called selected area diffraction (SAD) is formed (Fig. 3). The SAD operation is performed in the following way. The specimen is first examined in the image mode until a region of interest is found. The aperture is then inserted and positioned around this feature. The microscope is then switched into diffraction mode. SAD can be performed on regions of 10-4 cm diameter. If one needs to obtain SAD patterns in the nanometer scale it is necessary to use the CBED technique.

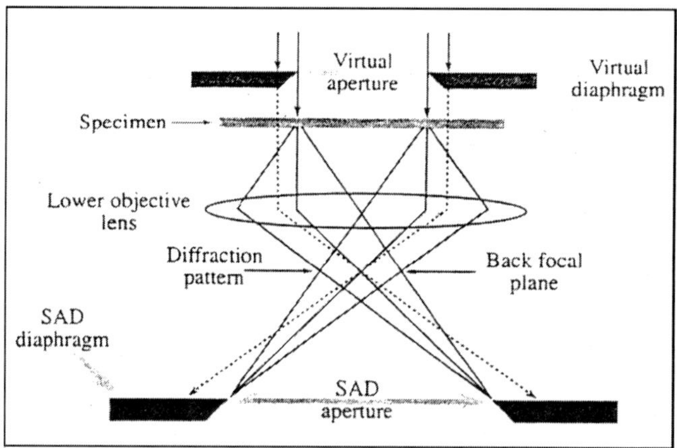

Fig. 3: Selected area diffraction in TEM

Bright-field and dark-field imaging

When one has a SAD pattern in the screen you can use it to form the two basic imaging operations in TEM: bright and dark field imaging. The SAD pattern contains a central spot, which corresponds to direct electrons, and some scattered electrons. In order to form bright and dark images we use the central spot or some or all of the scattered electrons respectively. In order to choose which electrons will finally form the image we inset the aperture in the back focal plane of the objective lens, blocking out most of the spots except for those that are visible through the aperture. When the direct beam is selected, the resultant image is a bright-field image and when we select scattered electrons of any form we call it a dark-field image.

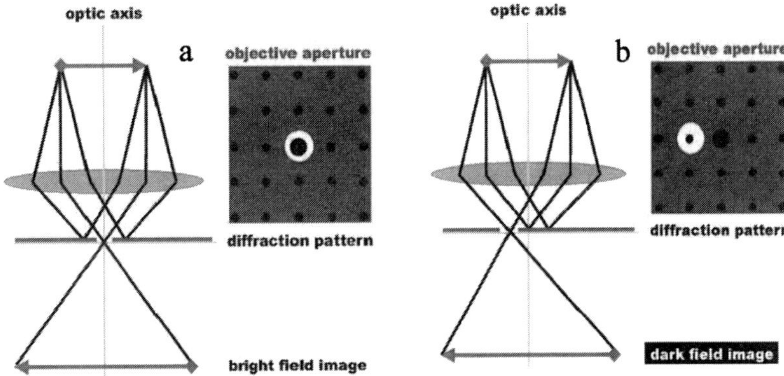

Fig. 4: (a) in the bright field (BF) mode, an aperture is placed in the back focal plane of the objective lens which allows only the direct beam to pass **(b)** In the dark field (DF) mode the direct beamis blocked by the aperture while one or more diffracted beams are allowed to pass the objective aperture.

Kikuchi Diffraction

When a specimen is thick enough a large number of scattered electrons will be generated, which will travel in all directions (diffusely scattered electrons). These electrons can then be Bragg diffracted by the planes forming the known Kikuchi diffraction. Let us consider that we have a diffraction pattern mode. Some of the electrons that have been in elastically scattered at a given point (marked with red) and scattered in all directions will travel at an angle èÂ to the hkl planes and then be Bragg rediffracted by the planes. Since the scattered electrons are travelling in all directions, the diffracted beam will lie on one or two cones, which are called Kossel cones. What we see in the diffraction pattern is the intersection of these two cones with the Ewald sphere and so they appear like parabolas in the viewing screen. If we consider region close to the optic axis those parabolas look like parallel lines well known as Kikuchi lines. If we draw a line in the halfway between the two Kikuchi lines this represents the trace of the plane (hkl).

The spacing of the pair of Kikuchi lines is the same as the spacing of the diffracted spots from the same plane. The position of the Kikuchi lines on the diffraction patterns are controlled only by the angle between the diffraction planes and the incident beam. As the crystal is slightly tilted, so the angle between the incident beam and these planes changes, Kikuchi lines will move as though rigidly attached to the crystal, while the position of the spots will remain stationary. Because all planes give rise to Kikuchi lines, the pattern of lines visible on the final screen changes as the specimen is tilted and the viewing screen is intersect by different sets of Kikuchi lines. As mentioned above, Kikuchi lines are sensitive to any difference in crystal orientation in respect to the incident beam. Therefore they allow us an accurate determination of crystallographic orientation, with accuracy to 0.1°.

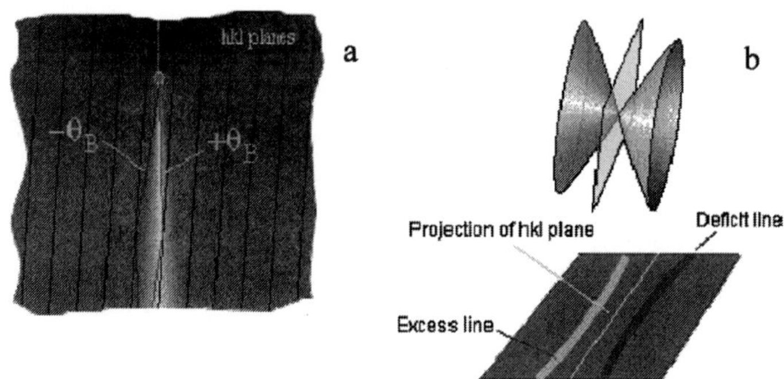

Fig. 5: (a) Some of the electrons are diffracted because they travel at a Bragg angle qB to centrain hkl planes **(b)** The diffracted electrons from the Kossel cones which create Kikuchi lines in the diffraction pattern

Contrast gormation mechanisms

The different imaging modes exploit different contrast formation mechanisms. The three main contrast formation mechanisms are mass-thickness, diffraction and phase contrast formation mechanisms. As electrons go through the specimen they are scattered off axis by elastic nuclear interactions. The mass-thickness contrast arises from incoherent scattered electrons. The interaction cross section or in other words the probability an electron with certain energy, interacting with the specimen, to be scattered in such a way depends strongly on (i) the atomic number Z, i.e. the mass or the density,ñ, as well as (ii) the thickness, t, of the specimen.

The mechanism by which differences in mass and thickness cause contrast is shown in (Fig. 6). Qualitatively, high-Z (i.e., high-mass) regions of a specimen scatter more electrons than the low-Z regions, given an equal thickness. Similarly, thicker regions will scatter more electrons than thinner regions of the same average Z, all other factors being constant. So, for a bright field image thicker and/or higher mass area will appear darker than thinner and/or lower-mass areas.

If an image is formed with electrons scattered at low angles (<~5°), mass thickness contrast dominates but competes with Bragg-diffraction contrast. At high angles (>~5°), where the coherent scattering is negligible low-intensity, incoherently scattered beams can be picked up. The intensity of these beams depends inversely on atomic number (Z) only. This is the so-called Z contrast and contains elemental information.

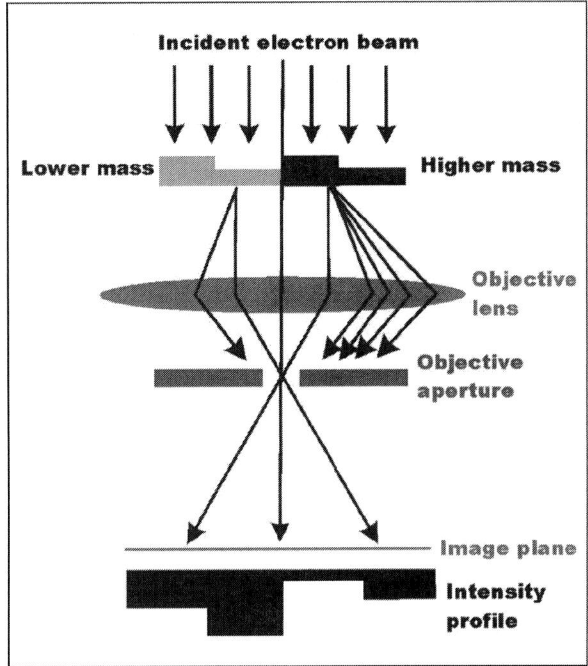

Fig. 6: Mechanisms in intensity profile

Mechanism of mass-thickness contrast in a BF image

Coherent elastic scattering produces diffraction contrast. There is one major difference between forming images to show mass-thickness contrast or diffraction contrast. Any scattered electrons can be selected, if one wants to take a dark field image, which shows mass-thickness contrast. In order to get strong diffraction contrast in both bright and dark field images, the specimen has to be tilted to the so called two-beam conditions, where only one diffracted beam is strong and the direct beam is the other strong spot.

The electron in the strongly excited hkl beam have been diffracted by a specific set of hkl planes and so the area that appears bright in the dark field image is the area where the hkl planes are at the Bragg conditions. Hence, the dark field image contains specific orientation information, not just general scattering information as is the case for the mass-thickness contrast. To produce good bright field diffraction contrast under two beam conditions one needs to tilt the specimen to the desired two-beam conditions (so that only one strong diffracted beam exists) and insert the objective aperture on axis.

A two beam dark field image is not so easily performed. One needs also to tilt the incident beam so that the strong hkl reflection moves onto the optic axis. If one does so, the hkl reflection will become weaker and the so called weak-beam image conditions will be set.

To set up a strong-beam dark field image, tilt in the hkl reflection which is initially weak, and it becomes strong as it moves on axis.

Contrast in TEM images can arise due to the difference in the phase of the electron waves scattered through the specimen. The phase contrast is used to image the atomic structure of thin specimens. In contrast to the above described TEM imaging approaches, the high resolution TEM phase-contrast image requires the selection of more than one diffracted beam. The more beams collected the higher the resolution of the image is.

Contrast in imperfect crystals – defect Imaging

To get best diffraction contrast from defects the specimen is advisory to be tilted so that we have two-beam condition with slight deviation from the Bragg conditions. This is measured by the vector s (K=g+s, where K is a diffraction vector and g is a reciprocal lattice vector), which is called the excitation error or deviation parameter. We say that s>0 if the excess Kikuchi line lies outside its corresponding diffraction spot (g). Alternatively we say that s<0 if the excess Kikuchi line lies inside it corresponding diffraction spot and we say that s=0 when the Kikuchi line runs exactly through its corresponding spot. The best possible strong-beam image contrast conditions for a defect imaging is small and positive deviation parameters.

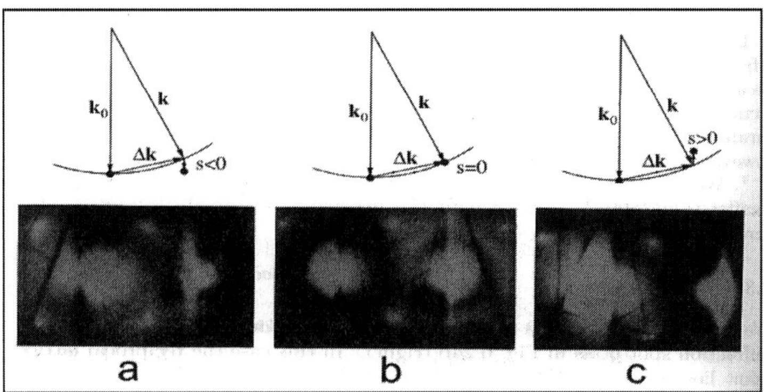

Fig. 7: Diffraction patterns for Kikuchi lines

Ewald sphere constructions and the diffraction patterns for one intense diffraction spot and its Kikuchi line from fcc Al. (a) s<a (b) s=0, and (c) s>.

Chemical analysis in TEM

There are a few physical processes by which electrons are scattered in elastically by the sample: photon creation, plasmon excitation and core electron excitation. Energy is conserved for all inelastic processes-the spectrum of energy gains by the sample is mirrored in the spectrum of energy losses of the high-energy electrons. Electrons undergoing energy losses due to crystal vibrations, by exciting an atomic electron from a core state, quantized as photons with E~10-2eV, are indistinguishable from elastically scattered electrons, given the present state of the art for electron energy-loss spectrometry (EELS) in a TEM. After a core electron has been excited from the atom, the remaining "core hole" decays quickly, often by the emission of a characteristic x-ray. Characteristic X-rays are useful for microchemical analysis by energy dispersive X-ray spectrometry (EDS). An EDS spectrum contains peaks at the energies of the characteristic X-rays from the elements in the material. Chemical mapping of element distributions in samples is possible also when EELS is performed in STEM mode. The electron beam is focused into small probe and EELS spectra are acquired from two dimensional grid of points across the sample.

Each "pixel" in the image can contain an entire EELS spectrum. The data set contains information of chemical variations across the specimen. Another method of chemical mapping is becoming popular. A conventional TEM uses all electrons that pass through the sample, but an instrument known as "energy filter" allows image formation with electrons that have undergone selected energy losses in the specimen. The technique of "Energy filtered TEM" (EFTEM), detects chemical contrast in specimens by adjusting an energy filter to pass electrons that have lost energy to core ionizations of selected elements. Under optimal conditions, this "energy filtered images" can reveal chemical contrast with sub nanometer spatial resolution. It is worth mentioning that EFTEM and EELS allow rapid, high-resolution spatial mapping and analysis of light elements (C,N,O etc) .

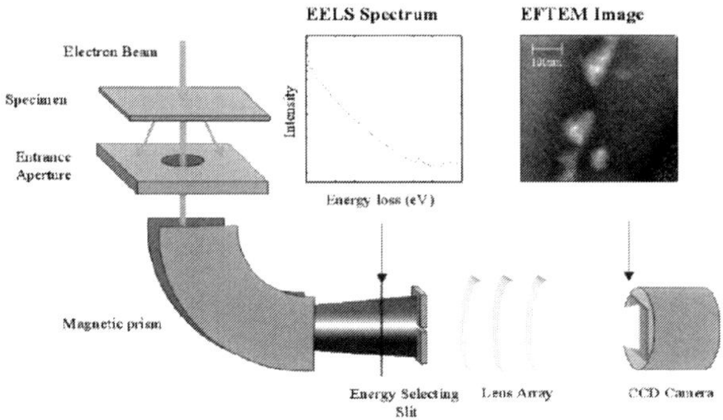

Fig. 8: A schematic of the EELS and EFTEM

References

Electron Beam Analysis of Materials, M.H. Loretto, Chapman and Hall, London New York 1984.

Transmission Electron Microscopy and Diffractometry of Materials, B.Fultz and J.M. Howe, Springer-Verlag Berlin Heidelberg New York 2001.

Transmission Electron Microscopy of Materials, Gareth Thomas and Michael J. Goringe, A Wiley-Interscience Publication, USA, 1979.

Transmission Electron Microscopy, D.B.Williams and C.B.Carter, Plenum Press, New York, 1996.

www.matter.org.uk

13

Atomic Force Microscopy

M. Kannan, S. Marimuthu, P. Meenakshisundaram

Introduction

Biological systems are complex and are full of mysteries. Understanding function in relation to structure is crucial in many biological systems. A microscope is used to magnify, resolve and visualize a substance that is impossible to see by naked eyes and plays a vital role in various biological studies. We can visualize macroscopic organisms like animals, plants, and insects by the unaided eye but not bacteria, viruses or individual proteins. Biological structures have sizes variable over such a wide range that is not possible for a single microscopic technique to analyze all of them. Therefore, different types of microscopes with varying magnification and resolving abilities were developed to unveil the structural complexities of biomolecules (Fig. 1). Over the last 100 years, several types of microscopes have been designed and used in various fields. The microscopes work on different principles, and have a broad range of applications.

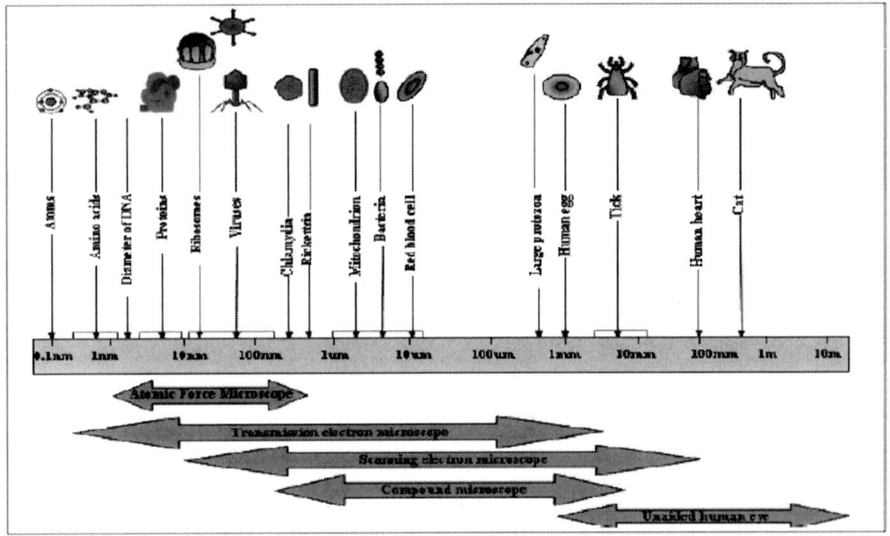

Fig. 1: Magnifying range of different microscopes

SPM and AFM

Scanning probe microscopes (SPM) define a broad group of instruments used to image and measure properties of material, chemical, and biological surfaces. SPM images are obtained by scanning a sharp probe across a surface while monitoring and compiling the tip–sample interactions to provide an image. The two primary forms of SPM are scanning tunneling microscopy (STM) and atomic force microscopy (AFM). STM was first developed in 1982 at IBM in Zurich (Binnig, et al., 1986). The invention of the scanning tunneling microscope (for which Binnig and Rohrer were awarded the Nobel Prize in Physics in 1986) has had a great impact on the technical community by providing a new and unique tool to advance fundamental science and technology. Although the ability of the STM to image and measure material surface morphology with atomic resolution has been well documented, only good electrical conductors are candidates for this technique. This significantly limits the materials that can be studied using STM and led to the development, in 1986, of the Atomic force microscope by Binnig, Quate, and Gerber. This enabled the detection of atomic scale features on a wide range of insulating surfaces that include ceramic materials, biological samples and polymers. AFM has significantly impacted the fields of materials science, chemistry, biology, physics and the specialized field of semiconductors.

Prior to the invention and commercial availability of SPMs, researchers traditionally used (and still use) a variety of microscopes to image surfaces and measure surface morphology on a microscale. Optical microscopes are the most common instrument available to image any sample that is not completely optically transparent. Resolution is limited to about 1 μm and only images and size measurements from features lying in the surface (*x*-*y*) plane are obtainable. Also, optical microscopy has a relatively small depth of field. A more advanced technique, scanning electron microscopy (SEM), has been widely used, since its inception in the mid-1900s, to image microscopic features on sample surfaces. SEMs provide much greater resolution (~5nm) than optical microscopes and have a relatively large depth of field. Because a beam of electrons must travel to the sample to provide an image, the samples must be vacuum compatible and either electrically conductive or coated with a conductive layer to avoid charge buildup.

Advantages of AFM over conventional microscopy techniques

1. High-resolution three-dimensional *x*, *y* and *z* (normal to the surface) images can be obtained. The resolution in the *x*–*y* plane ranges from 0.1 to 1.0 nm and in the *z* direction it is 0.001 nm (atomic resolution).
2. AFM images are free of any artifacts while images obtained by EM consist of a large amount of artifacts which in several cases result in misleading conclusions.
3. AFM requires neither a vacuum environment nor any special sample preparation. Instead, imaging can also be performed in liquid medium, which permits the samples to be analyzed in a near native condition.
4. Sample preparation of AFM is comparatively simple and less time consuming.

Principle of AFM

Both AFM and light microscope amplify the image of the sample but the major difference between them is that the former does not use visible light; instead it uses a cantilever made from silicon or silicon nitride having a very low spring constant to image a sample. At one end of the cantilever, a very sharp tip (around 100–200 nm long and 20–60 nm radius of curvature) is fabricated using semiconductor processing techniques. The cantilever scans above the surface of the sample by progressively moving backward and forward across the surface. A piezo-electric crystal raises or lowers the cantilever to maintain a constant bending of the cantilever.

The force exerted on the tip varies with the difference in the surface height and thus leads to the bending of the cantilever. A laser beam gets constantly reflected from the top of the cantilever towards a position-sensitive photo detector consisting of four side by- side photodiodes (Fig. 2). This laser beam detects the bend occurring in the cantilever and calculates the actual position of the cantilever. Thus, AFM records a three-dimensional image of the surface topography of the sample under a constant applied force (as low as nano Newton range), which provides a maximum resolution image without causing any damage to the sample surface. As the tip scans the surface of the sample, the force between the tip and the sample varies. This change in force is sensed by the tip attached to the flexible cantilever. The amount of force between the probe and the sample is dependent on the spring constant of the cantilever and the distance between the probe and the sample surface. According to Hooke's Law, this force can be described as

$$F = -k\,\ddot{A}x$$

Where F is the force, k is the spring constant and x is the cantilever deflection.

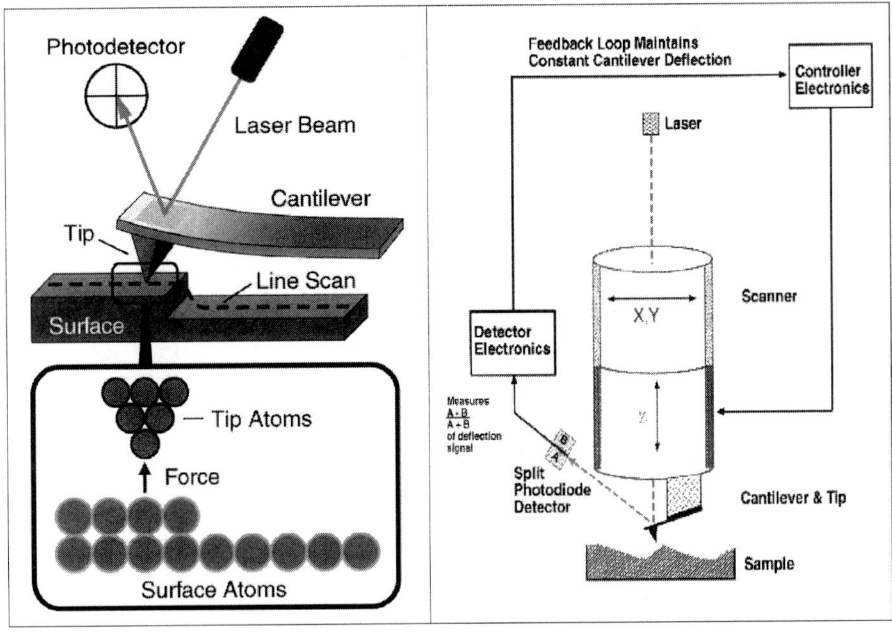

Fig. 2: Schematic diagram showing the operating principles of the AFM

The interactions between the tip of the cantilever and the sample surfaces are regulated by different types of forces, *viz.,* van der Waals forces, capillary and adhesive forces, and double layer forces. Among these, van der Waals force is most commonly associated with AFM. In the contact region, the distance between the tip of the cantilever and the sample surface is less than a few angstroms (Å). Thus the probe experiences repulsive van der Waals force. In the non-contact region, the tip is several tens to hundreds of angstroms away from the sample surface and hence experiences an attractive van der Waals force. Various scanning modes operate in different regions of the van der Waals force vs distance curve as represented in Figure 3. The contact mode operates in the repulsive region and the non-contact mode operates in the non-repulsive region of the curve, while the intermittent or tapping mode fluctuates between the two.

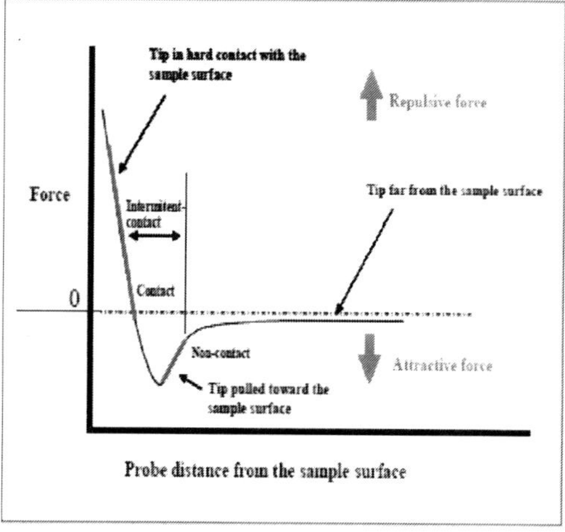

Fig. 3: Graphical representation of force–distance relationship

Operating Modes of AFM

Contact mode

Contact mode or repulsive mode is the simplest mode of operation for AFM. This mode of operation involves sideways scanning by the cantilever tip over the sample surface. The tip makes soft physical contact with the sample and the scanner gently traces the tip across the sample. However, this mode is also associated with disadvantages, viz., general integrity of the delicate biological samples can be lost. This drawback prevents the use of contact mode in studies pertaining to biological substances. In addition, contact mode yields very low resolution with large soft samples. As biological substances are very soft and delicate, the vertical and shear forces exerted by the tip can damage the sample. Therefore, contact mode is not ideal for biological systems. This lacuna can be overcome by applying tapping mode in fluid that reduces the shear forces and minimizes the damage to biological samples.

Tapping mode

Tapping mode or the intermittent contact AFM is the most preferred operating mode for high-resolution topographic imaging of sub cellular structures, and soft and delicate biological samples.

Tapping mode in air

In this operating mode, a small piezoelectric crystal mounted in the multimode AFM tip holder makes the cantilever oscillate up and down at or slightly below its resonance frequency. The tip oscillates vertically, alternately contacts the surface and lifts off. The amplitude of this oscillation typically ranges from 20nmto 100nm. The oscillating tip lightly touches or 'taps' on the sample surface during scanning. When the tip comes close to the sample surface, forces like van der Waals force, dipole-dipole interactions, electrostatic forces, etc., act on the cantilever and lead to a decrease in the amplitude of oscillation. Thus, the image is obtained by imaging the force of the oscillating contacts of the cantilever tip with the sample surface.

Tapping mode in fluids

Tapping mode operation in aqueous medium is a very useful tool for biologists because the samples are in a state that closely resembles the *in vivo* environment as compared to dehydrated samples.

Non-contact mode

This mode is preferentially used in the study of chromatin dynamics. The Non-Contact AFM (or NC-AFM) operates with increased tip–sample separation without being in contact with the samples under normal imaging conditions. The cantilever oscillates above the sample surface with small amplitude at a frequency larger than its resonance frequency. This results in increased sensitivity in comparison with tapping mode AFM. Outstanding spatial resolution can be achieved by performing NC-AFM in ultra high vacuum (UVH). In biological studies tapping mode and non-contact modes are most extensively used.

Applications of AFM

- **Compound semiconductor:** Imaging compound semiconductor surfaces on atomic scale.
- **Electronic materials:** AFM tip is used to write and erase nanoscale electronic structures.
- **Data storage:** AFM is nowadays used for data storage due to space and cost constraints.
- **Life science:** AFM measures surface structures of various biological substances, and is capable of visualizing biological objects from living cells down to single molecule levels.

- **Pharmaceuticals:** In pharmaceutical industries, AFM is used for drug crystallization study, particle characterization and tablet coatings.
- **Semiconductor:** AFM is used in characterizing semiconductors.
- **Optics:** AFM is used in optics for the metrological measurement, since the optical profilers are unable to image transparent specimens.
- **Polymers:** Polymers are generally insulators. High resolution images of the polymers are very difficult to obtain using SEM/TEM, as the samples need to be coated with a conductive layer. AFM however can generate images of polymers without sample preparation.

Limitation of AFM

1. The AFM can be used to study a wide variety of samples (i.e. plastic, metals, glasses, semiconductors, biological samples such as the walls of cells, bacteria).
2. Unlike STM or scanning electron microscopy it does not require a conductive sample. However there are limitations in achieving atomic resolution.
3. The physical probe used in AFM imaging is not ideally sharp.
4. As a consequence, an AFM image does not reflect the true sample topography, but rather represents the interaction of the probe with the sample surface (tip convolution).

Need of AFM in Biological Sciences

Biological samples have been studied by a variety of microscopic techniques, starting from light microscope, compound microscope and electron microscope, but each technique has its own limitations. Magnification achieved by compound microscope is not satisfactory, while electron microscope provides a lot of artifacts which could in turn convey wrong and misleading information. AFM has emerged as an excellent tool for biological studies, as the magnification and resolution achieved is quite satisfactory and with limited artifacts. Proper sample preparation could lead to a resolution as low as 2–3 nm. The AFM has been used to visualize different biologically important structures, ranging from DNA–protein complexes, viruses to sub-cellular components of eukaryotic cell. It has also enabled us to examine the dynamic structures arising from changes in different physiological conditions (e.g., change in the shapes of bacteria upon treatment with various drugs).

Study of Viruses with AFM

Viruses are sub-microscopic particles that are considered as entities between living and non-living because they cannot replicate outside the host cell. They are composed of an outer protein coat surrounding the genetic material that consists of either DNA or RNA. AFM was first used for virology in 1992 by Kolbe and colleagues for the study of T4 bacteriophages. AFM can be used to study intact virion, without causing any damage by fixation or staining and has the ability to analyze host–viral interaction in real-time.

Tobacco Mosaic Virus (TMV)

Tobacco mosaic virus (TMV) is a rod-shaped virus containing single-stranded RNA as its genetic material (Fig. 4a). TMV infects plants especially tobacco and other specific members of the family *Solanaceae.* It causes the mosaic disease in tobacco. AFM observations revealed that the width of TMV is in the range of 25–35nm while the height is in the range of 17–23nm in tapping mode and 19–23nm in the case of contact mode (Fig. 4). The structure of TMV is very interesting and artistic. The protein subunits are arranged in a helical fashion with three turns of viral helix consisting of 49 subunits. The single-stranded RNA follows the viral helix where each protein is attached to three nucleotides. One of the most intensively studied TMV proteins is the Coat Protein (CP). It forms the virus particle and in turn protects the viral RNA. CP assembles into the virus particle and this ability helps in the long distance movement of the virus. CP is essentially made up of two layers of helices oriented perpendicular to the axis of the virus particle, while the RNA remains deeply embedded in the CP.

Conclusion

The continuing development of AFM technology provides scientists with a powerful tool to characterize a variety of sample surfaces. Minimal sample preparation, use in ambient conditions, and the ability to image non-conducting specimens at the atomic scale (in some cases) make AFM an extremely versatile and useful form of microscopy. Recent advances in AFM have allowed the successful imaging of real time images of different biochemical reactions, living cells biological samples and the imaging of magnetic microstructures.

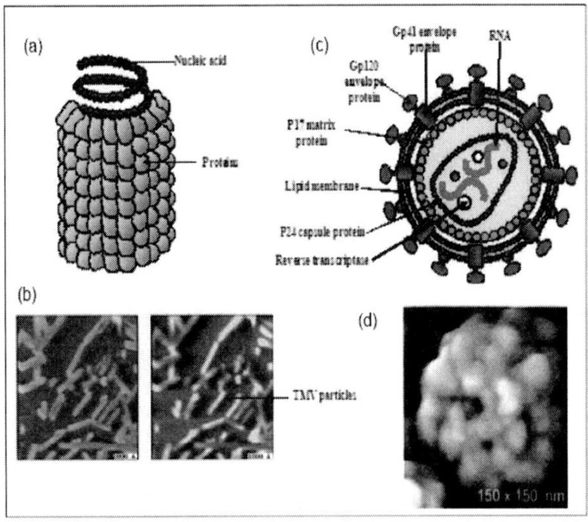

Fig. 4: Visualization of viruses by AFM: **(a)** Artistic representation of the Tobacco Mosaic Virus (TMV) **(b)** AFM image of TMV (From Internet). **(c)** Schematic representation of Human Immunodeficiency Virus (HIV) (d) AFM image of HIV

References

Binnig, G, Quate, C.F. and Gerber, C. 1986. Atomic force microscope, *Phys. Rev. Lett.*, **56(9):** 930-933.

Marti, O., V. Elings, M. Haugan, C.E. Bracker, J. Schneir, B. Drake, S. AGould, J. Gurley, L. Hellemans, K. Shaw, et al., 1988. Scanning probe microscopy of biological samples and other surfaces, *J. Microsc.*, **152**:803–809.

Yao, N., Z.L. Wang. 2000. Handbook of Microscopy for Nanotechnology. 2005. Veeco. Scanning Probe Microscopy Training Notebook. Version 3.0.

14

Gas Chromatography – Mass Spectrometer (GC-MS)

N.B. Nandakumar, S.K. Rajkishore and R. Sunitha

Introduction

Gas chromatography–mass spectrometry (GC-MS) is equipment that combines the features of gas-liquid chromatography and mass spectrometry to identify different substances within a test sample. Applications of GC-MS include drug detection, environmental analysis include pesticides residue in soil, water and plant crop produce, greenhouse gas quantification, food toxicants, assessment of antibiotic compounds, estimation of volatile organic compounds from biological systems besides identification of unknown compounds.

Instrumentation

The GC-MS is composed of two major components: the gas chromatograph and the mass spectrometer. The gas chromatograph utilizes a capillary column which depends on the column's dimensions (length, diameter, film thickness) as well as the phase properties (e.g. 5% phenyl polysiloxane). The difference in the chemical properties between different molecules in a mixture will separate the molecules as the sample travels the length of the column. The molecules are retained by the column and then elute (come off) from the column at different times (called the retention time), and this allows the mass spectrometer downstream to capture, ionize, accelerate, deflect, and detect the ionized molecules separately. The mass spectrometer does this by breaking each molecule into ionized fragments and detecting these fragments using their mass to charge ratio.

These two components, used together, allow a much finer degree of substance identification than either unit used separately. The mass spectrometry process normally requires a very pure sample while gas chromatography using a traditional detector (e.g. Flame ionization detector) cannot differentiate between multiple molecules that happen to take the same amount of time to travel through the column (i.e. have the same retention time), which results in two or more molecules that co-elute. Sometimes two different molecules can also have a similar pattern of ionized fragments in a mass spectrometer (mass spectrum). Combining the two processes reduces the possibility of error, as it is extremely unlikely that two different molecules will behave in the same way in both a gas chromatograph and a mass spectrometer. Therefore, when an identifying mass spectrum appears at a characteristic retention time in a GC-MS analysis, it typically increases certainty that the analyte of interest is in the sample.

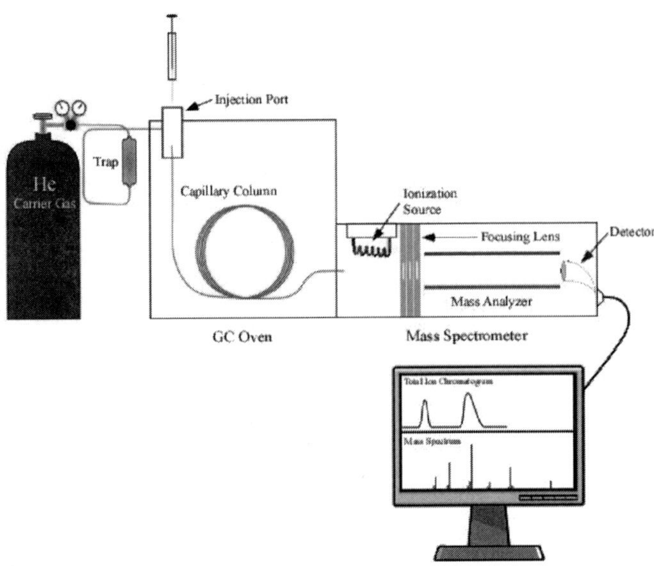

Fig. 1: Diagram of Gas chromatography–mass spectrometry

Gas Chromatograph

A Gas Chromatograph is used to detect the components based on the selective affinity of components towards the adsorbent materials. The sample is introduced in the liquid/gas form with the help of GC syringe into the injection port, it gets vaporized at injection port then passes through column with the help of continuously flowing carrier stream (mobile phase), mainly Helium, and gets separated/detected at the detection port with suitable temperature programming.

Different chemical constituents of the sample travel through the column at different rates depending upon,

1. Physical properties
2. Chemical properties, and
3. Interaction with a specific column filling (stationary phase).

As the chemicals exit the end of the column, they are detected and identified electronically. The function of the stationary phase in the column is to separate different components, causing each one to exit the column at a different time (*retention time*). Other parameters that can be used to alter the order or time of retention are the carrier gas flow rate, and the temperature.

Physical Components involve inlet port, Adsorption column, detector port, flow controller (to control the flow of carrier gas), etc.

Columns

Two types of columns are used in GC

Packed columns are 1.5-10 m in length and have an internal diameter of 2-4 mm. The tubing is usually made of stainless steel or glass and contains a *packing* of finely divided, inert, solid support material (eg. diatomaceous earth) that is coated with a liquid or solid stationary phase. The nature of the coating material determines what type of materials will be most strongly adsorbed.

Capillary columns have a very small internal diameter, on the order of a few tenths of millimeters, and lengths between 25-60 meters are common. The inner column walls are coated with the active materials (WCOT columns).

Some columns are quasi solid filled with many parallel micro pores (PLOT columns). Most capillary columns are made of fused silica with a polyimide outer coating. These columns are flexible, so a very long column can be wound into a small coil. Both physical (length, internal diameter, and stationary phase), and parametric (temperature and flow velocity) column variables affect separation process. To illustrate a wide range combinations to be considered when selecting GC capillary column, the overview of available internal diameters are shown in Table 1.

Table 1: Different Types of Capillary columns

Category	Column diameter range (mm)	Standard commercial column diameters (mm)	Max flow-rate (mL/min)
Megabore	≥0.5	0.53	≥660
Wide bore	≥0.3 to <0.5	0.32, 0.45	≥85 to <660
Narrow bore	≥0.2 to <0.3	0.20, 0.25, 0.28	≥17 to <86
Microbore	≥0.1 to <0.2	0.10, 0.15, 0.18	≥1 to <14
Sub-microbore	<0.1	Various	<1

Temperature dependence of molecular adsorption and of the rate of progression along the column necessitates a careful control of the column temperature to within a few tenths of a degree for precise work. Reducing the temperature produces the greatest level of separation, but can result in very long elution times.

Carrier gas

The choice of carrier gas (*mobile phase*) is important, with hydrogen being the most efficient and providing the best separation. However, helium has a larger range of flow rates that are comparable to hydrogen in efficiency, with the added advantage that helium is non-flammable, and works with a greater number of detectors. Therefore, helium is the most common carrier gas used.

Injectors

A number of options exists for GC inlet systems the most common being split/splitless, programmed temperature vaporiser and cold on-column injector. The choice of optimum sample introduction strategy depends besides of other aspects on the concentration range oftarget analytes (with special requirements in ultra trace analysis), their physico-chemical properties and on the occurrence of matrix co-extracts present in the sample. In any case, to generate accurate and reproducible data in the GC analysis of analytes occurring in foods at trace levels (like most of food toxicants), the cause of following adverse effects due to the injection process should be avoided in maximum feasible extent:

- Discrimination, *i.e.* changed composition of sample that enters the column as compared to that of injected sample
- Poor reproducibility of the amount of sample entering the column
- Poor reproducibility of retention times
- Thermal degradation, adsorption, rearrangement, and/or other changes of analyte
- Impairment of separation column performance.

Split/splitless injection

Due to an easy operation, split/splitless injection remains the dominating sample introduction technique in the analysis of GC MS.

In a *split injection* mode, small volume of sample extract (typically in a range of 0.1–2 μL) is rapidly delivered into a heated glass liner. Under these conditions, thanks narrow input band of analyte, even microbore capillary can be used supposing fast separation is an option. However, considering the loss of most of injected sample (depends on the setting of split ratio) this technique is obviously not suitable for trace analysis where very low detection limits are required Another problem associated with split injection is potential discrimination due to the heating of the syringe resulting in a change of relative abundances of sample components when the mixture of analytes largely differing in boiling points is analysed.

Now a days in trace quantitative analysis, hot *splitless injection* is the most commonly used injection technique since entire injected sample is introduced onto the GC capillary. As already discussed earlier, the major limitation of this inlet is that it suffers from the potential thermal degradation and/or adsorption of susceptible analytes what may result either in matrix-induced response enhancement or its diminishment. The extent of these effects depends on many factors like the inlet temperature, sample residence time (splitless period),the nature of sample matrix and/or liner material.

Detectors

A number of detectors are used in gas chromatography. The most common are the flame ionization detector (FID) and the thermal conductivity detector (TCD). While TCDs are essentially universal and can be used to detect any component other than the carrier gas (as long as their thermal conductivities are different than that of the carrier gas, at detector temperature), FIDs are sensitive primarily to hydrocarbons, and are more sensitive to them than TCD. Both detectors are also quite robust. Since TCD is non-destructive, it can be operated in-series before an FID (destructive), thus providing complementary detection of the same eluents.

General uses

Identification and quantitation of volatile and semivolatile organic compounds in complex mixtures. Determination of molecular weights and (sometimes) elemental compositions of unknown organic compounds in complex mixtures. Structural determination of unknown organic compounds in complex mixtures both by matching their spectra with reference spectra and by a priori spectral interpretation.

Analytes suitable for GC/MS

- In general, GC/MS is used to analyze non-polar compounds soluble in non-polar solvents such as the following:
- Hexane, Ethyl Acetate, Methanol, Acetonitrile, Acetone
- High-boiling solvents should be used only with appropriate methods
- Pyridine and Chlorinated Solvents MUST be avoided
- The compound(s) to be analyzed should be both volatile and chemically stable at the operation temperature of the injector (280°C)
- Thermally labile compounds may be pyrolyzed at the maximum temperature of injector and/or column (320°C)
- Water solutions, and organic solutions at pH lower than 2 and greater than 8 are not allowed for GC/MS analysis, as they may damage the column.

Mass Spectroscopy

Mass spectrometry (MS) is the science of displaying the spectra (singular spectrum) of the masses of the molecules comprising a sample of material. It is used for determining the elemental composition of a sample, the masses of particles and of molecules, and for elucidating the chemical structures of molecules, such as peptides and other chemical compounds. Mass spectrometry works by ionizing chemical compounds to generate charged molecules or molecule fragments and measuring their mass-to-charge ratios. In a typical MS procedure, a sample, which may be solid, liquid, or gas, is ionized. The ions are separated according to their mass-to-charge ratio. The ions are detected by a mechanism capable of detecting charged particles. The signal is processed into the spectra (singular spectrum) of the relative abundance of ions as a function of the mass-to-charge ratio. The atoms or molecules can be identified by correlating known masses by the identified masses or through a characteristic fragmentation pattern.

A mass spectrometer consists of three components: an ion source, a mass analyzer, and a detector. The ionizer converts some portion of the sample into ions. There are a wide variety of ionization techniques, depending on the phase (solid, liquid, gas) of the sample and the efficiency of various ionization mechanisms for the target species in question. An extraction system removes ions from the sample and gives them a trajectory that allows the mass analyzer to sort the ions by mass-to-charge. The detector measures the value of an indicator quantity and thus provides data for calculating the abundances of each ion present. Some detectors also give spatial information, e.g. a multi channel plate.

Types of mass spectrometer detectors

The most common type of mass spectrometer (MS) associated with a gas chromatograph (GC) is the quadru pole mass spectrometer. Another relatively common detector is the ion trap mass spectrometer.

Ionization

After the molecules travel the length of the column, pass through the transfer line and enter into the mass spectrometer they are ionized by various methods with typically only one method being used at any given time. Once the sample is fragmented it will then be detected, usually by an electron multiplier diode, which essentially turns the ionized mass fragment into an electrical signal that is then detected. The ionization technique chosen is independent of using full scan or SIM.

Electron ionization

By far the most common and perhaps standard form of ionization is electron ionization (EI). The molecules enter into the MS (the source is a quadrupole or the ion trap itself in an ion trap MS) where they are bombarded with free electrons emitted from a filament, not unlike the filament one would find in a standard light bulb. The electrons bombard the molecules, causing the molecule to fragment in a characteristic and reproducible way. This "hard ionization" technique results in the creation of more fragments of low mass to charge ratio (m/z) and few, if any, molecules approaching the molecular mass unit. Hard ionization is considered by mass spectrometrists as the employ of molecular electron bombardment, whereas "soft ionization" is charge by molecular collision with an introduced gas. The molecular fragmentation pattern is dependant upon the electron energy applied to the system, typically 70 eV (electron Volts). The use of 70 eV facilitates comparison of generated spectra with library spectra using manufacturer-supplied software or software developed by the National Institute of Standards (NIST-USA). Spectral library searches employ matching algorithms such as Probability Based Matching and dot-product matching that are used with methods of analysis written by many method standardization agencies. Sources of libraries include NIST, Wiley, the AAFS and instrument manufacturers.

Analysis

A mass spectrometer is typically utilized in one of two ways: full scan or selected ion monitoring (SIM). The typical GC-MS instrument is capable of performing both functions either individually or concomitantly, depending on the setup of the particular instrument.

The primary goal of instrument analysis is to quantify an amount of substance. This is done by comparing the relative concentrations among the atomic masses in the generated spectrum. Two kinds of analysis are possible, comparative and original. Comparative analysis essentially compares the given spectrum to a spectrum library to see if its characteristics are present for some sample in the library. This is best performed by a computer because there are a myriad of visual distortions that can take place due to variations in scale. Computers can also simultaneously correlate more data (such as the retention times identified by GC), to more accurately relate certain data.

Another method of analysis measures the peaks in relation to one another. In this method, the tallest peak is assigned 100% of the value, and the other peaks being assigned proportionate values. All values above 3% are assigned. The total mass of the unknown compound is normally indicated by the parent peak. The value of this parent peak can be used to fit with a chemical formula containing the various elements which are believed to be in the compound. The isotope pattern in the spectrum, which is unique for elements that have many isotopes, can also be used to identify the various elements present. Once a chemical formula has been matched to the spectrum, the molecular structure and bonding can be identified, and must be consistent with the characteristics recorded by GC-MS. Typically, this identification done automatically by programs which come with the instrument, given a list of the elements which could be present in the sample.

A "full spectrum" analysis considers all the "peaks" within a spectrum. Conversely, selective ion monitoring (SIM) only monitors selected ions associated with a specific substance. This is done on the assumption that at a given retention time, a set of ions is characteristic of a certain compound. This is a fast and efficient analysis, especially if the analyst has previous information about a sample or is only looking for a few specific substances. When the amount of information collected about the ions in a given gas chromatographic peak decreases, the sensitivity of the analysis increases. So, SIM analysis allows for a smaller quantity of a compound to be detected and measured, but the degree of certainty about the identity of that compound is reduced.

Full scan MS

When collecting data in the full scan mode, a target range of mass fragments is determined and put into the instrument's method. An example of a typical broad range of mass fragments to monitor would be m/z 50 to m/z 400. The determination of what range to use is largely dictated by what one anticipates being in the sample while being cognizant of the solvent and other possible interferences. A MS should not be set to look for mass fragments too low or else one may detect air (found as m/z 28 due to nitrogen), carbon dioxide (m/z 44) or other possible interferences. Additionally, if one is to use a large scan range then sensitivity of the instrument is decreased due to performing fewer scans per second since each scan will have to detect a wide range of mass fragments.

Full scan is useful in determining unknown compounds in a sample. It provides more information than SIM when it comes to confirming or resolving compounds in a sample. During instrument method development it may be common to first analyze test solutions in full scan mode to determine the retention time and the mass fragment fingerprint before moving to a SIM instrument method.

Selected ion monitoring

In selected ion monitoring (SIM) certain ion fragments are entered into the instrument method and only those mass fragments are detected by the mass spectrometer. The advantages of SIM are that the detection limit is lower since the instrument is only looking at a small number of fragments (e.g. three fragments) during each scan. More scans can take place each second. Since only a few mass fragments of interest are being monitored, matrix interferences are typically lower. To additionally confirm the likelihood of a potentially positive result, it is relatively important to be sure that the ion ratios of the various mass fragments are comparable to a known reference standard.

Components of mass spectroscopy

Vacuum system

Vacuum system components maintain the low pressure necessary for the ion source,quadrupole, and ion detector to operate properly. The vacuum manifold, which houses theion source, quadrupole, and ion detector, is pumped (or evacuated) by the high vacuum (turbo molecular) pump. All components, except the rotary-vane pump, are located around the vacuum manifold.

The vacuum system consists of the following components

- Fore pressure gauge
- High vacuum pump
- Ion gauge (upgrade)
- Rotary-vane pump
- Vacuum manifold

Ion Source

The ion source is the part of the MS where ions are formed. It is located inside and to the front of the vacuum manifold. The ion source has two main functions: to generate a beam of electrons and to provide a site for these electrons to interact with sample or reagent gas molecules to form ions. Once formed, the seions are then focused by the lenses into the prefilter and then the quadrupole mass filter.

The ion source uses an inter changeable ion volume so that three different ionization modescan be used (EI, PCI, or NCI). The ion volume sits inside the ion source, which is heated by three cartridge heaters. Three lenses at the back of the ion volume draw out either positive or negative ions from the ion volume and pass them into the prefilter.

The ion source consists of the following components:

- Filament
- Ion source
- Ion volume
- Lenses
- Magnets and magnet yoke

Prefilter and quadrupole

The quadrupole consists of the following components:

- Prefilter
- Entrance Lens
- Quadrupole Mass Filter
- Exit Lens

After ions leave the ion source lenses, they pass through the prefilter. The prefilter is mounted to the back side of the baffle wall and consists of four metal rods held in a square array. The prefilter is curved to prevent neutral species (like helium) from reaching the ion detector. Ions next pass through the entrance lens and into the quadrupole mass filter. The quadrupole massfilter, like the prefilter, is composed of four metal rods held in a square array. However, these rods are precision-aligned so the quadrupole should be handled with care. Finally, ions pass through the exit lens and to the ion detector.

The four quadrupole mass filter rods are charged with a combination of radio frequency (RF) and direct current (DC) voltages. The magnitudes of these voltages give stable oscillations to ions with a specific m/z ratio and unstable oscillations to all others. The unstable ions strike one of the rods, become neutralized, and are pumped away. The quadrupole mass filter be haves as a filter allowing only ions with a certain m/z ratio to reach the ion detector. As the RF and DC voltages are ramped, ions of different m/z ratios are transmitted through the exit lens to the ion detector. This produces the mass spectrum.

Ion detector

Ion detector consists of a conversion dynode and an electron multiplier. The ion detector is mounted inside the vacuum manifold directly behind the quadrupole.

Conversion dynode

The conversion dynode is a concave metal surface located at a right angle to the ion beam. A potential of +10 kV for negative ion detection or -10 kV for positive ion detection is applied to the conversion dynode. The conversion dynode increases signal and decreases noise. High voltage applied to the conversion dynode results in a high conversion efficiency and increased signal. That is, for each ion striking the conversion dynode, many secondary particles are produced. The increase in conversion efficiency is more pronounced for more massive ions than for less massive ions.

When an ion strikes the surface of the conversion dynode, one or more secondary particles are produced. These secondary particles can include positive ions, negative ions, protons, electrons, and neutrals. When positive ions strike a negatively charged conversion dynode, the secondary particles are negative ions and electrons. When negative ions strike a positively charged conversion dynode, the secondary particles are positive ions and protons. These secondary particles are focused by the curved surface of the conversion dynode and accelerated toward the electron multiplier. Because of the off-axis orientation of the ion detector assembly relative to the quadrupole assembly, neutral molecules from the mass analyzer tend not to strike the conversion dynode or electron multiplier. Thus, neutral noise is reduced.

Electron multiplier

An electron multiplier amplifies a current by generating an electron cascade from an oxide-coated surface of an electrode. The two components of an electron multiplier are thecathode and anode.

Cathodes serve as the electrode the electron multiplier uses to generate the electron cascade. It receives a potential of up to -2.5 kV from the high voltage ring. The anode is near ground potential. Secondary particles from the conversion dynode strike near the inner walls of the electron multiplier cathode with sufficient energy to eject electrons. The ejected electrons are accelerated farther into the cathode, drawn by the increasingly positive potential gradient. Due to the design of the eletron multiplier, the ejected electrons do not travel far before they again strike a surface, causing the emission of more electrons. Thus, a cascade of electrons is created that finally results in a measurable current at the anode. The current collected by the anode is proportional to the number of secondary particles striking the cathode. Typically, the electron multiplier has a gain of about 105. If the current of ions entering the electron multiplier from the conversion dynode is 10-12 A, and the gain of the electron multiplier is 105, then a current of 10-7 A leaves the electron multiplier through the anode. This current is converted to a voltage by the electrometer circuit and recorded by the data system.

Conclusion

GC-MS is a widely used analytical instrument for the assessment of volatile compounds derived from plants, soil, microbes and other organic substrates. This device can seprate organic compounds that are volatile in nature besides helping in qualitative and quantitative analysis. The application of GC-MS includeds pesticide residue analysis, plant volatiles, anti-biotic compounds from microbial substrates etc. Despite its wide applications, the technique has to be optimized case by case basis in order to exploit the potential of GC-MS principle.

References

Griffin, M.K., Burke, H.K., 2003. Compensation of hyperspectral data for atmosphericeffects. *Lincoln Laboratory Journal* **14 (1)**:29–54.

Kushalappa, A.C., Lui, L.H., Chen, C.R., Lee, B., 2002. Volatile fingerprinting (SPMEGCFID)to detect and discriminate diseases of potato tubers. *Plant Disease* **86:**131–137.

Lui, L., Vikram, A., Hamzehzarghani, H., Kushalappa, A.C., 2005. Discrimination ofthree fungal diseases of potato tubers based on volatile metabolic profiles developedusing GC/MS. Potato Research **48**:85–96.

Moalemiyan, M., Vikram, A., Kushalappa, A.C., 2007. Detection and discriminationof two fungal diseases of mango (cv. Keitt) fruits based on volatile metaboliteprofiles using GC/MS. Postharvest Biology and Technology **45**:117–125.

Moalemiyan, M., Vikram, A., Kushalappa, A.C., Yaylayan, V., 2006.Volatile metaboliteprofiling to detect and discriminate stem-end rot and anthracnose diseases ofmango fruits. *PlantPathology* **55**:792–802.

Prithiviraj, B., Vikram, A., Kushalappa, A.C., Yaylayam, V., 2004. Volatile metaboliteprofiling for the discrimination of onion bulbs infected by *Erwiniacarotovora*ssp. *carotovora*, *Fusariumoxysporum* and *Botrytis allii*. *European Journal of PlantPhysiology* **110**: 371–377.

Staudt, M., Lhoutellier, L., 2007. Volatile organic compound emission from holm oakinfested by gypsy moth larvae: evidence for distinct responses in damaged andundamaged leaves. *Tree Physiology* **27**:1433–1440.

Vuorinen, T., Nerg, A.M., Syrjälä, L., Peltonen, P., Holopainen, J.K., 2007.*Epirritaautumnata* induced VOC emission of silver birch differ from emission inducedby leaf fungal pathogen. Arthropod–Plant Interactions **1**:159–165.

Section 4: Applications

15

Nanotechnological Approaches in Seed Science

N. Natarajan, S. Kalaivani and S. Senthil Kumar

Introduction

Seed is the key input in agriculture deciding the fate of productivity of any crops. Indeed, seed is referred as a nature's "nano-gift" for agriculture. Seed is a smart delivery system where we shall make manipulation to achieve the desirable goals. The seed technologists in the country were trying to find a second or third generation technologies to address the issues of seed science. Nanotechnology is one of the cutting edge research areas that has enormous potentials to solve the unresolved field problems. The application of nanotechnology to the agricultural and food industries was first addressed by a United States Department of Agriculture roadmap published in September 2000. Nanotechnology in the agricultural front has been more useful in improving the existing crop management techniques. Most of the applications of nanotechnology spearhead around nano encapsulated agrochemicals, herbicides, nano coated gene delivery and nano fertilizers (Nair et al., 2010). The exploration of innovative ideas employing nano principles for seed hardening, seed invigouration and seed germination enhancement and plant establishment is now a new hope for seed technologists. Nanotechnological strategies for seed science includes 1. Quality enhancement, 2. Smart delivery of inputs, 3. High resolution imaging, 4. Quality control, and 5. Quick detection kits.

Seed quality enhancement

Seed invigouration implies an improvement in performance of seed by any post-harvest treatment resulting in good germinability, greater storability and better field performance. Pre-storage treatment of harvest fresh seeds is mainly given towards protection against deteriorative senescence (Basu, 1990). Basu (1994) observed that seed treatment with plant products *viz.,* Red chill fruit powder, Tamarind powder and *Trigonella* seed powder @ 2g kg^{-1} significantly slowed down the deterioration of seeds under various ageing conditions. The entry of the crude powder in the dry state into the dry seeds is through the cracks in the seed coat. Normally these crude powders are of macro-size and irrespective of its size; it gets entry into the seed. Under these circumstances, if nano sized plant powder is used; its efficiency of penetration and action will be more effective. As a result, the quantum of materials used for such treatment will be also minimal. Macro-sized particles can be reduced into nano size through ball milling procedure, which ultimately invigorate the seeds with consequent reduction in loss of vigour and viability during storage.

Utilization of nano TiO_2 by Zheng *et al.* (2005) on spinach for studying the germination and growth of naturally aged seeds had indicated that the nano TiO_2 treatment in proper concentration accelerates germination of aged seeds and increases its vigour. It also improves the growth of spinach and formation of chlorophyll and enhances the rubisco activity and the photosynthetic rate during the growth stage of spinach. This effect was found to be inversely proportional to the size of TiO_2 particles.

Scientists also found that nano-anatase titanium oxide (TiO_2) promoted antioxidant stress by decreasing the accumulation of superoxide radicals, hydrogen peroxide, malonyldialdehyde content and enhance the activities of superoxide dismutase, catalase, ascorbate peroxidase, guaiacol peroxidase and thereby increase the evolution of oxygen rate in spinach chloroplast under UV-B radiation (Lei *et al*., 2008). Entry of nano-TiO_2 into cells would induce oxidation-reduction reactions via the superoxide ion radical during germination in the dark, resulting in the quenching of free radicals in the germinating seeds (Zheng *et al*., 2005). In addition to TiO_2, nanoparticles of zinc oxides, iron oxide, platinum, Manganese etc are also found to have anti oxidant properties. If these particles are used to treat the seed, it can slow down the ageing process.

Metal nanoparticles influence on growth of *Lactuca* seeds was tested by measuring the length of the root and shoot of the plant after 15 days of incubation (Monica and Creonini, 2009). An increase in the shoot/root compared to that of control was evidenced.

Investigations were also made at the Department of Nano Science and Technology, Tamil Nadu Agricultural University, Coimbatore to study the effect of different dosage of ZnO, TiO_2 nanoparticles for the maintenance of vigour and viability of Black gram seeds. Upon accelerated ageing, the seed treated with nanoparticles (400 mg – 1000 g of seeds) maintained higher vigour and viability than the untreated check (Natarajan, 2011).

Lu *et al.* (2002) reported that a mixture of nano-SiO_2 and nano TiO_2 could increase the nitrate reductase in soybean, enhance its abilities of absorbing and utilizing water and fertilizer, stimulate its antioxidant system and apparently hasten its germination and growth. Study undertaken by Pandey (2010) utilizing ZnO nanoparticles on growth rate of *Cicer arietinum* indicated the presence of IAA in seeds which had a spurt in its concentration in the roots owing to the treatment of ZnO nanoparticles. Thus, either by utilizing the nano-formulations of growth regulators or by treating the seeds with ZnO nanoparticles, the level of growth regulator inside the seed can be altered to enhance its germination.

Carbon nanotubes (CNT) which could play an effective role in increasing the germination of the seeds through improved water penetration was testified by the scientists of University of Arkansas. Khodakovskaya et al. (2009) observed that addition of carbon nanotubes to the MS medium where the tomato seeds were placed for germination had accelerated the process of seed germination and significantly shortened the germination time. The germination percentage for seeds that were placed on regular medium averaged 32 % in 12 days and 71 % in 20 days while the germination percentage had a spurt for the seeds placed in MS medium with CNTs (74-82 % in 12 days ; 90 % in 20 days). It was also reported by them that seeds exposed to CNTs had a significantly higher level of moisture compared to the seeds that were not treated. With a start level moisture of 18.4 % content, CNT treated seeds accumulated 57.6 % while the non treated one had 38.9 % moisture.

Possible explanation for this phenomenon would be the role of CNT in penetrating the seed coat while supporting and allowing water inside the seeds. Mature seeds are relatively dry and need to take significant amounts of water before cellular metabolism. So the researchers hypothesized that the observed activation of germination is based on the role of CNTs in the process of water uptake inside the seed embryo. Since, this proven technology makes the seeds to emerge faster, it can be modified to coat the seeds with CNTs so that the treated seeds will germinate faster utilizing the available moisture once the seeds are sown in rainfed condition. Thus the ability of CNT to penetrate thick seed coat of tomato and supporting its germination in a hastened manner leaves a hope on its utilization in other crop seeds as well.

CNTs also enhanced root elongation in onion and cucumber (Canas *et al.*, 2008). Experiments reveal the positive effects of suspensions of Multiwalled CNTs (MWCNTs) on seed germination and root growth of six different crop species namely radish (*Raphanus sativus*), rape (*Brassica napus*), rye grass (*Lolium perenne*), lettuce (*Lactuca sativa*), corn (*Zea mays*) and cucumber (*Cucumis sativus*) (Lin and Xing, 2007). The study conducted by Nair *et al.* (2010) also supported the positive effects of carbon nanotubes on germination and growth of plants species. They studied the effects of both single walled CNT (SWCNTs) and MWCNTs on the germination of rice seeds and observed an enhanced germination for seeds germinated in the presence of nanotubes. Having studied the effect of forms of CNTs on various species, they concluded that the response of plants to nanomaterials varies with the type of plant species, their growth stages and the nature of nanomaterials since the studies showed contradictory effects of even the same nano material in different plants at different developmental stages.

Several polymers are available in the market of which water soluble polymers are of immense use for coating the seeds. These water soluble polymers at a threshold of moisture gets dissolved allowing the coated seed to absorb the moisture and to germinate and withstand. This technology if employed in the pre monsoon condition will help to ensure successful establishment of the treated seeds. When the seeds are coated with a specific water soluble nano polymer, they will break open only when the moisture content of the soil is above 45-50%. This will ensure continued growth of the germinating seedling for the successful establishment under dry land conditions.

Another type of polymer utilized for enhancing /delaying the seed germination is temperature responsive. These polymers exhibit temperature induced phase transitions that change the permeability properties of the polymer. Seeds treated with polymer were found to take less water at 10 ^{0}C compared to 25 ^{o}C. This facilitates the sowing of seeds coated with polymer during the cool months which will germinate once the time reaps up.

Landec Ag company has lauched three commercial product families utilizing the application of polymers. They are: 1) Early Plant® Corn coatings which allow the farmers to plant 3-4 weeks earlier than normal in the spring broadening the planting window and serving as a planting management tool. The seed coating protects the seed from imbibing moisture until soil temperatures are ideal for germination. This helps farmers plant all their crops on time, maximize yield potential, avoid late planting losses and lower post-harvest dry down costs, 2) Pollinator Plus® coatings which eliminate split planting operations and reduce weather related risks in seed production of hybrid corn and 3) Relay Crop

Soybean coating is a double cropping system that allows farmers to grow two crops in one season in a single field.

Most of these technologies were developed and utilized in the temperate countries especially in corn seed production and hybrid development. However, for the tropical countries like India, these technologies are to be tailored to suit our requirements wherein temperature dependent, water loving polymers to be developed and employed in coating with the seeds to make them germinate under favourable conditions. For example, thermosensitive nano polymer coated seeds if sown in high temperature regimes will not germinate and desiccate until the favourable condition for the germination set in owing to the lowered temperature. Similarly, hydrophilic nanopolymers can help in getting a better crop stand through uniform germination of the coated seed whenever the soil moisture attains an appreciable level.

Extension of this polymer coating technology includes polymer banding technique wherein two to three polymers can be coated on the seeds. Polymers may be either hydrophilic or hydrophobic, thermostable or thermolabile, nutri rich. Coating of multi layers of polymers will help in ensuring the uniform stand of the seed crop by allowing them to germinate under favourable condition, ensuring optimum growth and vigour.

Smart delivery of inputs

The use of silver nanoparticles as a coat in the seed has tremendous potential in effectively controlling the seed borne diseases. If the selected seeds are embedded / coated with silver nanoparticles, these particles can prevent the sporulation, penetration of fungus and the growth of bacteria which will help in saving the life of the seeds. Longevity of the seeds thus treated can be maintained without any deterioration in its seed quality and viability.

Nano formulation of Au metal was found to possess antibacterial and antifungal properties (Panacek *et al.*, 2009; Singh *et al.*, 2008; Jo *et al.*, 2009; Kim *et al.*, 2009). Silver nanoparticles were found to possess broad spectrum of antimicrobial activity which can reduce various plant diseases by controlling the spore producing fungal pathogens (Jo and Kim, 2009). Thus, the effectiveness of Ag NPs can be improved by applying them well before the penetration and colonization of fungal spores within the plant tissues. Exposure of fungal hyphae to Ag NPs caused severe damage by the separation of layers of hyphal wall and collapse of hyphae.

Recently, it has been found that electrospun fibre can be used as a smart delivery system to channelize insecticides and fungicides to deter the seed borne pathogens (Rameash, 2013). He developed electrospun fibre using synthetic polymer PLGA (Poly Lactic Glycolic Acid) and successfully fortified with imidacloprid (insecticide) and Tubuconazol (fungicide). The electrospun fibre was used to coat the cotton seeds that facilitated protection against seed borne pathogens.

High resolution imaging

Seeds, to be qualified as a silent living legend, had so many variations in its surface among the different genotypes of a particular species and innumerable variations among the different genus. Seed coat is a specific feature for each crop variety apart from protecting the embryo and plays a key role in the maintenance of viability particularly in Leguminacae. Hard seed coat aids in extending the seed longevity by preventing the moisture absorption from atmosphere (Yasseen et al., 1994).

The emerging nano world has options of studying these variations through new and novel means to investigate structures and systems, besides exploiting the well-known microscopic, diffraction and spectroscopic methods. Examination of seed coat under Scanning Electron Microscope (SEM) helps in-depth understanding of its structure for the following purpose:

1. Temporal changes in seed coat structure and its pattern during seed development and maturation will aid in devising suitable methods to protect the seeds.
2. Varietal/ species confirmation based on seed cot pattern and also confirmation of inter-specific and inter- generic hybrids.

Utilization of the images taken through Scanning Electron Microscope helps in understanding the mode of action of scarification treatment on seed coat. Since most of the times, effectiveness of pretreatments depends on seed characteristics; their application should be modulated with different period, concentration or degree of combinations for achieving maximum percent germination. Studies taken up by Jordan et al. (1985) brought out the role of cracks in the palisade cuticle of soy bean seed coat and its relationship to the permeability of water. Depression, small cracks appear in the cuticle covering palisade layer. The presence and absence of these cracks appears to control the permeability property of seed coat. SEM images helps in diagnosing the water uptake *vis-à-vis* seeds germinability. Seed coat architecture can be understood through SEM images which help in sharpening the knowledge on understanding seed germination (Fengshan et al., 2004). Thus SEM image rationalizes the seed scarification treatment based on its effectiveness.

A Scanning electron microscopy study of the seed and post-seminal development in *Angelonia salicariifolia* made by Fabiola et al., 2001 which allowed species identification at different stages of development under field conditions. It also facilitated the differentiation between very similar taxonomic groups. To establish the amount of intra- and inter- specific variation in seed and seedling characters, SEM images are more helpful.

Ultrastructural analysis of seed testa in developing *Lupinus pilosus* seeds helped in characterizing the development of 'rough-seed' during seed maturation. Roughness of seed testa surface starts from 26 days after flowering. At this point, seed dehydration starts quickly. Macrosclereid layer is basically responsible for seed testa roughness trait in *L. pilosus* (Agnieszka et al., 2008).

Effects of ageing on amylase activity and scutellar cell structure during imbibitions in wheat seed was studied by Ganguli and Sen (1993). In fresh seeds during later stages of germination, the scutellar cells develop finger like projections. These cells serve to absorb endospermic reserves hydrolysed by aleurone amylase. Aged non-germinating seed showed no finger like projection even after prolonged imbibitions. Observations on finger like projections appearing on the seeds helps in understanding the seed viability.

SEM studies helped in understanding the onion seed ageing phenomenon as described by Gorinstein et al. (2004). Microscopic studies help in understanding and visualizing structural changes and textural differences in protein fractions. This study helped to correlate electrophoretic pattern and microstructure of proteins which can exhibit close identity with each other. This method serves as an efficient alternate for studying the seed coat structural and textural changes. In general, seed coat micromorphology helps in classifying *Solanum* species. Studies undertaken by Junlakitjawat et al. (2010) had resulted in classifying eight species of *Solanum* into three groups based on presences of the spiral secondary thickening and fibrils or hair on the surface. The phylogenetic relationship among members of the genus indicated that possibilities of having variation in the seed surface in the genus can be elucidated.

Quality Control

Nano-barcodes are free standing, cylindrically shaped metal nano-particles having dimensions of 20-500 nm in diameter and 0.04-15 mm in length, which are encodable, machine – readable (Nicewarner-Pena et al., 2001). The particles are manufactured in a semi automated highly scalable process by electro-plating using inert materials such as gold, silver into templates defining particle diameter and then releasing the resulting stripped nano-rods from the templates.

The barcodes can be affixed on the seeds to determine the product traceability. If the company wants to know the whereabouts of the seed, the seed can be agitated in a solution to washout the barcodes and can be read using a barcode reader. Seed samples for the different seed lots can be nano bar coded differently. Information such as location, date of harvest, germination, purity can be downloaded easily and used. This will also help in IPR protection for seed companies seeking to enter highly competitive market of GM seeds. As far as legal issues are concerned, tracking seeds will help resolving disputes arising from use of IPR protected seeds.

Quick detection kits

Electronic nose

Seeds during storage release several volatile aldehydes that determine the degree of ageing. These gases are harmful to even other seeds. Such volatile aldehydes can be detected using E-nose and seeds showing signs of deterioration can be separated and invigorated prior to their use. Electronic nose (E-nose) is a device that mimics the operation of human nose in detecting an array of gases. This device contains several gas sensors to detect different types of odors. The main purpose is to identify the odorant, estimate the concentration of the odorant and find characteristic properties of the odor. The main component of e-nose is a gas sensor composed of Nano-particles (zinc oxide nanowires) whose resistance changes when a certain gas passes over it (Hossain et al., 2005; Sugunan et al., 2004). The change in resistance generates a change in electrical signal that forms the fingerprint for gas detection. The advantage of using nanoparticles is that they have improved uncontaminated surface area for better gas adsorption.

Genetic purity

Seed production generally requires growing of selected genotypes in isolation. Particularly for wind pollinated crops, isolation distance may run to several meters to avoid contamination of foreign pollen. Even some time after providing the required isolation, contamination of the seed crop due to invasion of foreign pollen occurs. Hence, detecting pollen load that will cause contamination is a sure method to ensure genetic purity. Usually pollen flight is determined by its weight, air temperature, humidity, and wind velocity and pollen production of the crop. Provision of micro electronic sensors specific to the seed crop pollen placed along the perimeter of the seed field at varied heights can help to detect the possible contamination. This technique has its own advantages like fixing up the isolation distance for the seed crop in lieu of the regular method which is now followed being tedious, for assessing the purity of the seed crop and to quantify the contaminants thereby taking steps to reduce the contamination. The same method can also be used to prevent pollen from Genetically Modified Crop from contaminating the natural crop.

Seed moisture detection

Seed moisture has a definite role in seed storage and longevity. High moisture seeds are poor storers. It is pertinent to reduce the moisture content of stored seeds to a safe limit prior to storage. Lithium chloride is currently used for doping in silica gel to indicate the hygroscopicity of silicon hydroxide. Dry, crystalline silica gel is colourless, upon exposure to moisture it swells, but is still colourless. Lithium chloride is blue when dry and upon hydration turns pink. The degree of pink colour is an indicator of its hydration.

Seeds can be coated with highly purified lithium chloride quantum dots and sprayed on seed so as to serve as indicator for monitoring moisture content during storage. Seed store managers can then use driers to dry the seed or plan to reduce seed moisture just by observing the colour changes in the quantum dots. This facilitates real time monitoring of moisture content in charge. This technology will be of immense in use in germplasm conservation.

References

Agnieszka, I., Piotrowicz-Cieslak, Barbara Adomas, Dariusz J. Michalczyk and Kamilla Górska. 2008. Ultrastructural analysis of seed testa in developing *Lupinus pilosus* seeds. Proceedings of the 12th International Lupin Conference, 14-18 Sept., Fremantle, Western Australia, New Zealand.

Avinash C. Pandey, Sharda S. Sanjay and Raghvendra S. Yadav. 2010. Application of ZnO nanaoparticlews in influenecing the growth rate of *Cicer arietinum*. *Journal of Experimental Nanoscience*. **5(6):** 488-497.

Basu, R.N. 1990. Seed invigouration for extended storability. Proc. of the Int. Conf. on Seed Sci. and Technol, New Delhi, 88.

Basu, R.N. 1994. An appraisal of research on wet and dry physiological seed treatments and their applicability with special reference to tropical and sub-tropical countries. *Seed Sci. and Technology*., **22**: 107-126.

Canas, J.E., M. Long, S. Nations, R. Vadan, L. Dai, M. Luo, R. Ambikapathi, E.H. Lee and D. Olszyk. 2008. Effects of functionalized and non-functionalized single-walled carbon nanotubes on root elongation of select crop species. *Environ. Toxicol. Chem*., **27**: 1922-1931.

Fabiola V. Moro., Ana C.R. Pinto, Jaime M. Dos Santos and Carlos F. Damiao Filho. 2001. A scanning electron microcopy study of the seed and post-seminal development in *Angelonia salicariifolia* Bonpl. (Scrophulariaceae). *Annals of Botany*. **88:** 499-506.

Fang Yang, Fashui Hong, Wenjuan You, Chao Liu, Fengqing Gao, Cheng Wu and Ping Yang. 2005. Influence of nano-anatase TiO_2 on the nitrogen metabolism of growing spinach. *Biological Trace Element Research*. **110**: 179-189.

Fengshan Ma., Ewa Cholewa, Tasneem Mohamed, Carola Peterson and Mark Gijzen. 2004. Carcks in the palisade cuticle of soybean seed coats correlate with their permeability to water. *Annals of Botany*. **94**: 213-228.

Ganguli, S and Swati Sen-Mandi. 1992. Effcets of ageing on amylase activity and scutellum cell structure during imbibitions in wheat seed. *Annals of Botany*. **71**: 411-416.

Gorinstein, S., Elke Pawelzik, Efren Delgado-Licon, Kazutaka Yamamoto, Shoichi Kobayashi, Hajime Taniguchi, Ratiporn Haruenkit, Yong-Seo Park, Soon-Teck Jung, Jerzy Drzewiecki and Simon Trakhtenberg. 2004. Use of scanning electrion microscopy to indicate the similoarities and differences in pseudocereal and cereal proteins. *International Journal of Food Science and Technology*. **39**:183-189.

Hossain, M.K., S.C. Ghosh, Y. Boontongkong, C. Thanachayanont and J. Dutta. 2005. Growth of zinc oxide nanowires and nanobelts for gas sensing application. Journal of Metastable and Nanocrystalline Materials. **23**: 27-30.

Jo, Y.K. and B.H. Kim. 2009. Antifungal activity of silver ions and nanoparticles on phytopathogenic fungi. *Plant Dis.,* **93**: 1037-1043.

Johnson, G.A. and D.H. Hicks. 1999. Use of temperature-responsive polymer seed coating to control seed germination. Proc. of the Int. Symp. Stand Establishment Seed. Eds. Liptay, Vavrina, Welbaum, Acta Hort., 504: ISHS, 229-236.

Jordan S. Lowell., James L. Jordan and Catalina M. Jordan. 1985. Changes induced by water on Euphobia supine seed coat structures. *Amer. J. Bot.*, **72(10):** 1530-1536.

Junlakitjawat, A., A. Thongpukdee and C. Thepsithar. 2010. Seed coat micromorphology of some *Solanum* species in Thailand. *Journal of the Microscopy Society of Thailand*. **24 (1):** 3-16.

Lei Zheng, Fashui Hong, Shipeng Lu and Chao Liu. 2005. Effect of nano-TiO_2 on strength of naturally aged seeds and growth of spinach. *Biological Trace Element Research.* 104:83-91.

Lin, D. and Xing. 2007. Phytotoxicity of nanoparticles: inhibition of seed germination and root growth. *Environ. Pollut.*, **150**: 243-250.

Lu, C.M., C.Y. Zhang, J.Q. Wen and G.R. Wu. 2002. Research of the effect of nanometer materials on germination and growth enhancement of *Glycine max* and its mechanism. *Soybean Sci.,* **21(3):** 168-171.

Mariya Khodakovskaya, Enkeleda Dervishi, Meena Mahmood, Yang Xu, Zhongrui Li, Frmiya Watanabe and Alexandru S. Biris. 2009. Carbon nanotubes are able to penetrate plant seed coat and dramatically affect seed germination and plant growth. *American Chemical Society*. 10(3): 3221-3227.

Min. J.S. *et al.,* 2009. Effects of colloidal silver nanoparticles on sclerotium-forming phytopathogenic fungi. *Plant Pathol. J.*, **25**: 376-380.

Mohamed Yaseen, Y., Sheryl A. Barringer, Walter E. Splittstosser and Suzanne Costanza. 1994. The role of seed coat in seed viability. *The Botanical Review.* **60**: 426-439.

Natarajan, N. 2011. Effect of ZnO nanoparticles on vigour and viability of blackgram. Unpublished data.

Nicewarner-Pena, S.R., R.G. Freeman, B.D. Reiss, L. He, D.J. Pena, I.D. Walton, R. Cromer, C.D. Keating and M.J. Natan. 2001. Submicrometer metallic barcodes. *Science*. **294 (5540):** 137-141.

Panacek, A., M. Kolar, R. Vecerova, R. Prucek, J. Soukupova, V. Krystof, P. Hamal, R. Zboril and L. Kvitek. 2009. Antifungal activity of silver nanoparticles against *Andiala spp. Biometals*. **30**: 6333-6340.

Park, H.J., S.H. Kim, H.J. Kim and S.H. Choi. 2006. A new composition of nanosized silica-silver for control of various plant diseases. *Plant Pathol. J.*, **22**: 295-302.

Remya Nair, Saino hanna Varghese, Baiju G. Nair, T. Maekawa, Y. Yoshida and D. Sakthi Kumar. 2010. Nanopartilces material delivery to plants. *Plant Science*. **179**: 154-163.

Roco, M.C., R.S. Williams and P. Alivasatos Eds. 1999. Nanotechnology Research Directions: IWGN workshop report, Kluwer Academy Pulishers: Norwell, MA.

Ruffini Castiglione Monica and Roberto Creonini. 2009. Nanoparticles and higher plants. *Caryologia*. **62(2)**: 161-165.

Singh, M., S. Singh, S. Prasad and I.S. Gambhir. 2008. Nanotechnology in medicine and antibacterial effect of silver nanaoparticles. *Digest J. Nanometer. Biostruct.* **3:** 115-122.

Sugunan, A., H.C. Warad, C. Thanachayanont, J. Dutta and H. Hofmann. Zinc oxide nanowires on non-epitaxial substrates from colloidal processing for gas sensing application, Nanostructured and Advanced Materials for Applications in Senson, Optoelectronic and Photovoltaic Technology, NATO-Advance Study Institute, Sozopol, Bulgaraia, 6-17th September, 2004.

Vanangamudi, K., K. Natarajan, T. Saravanan, R. Renuka, N. Natarajan, R. Umarani and A. Bharathi. Eds. 2006. Seed Hardening, Pelleting and Coating : Principles and Practices. Satish Serials Pub, 418 pages.

Willson, D.O. and M.B. McDonald. 1986. The lipid peroxidation model of seed deterioration. *Seed Sci. and Technol.*, **14**: 269-300.

Z. Lei *et al.*, 2008. Antioxidant stress is promoted by nano-anatase in spinach chloroplasts under UV-Beta radiation. *Biol. Trace Elem. Res.* **121**: 69-79.

16

Nano-Fertilizer for Balanced Crop Nutrition

K.S. Subramanian and C. Sharmila Rahale

Introduction

Fertilizers play an important role in improving soil fertility and productivity of crops regardless of nature of cropping sequence or environmental conditions. It has been unequivocally demonstrated that one third of crop productivity is dictated by fertilizers besides influencing use efficiencies of other agri-inputs. In the past four decades, nutrient use efficiency (NUE) of crops had hardly exceeded 35-50%, 18-20% and 30-35 % for N, P and K, respectively, despite our relentless efforts. The remaining nutrients stay in soil or enter into the aquatic environment causing eutrophication. In addition to the low nutrient efficiencies, Indian agriculture facing a problem of low organic matter, imbalanced fertilization and low fertilizer response that eventually caused crop yield stagnation (Biswas and sharma, 2008).The fertilizer response ratio in the irrigated areas of the country has dcreased from 13.4 kg grain / kg nutrient applied in 1970's to just 3.7 kg in 2005. In other words, more amounts of fertilizers is required to produce the same quantity of grain output. For instance, 27 kg NPK ha^{-1} was required to produce one tonne of grain in 1970 while the same level of production can be achieved by 109 kg NPK ha^{-1} in 2008. The optimal NPK fertilizer ratio of 4:2:1 is ideal for crop productivity while the current ratio is being maintained at 10: 2.7: 1 in India due to the excessive use of nitrogenous fertilisers. In order to achieve a target of 300 million tonnes of food grains and to feed the burgeoning population of 1.4 billion in the year 2025, the country will require 45 mt of nutrients as against a current consumption level of 23 Mt. The extent of multi-nutrient deficiencies are alarmingly increasing year by year which is closely associated with a crop loss of nearly 25-30%. The extent of nutrient deficiencies in the country are of the order of 89, 80, 50, 41, 49 and 33% for N, P, K, S, Zn and B, respectively. Thus from all sources, the country will be required to arrange for the supply of about 40-45mt of nutrients by 2025 (Subramanian and Tarafdar, 2009).

Nanotechnology is a bootom up approch in which atom-by-atom by manipulation is possible. This faciniating field of science has been exploited widely in engineering, health, electronics and material sciences and agricultural scientists have began to use it as a tool to improve the input use efficiencies by integrating nanotechnological approaches in the conventional production system. In this context, there would be greater importance of the information how to increase the nutrient use efficinecy of fertilisers by nanotechnology in the coming years. A pioneering effort has been taken at the Tamil Nadu Agricultural University, Coimbatore, in order to synthesize, characterize and apply nano-formulations in agricultural production system. In this review, current status of understanding of nano-fertilizer formulations and its associated effects on crop production systems has been narrated.

Nano-fertilizers

Nano-fertilizers are nutrient carriers of nano-dimensions ranging from 30-40 nm (10^{-9} m or one-billionth of a metre) and capable of holding bountiful of nutrient ions due to their high surface area and release it slowly and steadily that commensurate with crop demand. Subramanian et al. (2008) reported that nano-fertilizers and nano-composites can be used to control the release of nutrients from the fertilizer granules so as to improve the nutrient use efficiency while preventing the nutrient ions either get fixed or lost to the environment. Nano-fertilizers have high use efficiency and can be delivered in a timely manner to a rhizospheric target. There are slow-release and super sorbent nitrogenous and phosphatic fertilizers. Some new-generation fertilizers have applications to crop production on long-duration human missions to space exploration. (Lal, 2005).

According to the report of Iranian Nanotechnology Initiative Council (2009), Iranian researchers have produced the first nano-organic iron-chelated fertilizer in the world. Nano fertilizers have unique features like ultra high absorption, increase of 20 % to 200 % in production, rise in photosynthesis by 3.5 times and a 70% expansion in the leaves' surface area, Iranian Nanotechnology Initiative Council reported." While foreign samples cause an increase of up to 30 % in photosynthesis, the Iranian nano fertilizers are able to cause a 350% increase.

Moreover, these nano fertilizers are environmentally sustainable due to their organic base which makes them more suitable than foreign fertilizers that are hormone based.

Recently, Subramanian and Rahale (2009) have monitored the nutrient release pattern of nano-fertiliser formulations carrying fertilizer nitrogen. The data have shown the nano-clay based fertilizer formulations (zeolite and montmorillonite with a dimension of 30-40 nm) are capable of releasing the nutrients for a longer period of time (> 1000 hrs) than conventional fertilizers (< 500 hrs).

Subramanian and Tarafdar (2011) suggested that clay particles are adsorptive sites carrying reservoir of nutrient ions. Major portion of nutrient fixation occurs in the broken edges of the clay particles. Zero valence nano-particles can adsorb on to the clay lattice thereby preventing fixation of nutrient ions. Further, nano-particles prevent the freely mobile nutrient ions to get precipitated. These two processes assist in promoting the labile pool of nutrients that can be readily utilized by plants. Fertilizer particles can be coated with nano-membranes that facilitate in slow and steady release of nutrients. This process helps to reduce loss of nutrients while improving fertilizer use efficiency of crops.

The naturally occurring clay minerals and zeolites have been reduced to the size of nano dimensions using top down approach and nutrient at desirable proportion have been fortified in the clays after surface modification (Bansiwal, et al., 2006). The nanofertiliser formulations before and after loading with nutrients have been characterized using High Resolution Microscopes and Spectroscopy (Liu et al., 2006). Nano zeolites are capable of retaining nutrients due to its extensive surface area and release slowly and steadily for an extended period of 1000-1200 hrs while conventional fertilizers could release for about 300-400 hours (Subramanian and Rahale, 2010). The literature strongly suggests that nano technological applications to improve the NUE and productivity of crops without associated environmental hazard.

Nitrogen

About 90% of Indian soils are deficient in N and thus there is a universal response for the addition of nitrogenous fertilizers. Indiscriminate use of N fertilizers caused major detrimental impacts on the diversity and functioning of the non-agricultural ecosystems due to eutrophication of freshwater and marine water (Tilman, 1999; Beman et al., 2005). In addition, there can be gaseous emission of N reacting with the stratospheric ozone and the emission of toxic ammonia into the atmosphere (Ramos, 1996; Stulen et al., 1998). The nitrogen use efficiency (NUE) by crops is very low (30-35%) due to the loss of N to the tune of 50 - 70% by leaching, volatalisation and microbial mineralization (Salman et al., 1989). One of the attempts to increase NUE is slow release or controlled release fertilizers which releases N slowly in available form or to develop materials which control the release of N in available form slowly. The most important slow release fertilizer is the coating of conventional N fertilizer with sulphur, neem, lac or clay (Saratin, 2010). The idea was due to coating with these materials, urea comes into the soil solutions through diffusion process very slowly and in this way they supply nitrogen to the plants at a controlled rate or slow rate but for a longer period. But these attempts to increase the NUE were yielded with little success due to the mismatch between the nutrient release and crop demand. Nano-fertilizer may regulate the release of nutrients and deliver the correct quantity of nutrients required by the crops in suitable proportion and promote productivity while ensuring environmental safety (DeRosa, 2010).

Griffith and Streeter (1994) stated that NH_4^+ occupying the internal channels of zeolite should be slowly set free, allowing the progressive absorption by the crop which results in a higher dry matter production of the crop. Dwairi (1998) suggested that zeolite impregnated with urea can be used as slow release fertilizer carrying the slow and steady release of N from nano-zeolite. Perrin et al. (1998) demonstrated that amending sandy soil with ammonium-loaded zeolite can reduce N leaching while sustaining growth of sweet corn and increasing N-use efficiency compared to ammonium sulphate. The same result was also demonstrated by Hernandez et al., (1994) that the combination of zeolite and slow release N fertilizers would increase the N efficiency.

Rahale (2010) reported that nano-fertiliser increased the NUE up to 45 % over control. She also reported that the release of nitrate from nano-zeolite continued even after 1176 h, with concentrations ranging from 110 to 114 mmol L^{-1}. The results clearly demonstrated slow and steady release of N from nano-zeolite for more than 45 days while conventional fertilizer does it for only 8 days.

Phosphorus

Phosphorus (P) is a vital nutrient for plants because of its role in ribonucleic acid and function in energy transfers via ATP. It is often the limiting nutrient in agricultural ecosystems due to its low availability in soils. Phosphate ions in soil gets often fixed as Fe, Al and Ca in soils and a small fraction of phosphate in available form (Ozanne et al., 1980; Du et al. 2005; Li et al. 2005; Mcbeath et al. 2005). The efficiency of fertilizer P use by crops ranged from 18 to 20 % in the year that it is applied. The remaining 78 to 80% becomes part of the soil P pool which is released to the crop over the following months and years (Malhi et al., 2003).

Malhi et al. (2003) also reported that polymer coating of mono ammonium phosphate (MAP) improved plant recovery of fertilizer phosphorus (P) and provided a modest barley grain yield advantage relative to uncoated MAP. Coating P fertilizer could limit the contact of applied P with soil, possibly reducing its precipitation and/or adsorption on soil colloids, and increase its availability to developing plant roots. The development of thin polymer coatings has improved the opportunity to coat fertilizer granules and increased the predictability of when nutrients become available from the controlled-release product.

The release of P from fertilizer-loaded unmodified zeolite and surface modified zeolite (SMZ) and from solid KH_2PO_4 was performed using the constant flow percolation reactor. The results showed that the P supply from fertilizer-loaded SMZ was available even after 1080 h of continuous percolation, whereas P from KH_2PO_4 was exhausted within 264 h. The results indicated that SMZ is a good sorbent for PO_4^{3-}, and a slow release of P is achievable (Bansiwal et al., 2006). These properties suggested that SMZ has a great potential as the fertilizer carrier for slow release of P. Eberl (2008) reported that phosphate (H_2PO_4) can be released to plants from phosphate rock (P-rock) composed largely of the calcium phosphate mineral apatite by mixing the rock with zeolite having an exchange ion such as ammonium. The approximate reaction in soil solution is as follows:

$$(\text{P-rock}) + (NH_4 - \text{zeolite}) = (\text{Ca zeolite}) + (NH_4) + (H_2PO_4^-)$$

The zeolite takes Ca^{2+} from the phosphate rock, thereby releasing both phosphate and ammonium ions. Unlike the leaching of very soluble phosphate establish equilibrium. The fertilizers (for example, super phosphate), the controlled-release phosphate is released of a specific chemical reaction in soil. As phosphate is taken up by plants or by soil fixation, the chemical reaction releases more phosphate and ammonium in the attempt to reestablish equilibrium. The rate of phosphate release is controlled by varying the ratio of P-rock to zeolite. Phosphorus is also released from the rock by the lowering of soil pH as ammonium ions are converted to nitrate.

Potassium

The common potassium fertilizers are completely water-soluble and, in some cases, have a high salt index. Consequently, when placed too close to seed or transplants, they can decrease seed germination and plant survival. This fertilizer injury is most severe on sandy soils, under dry conditions, and with high rates of fertilization — especially nitrogen and potassium. Some crops such as soybeans, cotton, and peanuts are much more sensitive to fertilizer injury than corn. There is little information available about potassium (K) use efficiency. However, it is generally considered to have a higher use efficiency than N and P because it is immobile in most soils and is not subject to the gaseous losses that N is or the fixation reactions that affect P. First year recovery of applied K can range from 20% to 60% (Roberts, 2009). Improvement of nutrient efficiency in crops is an important issue in agriculture for reducing cost in agriculture production and for protecting the environment (Oosterhuis et al., 2008). Efficient use of nutrients is the relative ability of plant to produce maximal amounts of dry matter for each increment of nutrients accumulated or it is plant yield (productivity) per unit nutrient supply (Tawfik, et al., 2010). Information regarding slow-release K materials is lacking compared to information available on N materials.

According to Spark and Jardian (1984) an important term of Friendlich equation is the constant r, which is indicative of the adsorb rate of K^+. A relatively high r^2 values indicated that the model successfully describe the kinetics of K adsorption by nano-clays. Pino et al. (1995) provided data on the slow release effect of K from K-zeolite. The high CEC of the nanoclays is caused when silica (Si^{4+}) is substituted by aluminum (Al^{3+}) thereby raising a negative charge of the mineral lattice. This negative charge is balanced by cations such as ammonium, sodium, calcium, and potassium, which are exchangeable with other cations (Curkovic et al., 1997). According to Weatherly and Miladinovic (2004), Co-existed cations such as Na^+, K^+, Ca^{2+}, NH_4^+ and Mg^{2+} are typically presented with K^+ and NH_4^+, the presence of these competing cations could affect K^+ desorption on zeolite.

Zhou and Huang (2007) reported the slow and steady release of K from nano-zeolite. This may be due to the ion exchangeability of the zeolites with selected nutrient cations, zeolites can become an excellent plant growth medium for supplying plant roots with additional vital nutrient cations and anions. The nutrients are provided in a slow-release, plant root demand-driven fashion through the process of dissolution and ion exchange reactions. The adsorption of nutrients by plant roots drives the dissolution and ion exchange reactions, pulling away nutrients as needed. The zeolite is then "recharged" by the addition of more dissolved nutrients. Their selectivity of ion exchange on zeolite was determined in an order of $K^+ > NH_4^+ > Na^+ > Ca^{2+} > Mg^{2+}$ (Guo et al., 2008).

Micronutrients

Micronutrients are elements which are essential for plant growth, but are required in much smaller amounts that include Fe, Mn, Cu, Zn, B and Mo. Sheta et al. (2003) suggested that natural zeolites, particularly clintopillolite have a high potential for Zn and Fe sorption with a high capacity for slow release fertilizers. According to Broos et al. (2007) reported that the slow release of Zn is attributed to the sparingly solubility of minerals and sequestration effect of exchanger, there by releasing trace nutrients to zeolite exchange sites where they are more readily available for uptake by plants.

Eberl (2008) reported that zeolite in soil can aid in the release of some trace nutrients and in their uptake in plants. The release of cationic micronutrients has enhanced by the presence of zeolite in neutral soil. The concentration of Cu and Mn in sudangrass (in mg/kg) was significantly related to the zeolite/P-rock in experimental systems that used two different NH_4 saturated zeolites, two different soils and two different forms of P-rock. The concentrations of trace elements were also increased by 19% for iron (Fe^{2+}) and 10% for manganese (Mn^{2+}). Overall results indicated that soil amendment with zeolite could effectively ameliorate salinity stress and improve nutrient balance in a sandy soil (Al- Busaidi et al., 2008).

Lin and Xing (2008) reported that zinc oxide nano-particles were shown to enter the root tissue of ryegrass and improved the germination. Methods of visualization of carbon coated nano-tubes in plant cells using pumkin plants as the model were reported by Melendi et al. (2008)

Nanocomposites

Nano-composites have been developed in oder to supply wide range of nutrients in desirable properties. These compounds compounds are capable of regulating the inputs depending on the conditions of soil or requirement of crops. Zinc–aluminium layered double-hydroxide nano-composites have been used for the controlled release of nutrients that regulate plant growth (Hossain et al., 2002).

In soil niches, nano-materials are porous and hydrated and as such they control moisture retention, permeability, solute transport, and availability of plant nutrients in soils. These nano-materials materials also control exchange reactions of dissolved inorganic and organic species between the soil solution and colloidal surfaces. The physico-chemical properties in the surface of nano-composites provide much of reactivity to soil biological and abiotic processes (Navrotsky, 2004). Zhang et al. (2006) reported that results of measurements on nutrients of wheat plant showed that in treatments with the five kinds of bulk blended slow-release fertilizers, the contents of NPK absorbed by wheat in shooting period were higher than that of chemical fertilizer of equal amount of NPK, which was consistent with the rule of nutrients demanded by wheat.

Liu et al. (2005) reported that the organic material was intercalated in the layers of kaoline clays, and the natural kaoline exfoliated into nanometer sized layers. The organic agent and clays formed nano-composites through hydrogen bond combination. The SEM pictures of polystyrene-starch nano-subnanocomposites showed that many pores were present on the surface of film at sizes ranging from 10 to 20 nm. These nano-subnanocomposites were used as the cementing and coating materials of slow/controlled release fertiliser.

Liu et al. (2006) studied the nitrogen slow release behaviour of the super absorbent nitrogen fertilizer (SSNF) in water and water retention capacity of the soils with SSNF. They reported that the surface cross linked product not only had good slow release property but also excellent soil moisture preservation capacity, which could effectively improve the utilization of fertilizer and water resources simultaneously. They also obtained that the SSNF could be found an application in agriculture and horticulture, especially in drought prone areas where the availability of water is insufficient.

Qiang et al. (2008) reported that the wheat grain yield and protein were improved in some degree, but protein content was increased insignificantly, and soluble sugar content was decreased by slow/controlled release fertilizer coated and felted by nano-materials compared with NPK chemical fertilizer. It was effective to use slow/controlled release fertilizer coated by nano-materials to improve wheat yield and quality.

Effect of nano-composites on crop growth

Liu et al. (2005a) reported that the kaoline nano-subnanocomposite was prepared by the methods of organic material intercalation under certain temperature and pressure. This compound was used as the cementing and coating material of slow/controlled release fertilizer because of its strong adsorption and thickness to macronutrients and organic C.

Liu et al. (2005b) reported that the addition of nano-subnanocomposites benefits the soil, and raises the utilized efficiency of fertilizer because of its excellent characteristics. The physical adsorption and chemical combination occurred between nutrient elements and nanocomposites due to surface reaction and small size reaction of nano-composites. They formed the efficient multifunctional fertilizer, which heightened the adsorption of nutrient elements by plants, lowered the leaching in soil, and the fixation of fertilizer in the soil.

Liu and Zhang (2005) reported that the nano-subnanocomposites significantly affected or controlled the structure and penetrability of the soil, increased the organic mineral granule of the soil, improved fertilizer storage and water holding capability in the soil, promoted action of microorganisms, regulated the ratio of C/N, enhanced the fertility of the soil and so on. Improved yields have been claimed for fertilizers that are incorporated into co-chleate nano-tubes (rolled-up lipid bilayer sheets). The release of nitrogen by urea hydrolysis has been controlled through the insertion of urease enzymes into nanoporous silica (Hossain et al., 2008).

Eberl (2008) tested the controlled release fertilizers in greenhouse pot experiments with sorghum-sudan grass using NH_4 – saturated zeolite and P-rock with a phosphate application rate of 340 mg per kg soil, and zeolite/P-rock ratios ranging from 0 to 6. Total phosphate uptake and phosphate concentration measured for the grass were related linearly to the zeolite/P-rock ratio, and yields summed over four cuttings were as much as four times larger than control experiments.

Canas et al. (2008) reported that non functionalized carbon nano-tubes affected root length more than functionalized nano-tubes. Nonfunctionalized nano-tubes inhibited root elongation in tomato and enhanced root elongation in onion and cucumber. Functionalized nano-tubes inhibited root elongation in lettuce. Cabbage and carrots were not affected by either form of nano-tubes. Effects observed following exposure to carbon nano-tubes tended to be more pronounced at 24 h than at 48 h. Microscopy images showed the presence of nano-tube sheets on the root surfaces, but no visible uptake of nano-tubes was observed.

Carbon nano-tubes were found to penetrate tomato seeds and affect their germination and growth rates. The germination was found to be dramatically higher for seeds that germinated on medium containing CNTs compared to control. Analytical methods indicated that the CNTs are able to penetrate the thick seed coat and support water uptake inside seeds, a process which can affect seed germination and growth of tomato seedlings (Khodakovskaya, 2009). Sultan et al. (2009) reported that the development of functional nanoscale films and devices has the potential to produce significant gains in the NUE and crop production.

Liu et al. (2009) demonstrated the capability of single walled carbon nano-tubes (SWNTs) to penetrate the cell wall and cell membrane of tobacco cells.

Liu and Zhiming (2007) investigated the capability of single-walled carbon nano-tubes (SWNTs) to penetrate the cell wall and cell membrane of intact plant cells. Confocal fluorescence images revealed the cellular uptake of both SWNT/fluorescein isothiocyanate and SWNT/DNA conjugates, demonstrating that SWNTs also hold great promise as nano-transporters for walled plant cells. Moreover, the result suggested that SWNTs could deliver different cargoes into different plant cell organelles.

Srinivasan and Saraswathi (2010) reported that the seeds incubated in CNTs for two days possessed a moisture content of 57.6% compared to the value of

38.9% for the control. This is due to generation of new pores generated during penetration of seed coat by CNTs which aided better water permeation. Another possible cause reason could be the efficient gating of water channels by the CNTs in the seed coat.They also reported that, CNTs induces increased biomass, rapid germination and growth rates will intun yield agriculture production in a short duration. Furthermore, nano-silica particles absorbed by roots have been shown to form films at the cell walls, which can enhance the plant's resistance to stress and lead to improved yields. (DeRosa, 2010).

It is quite evidence that nanotechnology is an emerging field of science being exploited to derive solutions to highly complicated unresolved issues in engineering and biological sciences. In agriculture, to address the challenges of imbalanced fertilization, we should think of alternate technology such as " nanotechnology" to precisely detect and deliver the correct quantity of nutrients required crops in suitable proportion that and promote productivity while ensuring environmental safety. For this, zeolite can be used as the carrier for the nanofertiliser which releases nutrients in slow and steady state.

References

Ahmed ,O.H., A. Hussin., H. Mohd Hanif Ahmad., M. Boyie Jalloh., A. Abd Rahim and N. Muhamad Majid. 2008. Ammonia volatilization and ammonium accumulation from urea mixed with zeolite and triple superphosphate. Acta Agriculturae Scandinavica, Section B - *Plant Soil Sci.*, **58** : 182-186.

Ahmed ,O.H., A. Hussin., H. Mohd Hanif Ahmad., M. Boyie Jalloh., A. Abd Rahim and N. Muhamad Majid. 2009. Enhancing the Urea-N Use Efficiency in Maize (*Zea mays*) Cultivation on Acid Soils using Urea Amended with Zeolite and TSP. American *J. Applied Sci.*, **6** : 829-833.

Bansiwal, A.K., S. S. Rayalu., N. K. Labhasetwar., A. A. Juwarkar and S. Devotta.2006. Surfactant-Modified Zeolite as a Slow Release Fertilizer for Phosphorus. *J. Agric. Food Chem.*, **54**: 4773-4779.

Busaidi, A., T.Yamamoto., M. Inoue and A. Egrinya Eneji. 2008. Effects of Zeolite on Soil Nutrients and Growth of Barley Following Irrigation with Saline Water. The 3rd International Conference onWater Resources and Arid Environments and the 1st Arab Water Forum.

Beman, JM, K.Arrigo and PM. Matson. 2005. Agricultural runoff fuels large phytoplankton blooms in vulnerable areas of the ocean. *Nature.*, **434**:211–214.

Bernardi. 2009. Yield and nitrogen levels of silage corn fertilized with urea and zeolite. *The Proceedings of the International Plant Nutrition Colloquium* XVI, Department of Plant Sciences, UC Davis, USA.

Broos, K., J. Warne., D. A. Heemsbergen., D. Stevens., M.B. Barnes., R.L. Correll and M.J. Mclaughlin. 2007. Soil factors controlling the toxicity of Copper and Zinc to microbial processes in Australian Soils. *Environ. Toxi. and Chem.,* **26** (4): 583–590.

Biswas, P.P and S.P Sharma. 2008. Nutrient Management – Challenges and Options. *J. Indian Society of Soil Sci.*, **56** :22-25.

Buentello, S., DL. Persad.,EB. Court., DK.Martin., AS. Daar and A.Peter.2005. Nanotechnology and the Developing World. *Pl.Med.*,2:e97 doi:10.1371/journal.pmed.0020097.

Cañas, JE, M. Long., S. Nations., R .Vadan., L. Dai., M. Luo and R. Ambikapathi. 2008. Effects of functionalized and nonfunctionalized single-walled carbon nanotubes on root elongation of select crop species. *Environ. Toxicol. Chem.*,**27** :1922-31.

Caballero, P.R., J.Gil., C.Benitez and J.L.Gonzalez 2008. The Effect of Adding Zeolite to Soils in Order to Improve the N-K Nutrition of Olive Trees. Preliminary Results. *Am. J. Agricul. and Biol. Sci.*, **2: 3**21-324.

Curkovic, L., S. Cerjan-Stefanovic and T.Filipan. 1997. Metal ion exchange by natural and modified zeolites. *Water Res.*, **31**: 1379–1382.

Chaudhry, Q., L. Castle and R.Watkins. 2010. Nanotechnologies in Food, Royal Society of Chemistry Publishers (ISBN 978-0-85404-169-5).

DeRosa, M.C., C. Monreal., M. Schnitzer., R. Walsh and Y.Sultan.2010.Nanotechnology in fertilizers. *Nature nanotechnol.*, (5) February.

Dwairi, J. M. 1998. Renewable, controlled and environmentally safe phosphorous released in soil mixtures of NH_4^+ -phillipsite tuff and phosphate rock. *Environ. Geology.*, **34** :293-296.

Du, J., T. Durt., P. Zou., H. Li., L. C. Kwek., C. H. Lai, C. H. Oh, and Arturker. 2005. Experimental Quantum Cloning with Prior Partial Information. *Phys. Rev.Lett.*, **94:**040505.

Eeberl, D.D. 2008. Controlled Release Fertilisers Using zeolites. USGS Science for Changing world. *Tech. Transfer.*, 1-3.

Guo, X., L.Zenga., X. Li and H.Spark 2008. ammonium and potassium removal for anerobically digested wastewater using natural clinoptilolite followed by membrane pretreatment. *J. hazardous Mater.*, **151**: 125-133.

Hossain, K. Z., .M. Monreal and Sayari. 2002.Controlled release of a plant growth regulator, á naphthalene acetate from the lamella of Zn-Al layered double hydroxide complex. *J. Control. Release* **8 :** 417–427.

Griffith, S.M.,and D.J.Streeter. 1994. Nitrate and ammonium nutrition in ryegrass: changes in growth and chemical composition under hydroponic conditions. J. *Plant Nutri.*, **17**:71-81.

Lal, R. 2008. Soils and India's Food Security. *J. Indian Soc. Soil Sci.*, **56**: 129-138.

Liu, X and M. Zhang .2005.Characteristicts of nano-subnanocomposites and response of soil and plant nutrition to them. Ph.D dissertation. Chinese Academy of Agricultural Sciences.

Liu, X., F. Zhao-bin., Z. Fu-dao., Z.Shu-qing and HE. XU-sheng. 2006. Preparation and Testing of Cementing and Coating Nano-Subnanocomposites of slow controlled –release Fertiliser. *Agril. Sci. in India.*, **5** :700-706.

Liu, X., F. Zhao-bin., Z. Fu-dao., Z.Shu-qing and HE. XU-sheng. 2005a. Study on adsorption and desorption properties of nano-kaoline to nitrogen, phosphorus, potash and organic carbon. *Sci. Agricul.*, **38** :102-109.

Liu, X., F. Zhao-bin., Z. Fu-dao., Z.Shu-qing and HE.XU-sheng.2005b. Responses of peanut to Nano-calcium carbonate. *Pl. Nutr. and Ferti. Sci.,* **11**:385-389.

Lin, D. and B.Xing. 2008. Root uptake and Phytotosicity of ZnO nanoparticles. *Environ. Sci. Technol.* **42**: 5580–5585.

Melendi,G.,P.Fernandez-Pacheo.,R.Coronado., M.J.Corredor., E.Testillano., P.S. Risueno and M.C. Marquina. 2008. Nanoparticles as Smart treatment delivery systems in Plants: assessment of Different techniques of Microscopy for their visualization in Plant Tissues. *Ann.Bot.,* **101**:187-195.

Navrotsky, A. 2004. Environmental nanoparticles. In: Dekker Encyclopedia of *Nanoscience and Nanotechnol.*, **2:** 1147-1156.

McBeath, T. M.; R.D.Armstrong., E. Lombi., M. McLaughlin and R.e.Holloway. Optimizing the efficiency of phosphorus *Aust. J. Soil Res.*, **43**: 203-212.

Malhi, S.S., L.K.Haderlin., D.G.Pauly and A.M.Johnson. 2002. Improving Fertiliser Use Efficiency. *Better Crops.*, **86**:22-25.

Milosevic, T., and N. Milosevic.2009. The effect of zeolite, organic and inorganic fertilizers on soil chemical properties, growth and biomass yield of apple trees. *Plant Soil Environ.*, **55**: 528–535.

Millán, G., F Agosto., M. Vázquez., L. Botto., L Lombardi and Luciano Juan.2008. Use of clinoptilolite as a carrier for nitrogen fertilizers in soils of the Pampean regions of Argentina. *Cien. Inv. Agr.*, **35**: 245-254.

Ozanne, P.G. and A. Petch. 1980. In "Plant Nutrition 1978 Proc. 8th Int. Coll. Plant Analysis and Fertiliser Problems". p 361 (Eds. A.R. Ferguson, R.C. Bieleski, I.B. Ferguson). (Govt. Printer Wellington).

Oosterhuis, D.M and D. Howard. 2008. Evaluation of slow-release nitrogen and potassium fertilizers for cotton production. *African J. Agricul.. Res.*, **3** : 068-073.

Perrin T. S.; D.T.Drost., J.L.Boettinger and J. M .Norton.1998. Ammonium-loaded clinoptilolite : A slow-release nitrogen fertilizer for sweet corn. *J. Plant Nutri.*, **21**:515-530.

Hernandez, G., R. Diaz Diaz., JS. Notario del Pino and MM. Gonzalez Martin. 1994. NH_4^+ Na-exchange and NH_4^+ - release studies in natural phillipsite. *Appl Clay Sci.*, **9**: 29–137.

Khodakovskaya, M., E Dervishi., M.Mahmood., Y. Xu., Z. Li., F.Watanabe and A. S.Bris. 2009. Carbon Nanotubes are able to Penetrate plant seed coat and dramatically affect seed germination and plant growth. *ACS Nano* **3**:3221-3227.

Pino,N., J.S. Arteaga Padron., I.J. Gonzdlez Martin and J.E. Garcfa Herndndez.1995. Phosphorus and potassium release from phillipsite-based slow-release fertilizers. *J. of Controlled Release.*, **34**: 25-29.

Park, M. and S. Komarneni. 1998. Ammonium nitrate occlusion vs nitrate ion exchange in natural zeolite. *Soil Sci. Soc. Am. J.* **62**:1455-1459.

Qi., Y.E.2010. Nano diatomite and zeolite ceramic crystal powder. *Proceedings of the National Academy of Sciences*, USA **93**: 6000-6005.

Ramos C. 1996. Effect of agricultural practices on the nitrogen losses in the environment. In: Fertilizers and environment, The Netherlands: Kluwer Academic Publishers. 355–361.

Salman, O.A., G. Hovakeemian and N. Khraishi. 1989. Polyethylene-coated urea. Urea release as affected by coating material, soil type and temperature. *Ind. Eng. Chem. Res.*, **28**: 633–638.

Sartain, J.B.2010 Food for turf: Slow-release nitrogen. *Grounds maintenance.*, **2:**6-4

Sheta, A.S., A.M. Falatah., M.S. Al-Sewailem., E.M. Khaled and A.S.H. Sallam.2003. Sorption characteristics of zinc and iron by natural zeolite and bentonite. *Microporous and Mesoporous Mat.*, **61:**127-136.

Sultan, Y., R.Walsh., C. Monreal and M.C. DeRosa.2009.Preparation of functional aptamer films by layer layer self assembly. *Biomacromolecules.,* **10**: 1149–1154.

Srinivasan, C. and R. Saraswathi. 2010 August. Nano-agriculture – carbon nanotubes enhance tomato seed germination and plant growth. *Current Sci.,* **99**: 274-275.

Stulen, I., M. Perez-Soba., L.J. De Kok., L. Van Der Eerden 1998. Impact of gaseous nitrogen deposition on plant functioning. *New Phytologist.*, :**139**:61–70.

Subramanian, K.S.,C. Paulraj and S.Natarajan, S. 2008. Nanotechnological approaches in Nutrient Management. In: Nanotechnology Applications in Agriculture. ISBN :978-81-904337-3-0 PP :37-42.

Subramanian K S & Tarafdar J C. 2011. Prospects of nanotechnology in Indian farming, Indian Journal of Agricultural Sciences **81**:887-893.

Subramanian, K.S. and S. Rahale. 2009. Synthesis of nanofertiliser formulations for balanced nutrition. Proceedings of the Indian society of Soil Science-Platinum Jubilee celebration. December 22-25, IARI, Campus, New Delhi. PP :85.

Tilman, D.1999. Global environmental impacts of agriculture expansion; the need for sustainable and efficient practices. Proceedings of the National Academy of Sciences., USA 1999, **96**:5995–6000.

Tawfik, M.M.,I.Mirvat., Gobarah and M.H. Mohamed. 2010. Management Practice For Increasing Potassium Fertilizer Efficiency of Sugar Beet in North Delta, Egypt. *International J. Academic Res.*, **2** :220-225.

Zhou J.M. and P.M. Huang, 2007. Kinetics of potassium release from illite as inf;uenced by different phosphates. *Geoderma*, **138**:221-228.

Zhang, J.P., L.Q Sheng., P. Chen. 2003. Synthesis of Various Types of Silver Nanoparticles Used as Physical Developing Nuclei in Photographic Science. *Chinese Chemical Letters,* **14**:645-648.

17

Nano Herbicides for Rainfed Agriculture

C.R. Chinnamuthu, N. Sunil Kumar and K. Brindha

Introduction

Weeds are unwanted plants in agricultural production causing huge losses in crop production system particularly rainfed agriculture. Of the total estimated losses caused in crop production by pests, diseases and weeds in the world, the weeds alone are responsible for one-third of it. Weeds interfere with crop growth and cause an yield loss to the tune of 10% of crop production which amounts to a total loss of about 20 million MT of food grains, valued at over Rs.10,000 crores, at current prices in India. Besides, losses of similar magnitude are experienced in vegetable, fruit and plantation crops. The losses inflicted on biodiversity, environment and health are considerable and have not been quantified scientifically.

Nano Herbicides

At present thousands of herbicide formulations are available in the market to combat weed plants under diverse situation. Although herbicides will continue to be the dominant technology in weed management programs, several problems have arisen from reliance on herbicides including herbicide movement to non-target areas, environmental contamination, and development of herbicide-resistant weeds. Continuous use of same herbicides or herbicides belonging to a similar class is believed to be the chief reason for development of herbicide resistance in weeds. International survey of herbicide resistant weeds in 2006 recorded 304 resistant weed biotypes in 42 countries. In India extensive use of isoproturon for over 20 years in rice-wheat ecosystem led to development of resistance in *Phalaris minor* a grass weed that resembles wheat crop. Development of herbicide resistant crops like "Roundup Ready" in soybean poses a threat of becoming a *"Super Weed"* in the following crops. Various non-chemical approaches such as cover crops, mechanical cultivation, competitive cultivars and biological control agents have been found to provide various levels of weed suppression but often they are inadequate to provide acceptable and consistent control of weeds by themselves.

Herbicides available in the market are designed to control or kill the growing above ground part of the weed plants. None of the herbicides are inhibiting activity of viable underground plant parts like rhizome or tubers, which act as a source for new plants in the ensuing season. Since no soil applied herbicides are available to kill the underground propagating material of perennial weeds, one must wait for the appearance of plants and apply the foliar herbicides. Lack of selective herbicides for perennial weeds, herbicides should be applied in the absence of crops or keep field without crop for season or less by compromising the growing season. Tilling operation many a times worsen problem of perennial weed population build. Amidst this situation, the new science, nanotechnology throws rays of hopes for the development of new nanoherbicides with highly specific, controlled release and increased efficiency to circumvent the weed competition under different ecosystem of crop production. In Tamil Nadu Agricultural University, Coimbatore, efforts are being undertaken to synthesize encapsulated form of herbicide that gets released at a specific moisture regime that is highly useful for rainfed agriculture.

Management of problem weeds

Generally, weed scientists are aiming to control the weeds belongs to communities with a single herbicide molecule. The multi-species approach in the cropped environment resulted in poor control and lead to development of herbicide resistance. Continuous exposure of plant community having mild susceptibility to an herbicide in one season and different herbicide in another season develops resistance to all the chemicals in due course and become uncontrollable through chemicals. The target domains of the present day herbicide in a plant cell are destruction of structure and function of the plant-specific chloroplast, inhibition of lipid biosynthesis and interference with cell-division by disrupting the mitotic sequence or inhibiting the mitotic entry, inhibition of cellulose biosynthesis and deregulation of auxin-induced cell growth (Ko Wakabayashi and Peter Böger, 2004).

Although molecular mechanisms of action are not yet completely understood even for some commercially available herbicides, about 60% of conventional herbicides interfere with the PET system of the chloroplast. The PET is embedded in thylakoids of chloroplast and its function is to convert light energy into chemical energy, namely NADPH and ATP. The chemical energy is used to reduce CO_2 and produce sugars, the basic blocks of plant biomass. The PET consists of photosystems I (PS-I) and II (PS-II), being combined by the cytochrome-b/f complex. The PS-II is connected with the O_2-evolving complex. At present, no commercial herbicides exist that interfere with CO_2-fixation and sugar production. Some of the herbicides affecting photosynthesis inhibit the biosynthesis of photosynthetic pigments (i.e. chlorophylls or carotenoids), causing bleaching (1); interfere with the photosynthetic electron flow as electron-transport inhibitors by binding to the D1-protein of PS-II (2); produce superoxide by dragging off electrons at the end of PS-I like paraquat, inducing radical formation that results in peroxidation

Nano-herbicide to exhaust the weed seed bank

Weed seed banks in soil is a reserve for viable and notorious weed species. The seed bank is an indicator of past and present weed populations. For example, the seeds of *Striga spp* produce hundred and thousands of seeds per plant per season and remain viable in the soil for more than 20 years. The seeds will germinate when the weather factors are favourable. The tubers and rhizomes of the sedges are dormant and viable during unfavourable seasons. The easiest way of reducing the weed incidence is exhausting the weed seed bank, the source for weeds over generations. Existing stale seedbed technique, a fallow period cultural weed management method often practiced during summer to reduce the weed seed bank. It involves frequent tilling and irrigation which adds more to weed management cost. Molecular characterization of problem weed seed coat will help us to identify the receptor having specific binding property with nano-herbicide molecules. Developing receptor based herbicides tagged with nanoparticles like carbon nanotubes for delivery will destroy the specific weed species completely from the soil.

Nano-herbicides to eradicate the perennial weed

The task sounds simple but it remains unsolved over several decades. A perennial weed propagates/survive through underground structures like rhizomes and stolon (*Cynodan dactylon*), tubers (*Cyperus spp*) and deep root (*Solanum elaeagnifolium*). Cultural practices like field plough, hand weeding, hoeing through implements increase the infestation of these perennial weeds rather than controlling. The plant is difficult to control with herbicides because the root system is widespread and connects to adjacent above-ground growth. Studies indicated that repeated application of herbicide like glyphosate and picloram helps to reduce the intensity for a current season. Alterative soil fumigation methyl bromide which is too banned because of its mammalian toxicity, can be employed certain extent for eradication of small infestations.

The present day herbicide target domains in a plant cell are destruction of structure and function of the plant-specific chloroplast, inhibition of lipid biosynthesis, and interference with cell-division by disrupting the mitotic sequence or inhibiting the mitotic entry, inhibition of cellulose biosynthesis and deregulation of auxin-induced cell growth (Böger 1989; Wakabayashi 1989; Wakabayashi & Sato 1992 a & b; Wakabayashi 1993).

Compared to foliar absorption, root absorption is a simpler process. Roots do not have cuticles like leaves; although, mature roots may be covered by a suberized layer. This means that there are few barriers to herbicide absorption by plant roots. Since roots are essentially lipophilic, lipophilic herbicides will be readily absorbed. In fact, herbicides log K_{ow} values are good predictors of root absorption and xylem translocation. Theoretically, absorption could occur anywhere the root system comes in contact with the herbicide; however, there is evidence to suggest that most herbicide absorption occurs in the area a few millimeters behind the root tip. This is the area where most water and nutrient absorption occurs and is characterized by a profusion of root hairs intended to increase the surface area of the root. The Casparian strip is also less developed in this area. If we assume that herbicide absorption is primarily due to mass flow in the soil solution and diffusion in response to concentration gradients, then this area of the root is the likely location of most herbicide absorption. Further characterization of underground plant parts for a new target domain and developing a receptor based herbicide molecule expected to kill the viable underground propagules. Selectivity of such herbicides can be increased by smart delivery mechanism with the help of nanoparticles.

Conditional release of nanoherbicides

Water is the most important critical inputs deciding the success or failure of the crop. Weeds with well developed root system and more efficient in extracting moisture become a serious threat to crop production especially in rainfed areas. In this context, developing a new herbicide molecule, remain in the soil broadcasted along with seeds without loss and available whenever rain receives fulfill the objective of weed control in the rainfed agriculture. To accomplish this objective, development of a new herbicide molecule encapsulated with water soluble nanoparticle programmed to release based on the receipt of rainfall will be considered as suitable way to move forward.

Unmodified starch is an effective matrix for encapsulating solid and or liquid active ingredients. Formulations encapsulated with starch have been widely used for herbicides such as Atrazine, Alachlor and Metolachlor (Mills and Thurman, 1994). Sopena et al. (2007) found that herbicide liberation from the microencapsulated formulations was slow and controlled manner to the soil solution, which allows effective control of harmful weeds for a longer period of time. Sopena et al. (2007) studied the effect of microencapsulation on herbicide release and concluded that Ethylcellulose microencapsulated formulations of alachlor reduced the total leaching losses to the soil columns from 98% to 59%, minimizing the risk of groundwater contamination, maintaining the desired concentration of the herbicide in the top soil layer obtaining longer period of weed control.

Co-precipitation using a supercritical anti-solvent (SAS) technique can be exploited for the controlled herbicide delivery system. The preparation of a herbicide (diuron) loaded in amorphous microparticles of a biodegradable polymer (L-polylactic acid, L-PLA). Spherical particles of L-PLA entrapping diuron concentration, (3 wt.% and 0.1 wt.%). The concentrations of the two components had a major influence upon morphology and shape of the final product. The precipitation time being very short during the SAS process, the growth rates observed are 10000 times higher than the growth rates of diuron in aqueous solution (Taki et al., 2001). Indirect encapsulation of pendimethalin and metolachlor, pre-emergence herbicides for controlled release was done by synthesizing core, core-shell and hallow-shell with the water soluble polyelectrolyte nanoparticle using PAH (polyalanine hydrochloride) (PSS (poly styrene sulfonate) (Layer by layer method) (Chinnamuthu et al., 2010). This method is a lengthy process and it carries less loading efficiency of herbicide (15-20%).

Slow release nanoherbicides

The performance of herbicides in tropical environments can sometimes be erratic and inefficient. This is particularly true for soil-applied herbicides where high temperatures, intense rainfall, low soil organic matter and microbial activity results in rapid breakdown and loss through leaching. The climatic variables involved in herbicide breakdown are moisture, temperature, and sunlight. Herbicide degradation rates generally increase as temperature and soil moisture increases, since both chemical and microbial decompsition rates increases with higher temperature and moisture levels. Further, the irrigation process reduces the herbicide concentration which leads to reduced weed control efficiency coupled with leaching and potential ground water pollution. Thus the half life period for many soil herbicides remains very short period of time ranging from few hours to couple of weeks. Once the concentration of soil applied herbicide reduced to 50 per cent of its original strength, it loses its weed control efficiency correspondingly.

An effective herbicide should control weeds with reasonable doses remain selectively non toxic to crops, remain in the area where applied, persist throughout the growing season taking care of frequently germinating weeds and then leaving no residue at the end of the season permitting subsequent crop in the sequence. Systems capable of releasing a herbicide to the desired site at a controlled rate offer the potential of overcoming some of these problems and of allowing more efficient utilization of existing herbicides. Effective controlled release formulations of herbicides should allow extended periods of activity, while minimizing adverse environmental effect related to movement of the herbicides from the application site. Nanoencapsulating of herbicide active molecules with the polymers capable of releasing herbicides partially or fragmentally during the crop stages provides weed free environment till harvest of the crop.

Nanoparticles to remediate herbicide residues

Nanoscale particles may hold the potential to remediate the herbicide contaminated soils (Masciangioli and Zhang, 2003; EPA, 2003a, 2003b). The nanoparticles are extremely small particle sizes (1–100 nm) and can be easily leached out. It can also remain in suspension for extended periods of time to establish an in situ treatment zone. Some pesticides that are persistent in aerobic environments are more readily degraded under reducing conditions (Comfort et al., 2001). One application of this technique uses Zero Valent Iron (ZVI) as a chemical reductant. Treatment with ZVI can promote rapid abiotic degradation via reductive dechlorination. When halogenated organic pollutants are treated with ZVI, oxidation of ZVI and Fe(II) provides electrons for dechlorination. ZVI has been successfully used to transform chlorinated solvents (Doong and Lai, 2006), pesticides and herbicides (Sayles et al., 1997; Eykholt and Davenport, 1998; Singh et al., 1998; Satapanajaru et al., 2003; Shea et al., 2004; Satapanajaru et al., 2006) explosives (Agrawal and Tratnyek, 1996; Park et al., 2005), nitrates (Till et al., 1998; Huang et al., 2003) and metals (Blowes et al., 1997; Fiedor et al., 1998).

Joo and Zhao (2008) studied the effects of catalyst and stabilizer on the destruction of lindane and atrazine using stabilized iron nanoparticles under aerobic and anaerobic conditions. Satapanjaru et al. (2008) reported the remediation of atrazine-contaminated soil and water by Nano Zero Valent Iron. Zhang et al. (1998) proposed the treatment of chlorinated organic contaminants with nanoscale bimetallic particles. Cinthia et al. (2009) has reported that activated carbon/iron oxide composites can be used for the removal of atrazine from aqueous medium. Susha et al., 2008 reported that the use of bimetallic nanoparticle (Fe_3O_4@Ag-CMC) could able to degrade the atrazine herbicide effectively equally as that of zero valent iron nanoparticle under controlled environmental condition.

Overall, the reported literatures suggest that weed management can be effectively done by exploiting nanotechnological approaches especially in the area of smart delivery, eradication of weed seed bank and remediation of herbicide contaminated soil.

References

Agrawal, A. and P.G. Tratnyek. 1996. Reduction of nitro aromatic compounds by zero-valent iron metal. *Environ. Sci. Technol.*, **30**: 153–160.

Blowes, D.W., C.J. Ptacek and J.L. Jambor. 1997. In situ remediation of Cr (Vr)-contaminated groundwater using permeable reactive walls: Laboratory studies. *Environ. Sci. Technol.*, **3:** 3348–3357.

Böger P. 1989. New plant-specific targets for future herbicides. In: *Target Sites of Herbicide Action* (ed. by Böger P. and Sandmann G.). CRC Press, Boca Raton, 247–282.

Chinnamuthu, C.R., V.Kanimozhi and V.P.Dinesh.2010. Synthesis and Characterization of $MnCO_3$ core-shell nanoparticles for active loading of herbicides. Chapter in the "Biomedical Applications of Nanostructured Materials" Ed: V.Rajendran, B.Hillebrands, P.Prabu and K.E.Geckeler. Published by Macmillan Publishers India Ltd. (ISBN:10:0230-33201-0) pp411-416.

Cinthia, S. Castro, Mario C.Guerreiro, Maraisa Gonclaves, Luiz C.A.Oliveira, Alexander S.Anastacio. 2009. Activated carbon or iron oxide composites for the removal of atrazine from aqueous medium. *Journal of hazardous materials*, **164**: 609-614.

Comfort, S.D., P.J. Shea, T.A. Machacek, H. Gaber and B.T .Oh. 2001. Field scale remediation of a metolachlorcontaminated spill site using zerovalent iron. *Journal of environmental Quality*, **30**: 1636-1643.

Doong, R., and Y. Lai. 2006. Effect of metal ions and humic acid on the dechlorination of tetrachloroethylene by zerovalent iron. *Chemosphere*, **64**: 371–378.

EPA (US Environmental Protection Agency), 2003a. Technology Innovation Office, Permeable Reactive Barriers. http:// clu-in.org/.

Eykholt, G.R. and D.T. Davenport. 1998. Dechlorination of the chloroacetanilide herbicide alachlor and metolachlor by iron metal. *Environ. Sci. Technol.*, **32**: 1481–1487.

Fiedor, J.N., W.D. Bostick, R.J. Jarabek and J. Farrel. 1998. Understanding the mechanism of uranium removal from groundwater by zero-valent iron using X-ray photoelectron spectroscopy. *Environ. Sci. Technol.*, **32**: 1466–1473.

Huang, Y. H., T.C. Zhang, P.J. Shea and S.D. Comfort. 2003. Effects of oxide coating and selected cations on nitrate reduction by iron metal. *Journal of Environmental Quality*, **32**: 1306–1315.

Joo, S.H. and D. Zhao. 2008. Destruction of lindane and atrazine using stabilized iron nanoparticles under aerobic and anaerobic conditions: effects of catalyst and stabilizer. *Chemosphere*, **70**: 418–425.

Ko Wakabayashi and Peter Böger, 2004. Phytotoxic sites of action for molecular design of modern herbicides (Part 1): The photosynthetic electron transport system. *Weed Biology and Management* **4**: 8–18 .

Masciangioli, T. and W. Zhang. 2003. Environmental nanotechnology: Potential and pitfalls. *Environ. Sci. Technol.,* **37**: 102A–108A.

Mills, M.S., Thurman, E.M., 1994. Preferential dealkylation reactions of s-triazine herbicides in the unsaturated zone. *Environ. Sci. Technol.* **28**: 600-605.

Park, J., S.D. Comfort, P.J. Shea and J.S. Kim. 2005. Increasing Fe^0-mediated HMX destruction in highly contaminated soil with didecyldimethylaammonium bromide surfactant. *Environ. Sci. Technol.,* **39(24)**: 9683-9688.

Satapanajaru, T., P. Anurakpongsatorn and P. Pengthamkeerati. 2006. Remediation of DDT-contaminated water and soil by using pretreated iron byproducts from the automotive industry. *Journal of Environmental Science and Health,* Part B, **41**: 1291–1303.

Satapanajaru, T., P. Anurakpongsatorn, P. Pengthamkeerati and H. Boparai. 2008 Remediation of atrazine-contaminated soil and water by nano zerovalent iron. *Water Air Soil Pollut.,* **192**: 349–359.

Satapanajaru, T., S.D. Comfort and P.J. Shea. 2003. Enhancing metolachlor destruction rates with aluminum and iron salts during zerovalent iron treatment. *Journal of Environmental Quality,* **32:** 1726–1734.

Sayles, D.G., G. You, M. Wang and M.J. Kupferle. 1997. DDT, DDD and DDE dechlorination by zerovalent iron. *Environ. Sci. Technol.,* **31**: 3448–3454.

Shea, P.A., T.A. Machacek, and S.D. Comfort. 2004. Accelerated remediation of pesticide-contaminated soil with zerovalent iron. *Environmental Pollution,* **132**: 183–188.

Singh, J., P.J. Shea, L.S. Hundal, S.D. Comfort, T.C. Zhang and D.S. Hage. 1998. Iron-enhanced remediation of water and soil containing atrazine. *Weed Science,* **46**: 381–388.

Sopena, F., C. Maqueda and E. Morillo. 2007. Norfl urazon mobility, dissipation, activity and persistence in a sandy soil as influenced by formulation. *J. Agric. Food Chem.,* **55**:3561-3567.

Susha.V.S, C.R.Chinnamuthu, J. Stella Winnarasi, K.Pandian. 2008. "Mitigating Atrazine, 2-chloro-4-(ethyl amine)-6-(isopropylamine)-s-triazine herbicide residue in Soil using Pd and Ag Nanoparticles doped Iron Oxide Nanoparticles with different Stabilizing Agents" presented at the International Conference on "Magnetic Materials and their Applications for 21st Century (MMA-21) October 21-23, 2008. Organized by NPL, New Delhi and Magnetic Society of India (MSI).

Taki, S., E Badens· and G Charbit Controlled release system formed by supercritical anti-solvent coprecipitation of a herbicide and a biodegradable polymer. 2001. *The Journal of Supercritical Fluids,* **21(1):** 61-70.

Till, B.A., L.J. Weathers and P.J.J. Alvarez. 1998. Fe(0)- supported autotrophic denitrification. *Environ. Sci. Technol.,* **32**: 634–639.

Wakabayashi K. 1989. Herbicidal targets in photosynthetic light reaction and molecular design of new herbicides. *Plant Protection* **43**:575–584 (in Japanese).

Wakabayashi K. 1993. Targets of herbicides: mechanism of action of herbicides. In: *Pharmaceutical Research and Development (2nd Series); Management and Development of Agrochemicals,* Vol. 18 (ed. by Yajima H., Iwamura H., Ueno T. and Kamoshita K.). Hirokawa Publishing Co., Tokyo, 393–420 (in Japanese).

Wakabayashi K. and Sato Y. 1992a. Herbicides affecting the structure and function of chloroplasts (1). *Japan Pesticide Information* **60**:25–31.

Wakabayashi K. and Sato Y. 1992b. Herbicides affecting the structure and function of chloroplasts (2). *Japan Pesticide Information* **61:**14–19.

Zhang, W.-X, C. Wang and H. Lien. 1998. Treatment of chlorinated organic contaminants with nanoscale bimetallic particles. *Catalysis Today,* **40**: 387–395.

18

Nano-based Smart Delivery Systems in Agriculture

K.S. Subramanian, S. Manikandan and M. Praghadeesh

Nanotechnology is an emerging of field of science which is being highly exploited in the sphere of medicine where "nano" plays a vital role in delivery of drugs or chemical molecules to the point it is required in précised quantities in right proportions. Nanoscience infuses intelligence to the truck load of chemical constituents that are to be delivered at appropriate locations and cleave from the site after the task is complete. Such process will likely to reduce the cost besides ensuring environmental safety. Nanoscale devices with its unique properties make the agricultural system more smart and effective; such devices are capable of responding to different situations by themselves, thus taking appropriate remedial action with the need of external directions from humans. In short, these devices act as a detectors and if need arises serve as a solution/ remedy for the particular issue. These smart delivery systems of chemicals in controlled and targeted manner can be considered synonymous to the proposed nano-drug delivery system in human (Patolsky et al., 2006).

Smart delivery system in agriculture is yet to be investigated. However, foliar feeding of nutrients and seed coating of agricultural inputs are considered "smart" delivery system as this technique will deliver the inputs directly to the plants and the site of action. But, the nano-based smart delivery systems should go beyond the boundaries of foliar feeding. The smart delivery system should consider the factors or combinations of factors such as time controlled, specifically targeted, highly controlled, remotely regulated / preprogrammed and multifunctional characteristics to avoid biological barriers for successful targeted release (Roco, 2003). These nano-formulations of various agrochemicals proved to be advantageous than conventional formulations applied to crops by spraying and broadcasting. According to Boehm et al., 2003, in conventional method of applying agro-chemicals, only a very low concentration of active ingredient, which is much below the minimum effective required concentration, has reached the target site of crops due to problems such as leaching, degradation by photolysis, hydrolysis and bio-instability in atmosphere. This in turn warrants the need of repeated application for the effective control of pests or weeds resulting in higher cost of cultivation and environmental pollution (Remya et al., 2010). Nano-encapsulated agrochemicals should be designed in such a way that they possess all necessary properties (effective concentration, stability and solubility) time controlled release in response to certain stimuli, enhanced targeted activity and less eco-toxicity with safe and easy mode of delivery thus avoiding repeated applications.

Smart nano herbicides

Herbicides are chemicals that are used to kill the weeds in a wide array of agro-climatic systems particularly rainfed where it is considered menace. Under rainfed conditions, there is no guarantee for moisture availability and thus herbicides are to be designed and fabricated to release the active ingredient only when the soil receives a short spell of rainfall. The control of parasitic weeds with nano-capsulated herbicides thereby reducing the phytotoxicity of herbicides on crops best explains the benefits of smart delivery system in agriculture. Properly functionalized nano-capsules provide better penetration through cuticle and allow slow and controlled release of active ingredients on reaching the target weed. Nano encapsulation of chemicals with biodegradable materials makes safer and easy to handle by the growers. Efforts are underway to kill the notorious weeds like *Cyprus rotundus* through smart delivery system. This weed produces tubers rich in starch that has to be exhausted through a suitable smart delivery system (Perez-de-luque and Diego, 2009; Chinnamuthu and Kanimozhi, 2012).

Nano insecticides

Similarly to nano-herbicides, the efficacy of commercially available insecticides surface modified hydrophobic nanosilica has been successfully used to control a range of agricultural pests (Barik et al., 2008). This functionalized lipophilic nanosilica gets absorbed into the circular lipids of insects by physio-sorption and damages the protective wax layer and induces death by desiccation. The use of such nano-biopesticide is more acceptable since they are safe for plants and cause less environmental pollution in comparison to conventional chemical pesticides (Rahman et al., 2009).

Nano fungicide

The successful use of silver nanoparticles (Ag NPs) in diverse medical streams as antifungal and antibacterial agents has led to their applications in controlling phytopathogens. Ag NPs with broad spectrum of antimicrobial activity reduce various plant diseases caused by spore producing fungal pathogens. The effectiveness of Ag NPs can be improved by applying them well before the penetration and colonization of fungal spores within the plant tissues (Singh et al., 2001). The small size of the active ingredient effectively controls fungal diseases like powdery mildew. However it was also observed that a very high concentration of nano-silica-silver produced some chemical injuries on the cucumber. The use of Ag NPs as an alternative to pesticides for the control of sclerotium forming phytopathogenic fungi was also investigated. Exposure of fungal hyphae to Ag NPs caused severe damage by the separation of layers of hyphal wall and collapse of hyphae. The efficacy of Ag NPs in extending the vase life of gerbera flowers was also studied and the results show inhibited microbial growth and reduced vascular blockage which increased the water uptake and maintained the turgidity of gerbera flowers (Solgi et al., 2009). Apparently, the use of biocide containing polymeric nanoparticles for introducing organic wood preservatives and fungicides into wood products thereby reducing the wood decay was also studied (Liu et al., 2001).

Choice of polymer and its role in smart delivery

Crop protection agents (CPAs) include chemicals (fungicides, insecticides, herbicides), biochemical (antibiotics, RNA-based vaccines), semiochemicals (pheromones, repellents, allomones), microbial pesticides, growth regulators (insect or plant), micronutrients, all with beneficial effects on the protection of crops. Also, we recommend the most beneficial smart polymer-based methods for plants or crops and also offer newly envisioned "intelligent" polymer-based, nano-encapsulated CPAs, able to be self-activated by plant-specific stimuli (Roy et al., 2010).

Polymers and hydrogels have become established matrices for the responsive or controlled release of countermeasures, therapeutic agents, or active ingredients in general. Yet, their application in agriculture, for the delivery of CPAs, has been severely limited until quite recently. However, once micro- and nanotechnology were established and their benefits for agriculture became undeniable, then the polymer-based CPA responsive delivery at micro- or nano-scale began developing as enabling technologies, as was already the case in medical, veterinary or pharmaceutical practice.

The advantages of the responsive polymer and hydrogel based CPA includes (Zhang et al., 2007; Peteu, 2007, Remya et al., 2010):

- Stabilization against environmental degradation by light, air, humidity, or microorganisms
- Decreased dosage, evaporation, leaching, leading to reduced environmental pollution and drift
- Reduced irritation of the human mucous-membrane and a lower phytotoxicity
- Increase in the number of target organisms and ease of handling of harmful CPAs
- Longer application intervals, actuated by CPA's controlled release from responsive polymers
- In addition, nano-encapsulation offers the *in situ* avenue of CPA release; Added benefits when employing specifically conductive polymers, as responsive materials.

While the interest in responsive polymers research and development generated a notable number of publications, it is apparent that the majority of literature to date accounts mostly for two of the stimuli: pH and temperature. Thus, our review does not cover these two classes of stimuli. Existing CPA delivery methods of agricultural polymers occur typically through passive diffusion release, erosion of the capsule wall, or active diffusion via osmotic pressure. By contrast, the human and veterinary medicine has inspired the development of new and advanced stimuli responsive polymers, with positive impact for improving patient health and quality of life. One objective of this review is to stimulate interest in the transfer of the methods, configurations and designs of smart polymers—from medicine towards agricultural uses in plant protection. Based on emerging developments in Nano-Science and Plant Science, yet another goal herein is to put forward plant specific mechanisms as new stimuli for the polymer-based, intelligent, self-regulated CPA release for future, sustainable crop protection (Zhang et al., 2007).

Classes of stimuli for different responsive polymer (Peteu et al., 2010)

- Ultrasound responsive polymers
- Light responsive polymers
- Redox/thiol responsive polymers
- Magnetic field responsive polymers

- Enzyme responsive polymers
- Antigen-Antibody responsive polymers
- Electric field responsive polymers

Intelligent responsive polymers with integrated sensing-release dual function

The advanced use of conducting (also called electroactive) polymers in biological sensors and actuators has grown over the past decade, also due to their compliance and compatibility with nanofabrication. Electroactive polymers (EAPs) such as polypyrrole, polyaniline, polyethylene dioxythiophene (PEDOT), exhibit a high conductivity, mediate rapid electron transfer and can be synthesized under mild conditions through simple deposition onto conductive surfaces from mono-mersolutions with precise electrochemical control (Peteu et al., 2010b). Furthermore, the synergistic integration of similar-size biomolecules and nanomaterials resulted recently in novel cross-bred bio-nanomaterials, displaying an unusual set of electronic, photonic, catalytic and recognitive properties and functionalities. More specifically, the nano-EAPs integrated with other (non) polymeric nano entities, and receptors, enzymes, antibodies, whole cells, and nucleicacids in their matrix, lead to uniquely advanced smart or so-called intelligent materials, due to their ability to self-regulate in response to a given agent or challenge. These EAP-based intelligent materials were proven to enhance biosensing or to deliver actuation, both at nanoscale (Perez-de-Luque and Rubiales, 2009). This is intrinsically similar to multi-level hierarchical functions found in natural living systems that bio-inspired engineers are now trying to mimic on a finer level, with newly-found enthusiasm.

Mode of entry

Plant cell wall acts as a barrier for easy entry of any external agents including nanoparticles into the plant cells. The sieving properties are determined by pore diameter of cell wall ranging from 5 to 20 nm (Fleischer et al., 1999). Hence, only nanoparticles with diameter less than the pore diameter of the cell wall could easily pass through and reach the plasma membrane. There is also a chance for enlargement of pores or induction of new cell wall pores upon interaction with engineered nanoparticles which in turn enhance nanoparticle uptake. Further internalization occurs during endocytosis with the help of a cavity like structure that form around the nanoparticles by plasma membrane. They may also cross the membrane using embedded transport carrier proteins or through ion channels. In the cytoplasm the nanoparticles are applied on leaf surfaces, they enter through the stomatal openings or through the base of the trichomes and then translocated to various tissues. However, accumulation of nanoparticles on photo synthetic surface cause foliar heating which results in alterations of gas exchange due to stomatal obstructions that produce changes in various physiological and cellular functions of plants (Fernandez and Eichert, 2009).

Nanocapsules enter plants through stomata orifices and prevent infection

The CPAs are nano encapsulated and the resulted polymer nanocapsules are sprayed onto the leaf tissue. These nanocapsules enter the plant through the stomata orifices. The nanocapsule's chemical bonds of the polymer wall can be weakened or broken by a critical amount of stress enzymes present. Plant cell stress enzymes are activated by mechanical, thermal and chemical or biological stress. This stress sensitizes the plant during an attack and infection from fungi and bacteria. These polymer based CPA nanocapsules sprays are able to prevent this infection. In this case, the plant cell stress enzymes are the stimuli tiriggering the CPA release (Uzu et al., 2010).

Microelements enter plant through root-hairs and deliver nutrients

The step-by-step method includes the following: the micro-elements, such as Ca, Mg, Fe, S and Zn are encapsulated into microspheres and these polymer microcapsules are, with time, incorporated and dissolved into the soil. Once close to the root, the chemical bonds of the microcapsule's wall polymer are broken down by the organic acids or phenolic substances from the root exudates. These root exudates are typically released to enhance plant feeding during the plant growth process and represent the stimuli activating CPA release (Corredor et al., 2009).

The possibility of targeting the movement of nanoparticles to specific sites of an organism paves way for the use of nano-biotechnology in the treatment of plant diseases that affect specific parts of the plant. Different procedures have made use of nanoparticles in plants, such as the controlled release of bioactive substances in solid wood and plant transformation through bombardment with gold or tungsten particles coated with DNA. In recent years, a breakthrough has been made as a result of Torney et al., who were able to control the intracellular release of substances into protoplasts using mesoporous silica nanoparticles. Despite these advances, the delivery of nanoparticles into plant tissues has been limited to methods involving bombardment, a methodology that does not allow massive application of particles in large number of plants, thus being less exploited in crop improvement programs till date (Torney, 2007).

Copper bio-mineralisation with some wetland plants that transform copper into metallic nanoparticles at soil-root interface with the help of some endomycorrhizal fungi were reported that could reduce copper toxicity in the contaminated soils (Manceau et al., 2008).

Increased applications of engineered carbon-based nanomaterial secrete concerns about their toxicity to humans and animals. Such carbon nanomaterials (such as single-walled carbonnanotubes (SWCNTs), multi-walled carbon nanotube (MWCNTs), carbon buckyballs, etc.) also found increased applications in the field of agriculture and food. This raised questions regarding the safety of using such nanomaterials with crops. Various studies showed contradictory results on the phyto-toxicity of carbon nanomaterials in plants. The effects of functionalized SWCNTs (fCNTs; functionalized with poly-3-aminobenzenesulfonic for high dispersibility) and non-functionalized SWCNTs (CNTs) on root elongation of 6 different crop species namely cabbage (*Brassica oleracea*),carrot (*Daucus carota*), cucumber (*Cucumis sativus*), lettuce (*Lactuca sativa*), onion (*Allium cepa*) and tomato (*Solanum lycopersicum*)) were studied to understand their toxicity to crops (Canas et al., 2008). CNTs enhanced root elongation in onion and cucumber and nanotubes sheets were formed by both fCNTs and CNTs on cucumber root surface due to their interaction with root surface; however none entered into the roots. Cabbage and carrot were not affected by either form of nanotubes. Root elongation in lettuce was inhibited with fCNTs and tomato was found to be most sensitive for CNTs with significant root length reduction. However a very recent work reported on the effects of MWCNTs on the seed germination and growth of tomato plants showed a positive response upon interaction with nanotubes. Their results showed an increased water uptake by seeds in the presence of MWCNTs which enhanced the germination process. Some other studies also support the positive effects of suspensions of MWCNTs on seed germination and root growth of six different crop species (radish (*Raphanus sativus*), rape (*Brassica napus*), rye grass (*Loliumperenne*), lettuce (*Lactuca sativa*),corn (*Zea mays*) and cucumber (*Cucumis sativus*)) (Lin and Xing, 2007).

Effects of magnetic nanoparticles

The main advantage of using magnetic nanoparticles is that they allow a very specific localization of the particles to release their load, which is of great interest in the study of nanoparticulate delivery for plants. A few works have been reported regarding the uptake, translocation and specific localization of magnetic nanoparticles (less than 50 nm) in pumpkin plants. Magnetization signals of various strengths were observed from different portions (ranging from roots to leaves) of the treated plants which clearly indicates the successful translocation of nanoparticles in the entire plant system irrespective of the area of application. No toxicity on plant growth had been detected thus suggesting the safe use of such nanoparticles for nanoparticulate delivery in plants. Magnetic nanoparticles could be used as capping agents for porous carriers of various chemicals so that their specific localization and uncapping process has been carried out using external magnets, hence releasing the payload at target site. Such methods are possible for specific treatments in fruit trees or high-input crops under greenhouse conditions since it is easy to provide an external magnetic exposure that triggers the payload release (Zhu et al., 2008; Corredor et al., 2008).

In recent years, great attention has been paid to genetic effects of ferrofluids; that lead to chromosomal aberrations in young plants. The influence of magnetic nanoparticles coated with tetramethylammonium hydroxide (TMA-OH) (as a stabilizingagent) on the growth of maize plants in early growth stages was studied. 'Chlorophyll a' level was increased at low ferrofluid concentrations while at higher concentrations it was inhibited. In another work, water based magnetic fluid obtained by coating magnetic nanoparticles with perchloric acid was added to germinated maize seeds. A slight inhibitory effect was observed on the growth of the plantlets that led to brown spots on leaves at higher volume fractions of magnetic fluid. The excess iron treatment generated some oxidative stress in leaf cells which in turn affected photosynthesis and led to decreased metabolic process rates. They also studied the oxidative stress induced by the ferrofluid concentration on living plant tissues (Gonzalez-Melendi et al., 2008). Maize seeds were germinated in the presence of magnetic fluid followed by exposure to electromagneticfield called, LM-EMF samples. For LM-EMF samples, a decrease in assimilatory pigments was observed with increased volume fraction of magnetic fluid solution. Exposure to electromagnetic field produced some local heating on the applied region due to the electromagnetic field energy absorbed by internalized magnetic nanoparticles in plant tissues that in turn affect the redox reactions involved in photosynthesis process.

Effects of zinc based nanoparticles

The use of metal based nanoparticles like ZnO for increased permeability and creation of new holes in bacterial cell wall has paved way for the use of such nanoparticles for studying their cell internalization and further translocation in plant cells. The seed germination and root growth study of zucchini seeds in hydroponic solution containing ZnO nanoparticles showed no negative effects whereas the seed germination of rye grass and corn was inhibited by nanoscale zinc (nano-Zn (35 nm)) and zinc oxide (nano-ZnO (15–25 nm)) respectively (Lin and Xing, 2007; Stampoulis et al., 2009).

Nano-genetic manipulation of agricultural crops

Nanobiotechnology offers a new set of tools to manipulate the genes using nanoparticles, nanofibres and nanocapsules. Properly functionalized nanomaterials serve as a platform to transport large number of genes as well as chemicals that trigger gene expression in plants.

Exploring nanotechnology for delivering genetic materials into plants

Nanofibre arrays which can deliver genetic material to cells quickly and efficiently have potential applications in drug delivery, crop engineering and environmental monitoring. The successful delivery and integration of plasmid DNA was confirmed from the gene expression. This process has similarity with micro-injection method of gene delivery and hence possible with the plant cells in which the treated cells could be regenerated into whole plant that would express the introduced trait. It is possible to make DNA tethered on carbon nanofibers without allowing them to integrate into host genome but still allowing some transcriptions of the tethered genes and hence this technique does not pass modified traits to further generations. This differ it from the existing genetic engineering methods and a onetime modification of the cells is possible (McKnight et al., 2004; Remya et al., 2010).

The application of fluorescent labeled starch-nanoparticles as plant transgenic vehicle was reported in which the nanoparticle biomaterial was designed in such a way that it bind and transport genes across the cell wall of plant cells by inducing instantaneous pore channels in cell wall, cell membrane and nuclear membrane with the help of ultrasound. It is possible to integrate different genes on the nanoparticle at the same time and the imaging of fluorescent nanoparticle is possible with fluorescence microscope thus understanding the movement of exterior genes along with the expression of transferred genes. Hence successful generation of pores on cell wall and cell membrane by suitable agents help in nanoparticle mediated DNA transfer that might be more successful in regenerative calli and soft tissues.

The ability of surface functionalized mesoporous silica nanoparticles (MSNs) to penetrate plant cell walls also opens up new ways to precisely manipulate gene expression at single cell level by delivering DNA and its activators in a controlled fashion. It was reported that a honeycomb MSN system with 3-nm pores transport DNA and chemicals into isolated plant cells and intact leaves. MSNs were loaded with gene and its chemical inducer and the ends were capped with gold nanoparticles to protect the molecules from leaching out (Torney et al., 2007). Uncapping of gold nanoparticles enabled release of chemicals and triggered gene expression in plants under controlled-release conditions. In preliminary experiments protoplasts were incubated with fluorescently labeled MSNs. It was found that surface modification of MSNs with triethylene glycol was necessary to penetrate the cells. Such modification of MSNs also allowed plasmid DNA to adsorb on MSN surface. After entering the protoplasts, the plasmid DNA was released from the MSNs and the green fluorescent protein (GFP) marker encoded in the DNA was expressed in the cell which was detected by microscopy. In this method, the minimum amount of DNA that is required to detect marker expression was 1000-fold, less than that required for the conventional delivery method. This efficient delivery method has pronounced applications in various protoplast-based gene expression studies. Now a days gene gun or particle bombardment is one of the popular tools to deliver DNA into intact plant cells. Particles used for bombardment are typically made of gold since they readily adsorb DNA and are non-toxic to cells. Since, MSNs are too light, it is difficult for delivering foreign DNA attached on MSNs by gene gun method. This problem was solved by capping MSNs withgold nanoparticles which increased their momentum after acceleration by the gene gun. Experiments showed that the plasmid DNA transferred by gene gun method using gold-capped MSNs was successfully expressed in intact tobacco and maize tissues. The major advantage is the simultaneous delivery of both DNA and effector molecules to the specific sites that results in site targeted delivery and expression of chemicals and genes, respectively. This is how the nanoparticle mediated plant transformation differs from the conventional genetic engineering methods (like electroporation, microinjection, etc) (Klein et al., 1989; Deng et al., 2001).

Conclusion

The smart delivery systems are well established in medical sciences which made breakthroughs in finding solution to serious human diseases without associated side effects. The data from medical sciences serve as guiding tools that can be exploited in agricultural production systems. In this book chapter, the literature review made has clearly suggest that there is an abundance of scope to exploit smart delivery systems in agriculture which facilitate enhanced use efficiency of inputs besides environmental protection.

It is time that agricultural scientists should undertake research in the fascinating field of nano-based smart delivery systems so as to achieve the targeted delivery of inputs that enhances the crop productivity of crops with minimal use agri-inputs.

References

Barik, T. K., Sahu, B. and Swain, V. 2008. Nanosilica—from medicine to pest control. *Parasitol. Res.* **103**: 253–258.

Boehm, A.L., Martinon, I., Zerrouk, R., Rump, E. and Fessi, H. 2003. Nanoprecipitation technique for the encapsulation of agrochemical active ingredients. *J. Microencapsul.* **20**:433–441.

Canas, J.E., Long, M., Nations, S., Vadan, R., Dai, L., Luo, M., Ambikapathi, R., Lee, E.H. and Olszyk, D. 2008. Effects of functionalized and non- functionalized single-walled carbon nanotubes on root elongation of select crop species. *Environ. Toxicol. Chem.* **27**: 1922–1931.

Kanimozhi, V and Chinnamuthu, C. R. 2012. Engineering Core/hallow Shell Nanomaterials to Load Herbicide Active Ingredient for Controlled Release. *Research Journal of Nanoscience and Nanotechnology.* **2:** 58-69.

Corredor, E., Testillano, P.S., Coronado, M., Gonzalez-Meledni, P., Fernandez-Pacheco, R., Marquina, C., Ibarra, M. R., Rubiales, D., Perez-de-Luque, A. and Risueno, M.C. (2009). Nanoparticle penetration and transport in living pumpkin plants: *in situ* subcellular identification. *Plant Biology.* **9**:45.

Deng, X.Y., Wei, Z.M. and AN, H.L. 2001. Transgenic peanut plants obtained by particle bombardment via somatic embryogenesis regeneration system. *Cell Res.* **11**:156–160.

Fernandez, V. and Eichert, T. 2009. Uptake of hydrophilic solutes through plant leaves: current state of knowledge and perspectives of foliar fertilization. *Crit. Rev. Plant Sci.* 28: 36–68.

Fleischer, M.A. O'Neill, R. and Ehwald, R. 1999. The pore size of non-graminaceous plant cell wall is rapidly decreased by borate ester cross-linking of the pectic polysaccharide rhamnogalacturon II. *Plant Physiol.* **121**: 829–838.

Gonzalez-Melendi, Fernandez-Pacheco, R., Coronado, M.J., Corredor, E., Testillano, P.S., Risueno, M.C., Marquina, C., Ibarra, M.R., Rubiales, D. and Perez-De-Luque, A. (2008), Nanoparticles as smart treatment delivery systems in plants: assessment of different techniques of microscopy for their visualization in plant tissues, Ann. Bot. 101: 187–195.

Klein, T.M., Kornstein, L., Stanford, J.C. and Fromm, M.E. 1989. Genetic transformation of maize cells by particle bombardment. *Plant Physiol.* **91**: 440–444.

Lin, D. and Xing, B. 2007. Phytotoxicity of nanoparticles: inhibition of seed germination and root growth. *Environ. Pollut.* **150**: 243–250.

Liu, Y., Yan, L., Heiden, P. and Laks, P. (2001). Use of nanoparticles for controlled release of biocides in solid wood. *J. Appl. Polym. Sci.* **79**: 458–465.

Manceau, A., Nagy, K. L., Marcus, M. A., Lanson, M., Geoffroy, N., Jacquet, T., and Kirpichtchikova, T. 2008. Formation of Metallic Copper Nanoparticles at the Soil"Root Interface. *Environmental Science & Technology.* **42(5):** 1766-1772.

McKnight, T.E., Melechko, A.V., Griffin, G.D., Guillorn, M.A., Merkulov, V.I., Serna, F., Hensley, D.K., Doktycz, M.J., Lowndes, D.H. and Simpson, M.L. 2003. Intracellular integration of synthetic nanostructures with viable cells for controlled biochemical manipulation. *Nanotechnology*. **14:** 551–556.

Patolsky, F., Zheng, G, and Lieber, C.M. 2006. Nanowire sensors for medicine and life sciences, *Nanomedicine*. **1:** 51–65.

Perez-de-Luque, A. and Diego, R. 2009. Nanotechnology for parasitic plant control. *Pest Manag. Sci.* **65:** 540–545.

Perez-de-Luque, A. and Rubiales, D. 2009. Nanotechnology for parasitic plant control. *Pest Manag. Sci.* **65**: 540–545.

Peteu, S.F. 2007. Responsive materials configured for micro- and nano-actuation. *J. Int. Mater. Syst. Struct.* **18:**147–152.

Peteu, S.F., Gebremichael, E., Peiris, P. and Bayachou, M. 2010. Nanostructured poly(ethylenedioxythiophene)-metalloporphyrin catalytic platform for the detection of peroxynitrite. *Biosens. Bioelectron.* **25**: 1914–1921.

Peteu, S.F., Oancea, F., Sicuia, A., Constantinescu, F. and Sorina, D. 2010. Responsive polymers for Crop Protection. *Polymers.* **2**:229-251.

Rahman, A., Seth, D., Mukhopadhyaya, S.K., Brahmachary, R.L., Ulrichs, C. and Goswami, A. 2009. Surface functionalized amorphous nanosilica and microsilica with nanopores as promising tools in biomedicine. *Naturwissenschaften* **96:** 31–38.

Remya N, Saino N. H., Baiju, G. N., Maekawa, T., Yoshida, Y, and Sakthi, K.D. 2010. Nanoparticle material delivery to plants. *Plant Science.* **179:** 154-163.

Roco, M.C. 2003. Nanotechnology convergence with modern biology and medicine. *Curr. Opin. Biotechnol.* **14:**337–346.

Roy, D., Cambre, J.N. and Sumerlin, B.S. 2010. Future perspectives and recent advances in stimuli responsive materials. *J. Prog. Polym. Sci.* **35:** 278–301.

Singh, M., Singh, S., Prasad, S. and Gambhir, I.S. 2008. Nanotechnology in medicine and antibacterial effect of silver nanoparticles. *Digest J. Nanomater. Biostruct.* **3:** 115–122.

Solgi, M., Kafi, M., Taghavi, T.S. and Naderi, R. 2009. Essential oils and silver nanoparticles (SNP) as novel agents to extend vase-life of gerbera (Gerbera jamesonii cv. 'Dune') flowers. *Postharvest. Biol. Technol.* **53:** 155–158.

Stampoulis, D., Sinha, S. K. and White, J.C. 2009. Assay-dependent phytotoxicity of nanoparticles to plants. *Environ. Sci. Technol.* **43**:9473–9479.

Torney, F., Trewyn, B.G., Lin., S.Y. and Wang, K. 2007. Mesoporous silica nanoparticles deliver DNA and chemicals into plants. *Nat. Nanotechnol.* **2:**295– 300.

Uzu, G., Sobanska, S., Sarret, G., Munoz, M. and Dumat, C. (2010). Foliar lead uptake by lettuce exposed to atmospheric pollution. *Environ. Sci. Technol.* **44:** 1036–1042.

Zhang, R., Bowyer, A., Eisenthal, R. and Hubble, J. 2007. A smart membrane based on an antigen responsive hydrogel. *Biotechnol. Bioeng.* **97:** 976–984.

Zhu, H., Han, L., Xiao, J.Q. and Jin, Y. 2008. Uptake, translocation and accumulation of manufactured iron oxide nanoparticles by pumpkin plants. *J. Environ. Monit.* **10:**713-717.

19

Nano-fibre as a Smart Delivery System to Contain Seed Borne Pests and Diseases

K. Ramaesh

Nanofibres are an exciting new class of material used for several value added applications in the field of such as medical, filtration as barrier in composites, garments, wipes, insulation, and energy storage. Special properties of nanofibres make them suitable for a wide range of applications from medicine to consumer products and industrial to high-tech applications. The US - National Science Foundation (NSF) defines nanofibres as having at least one dimension of 100 nm or less (Hegde, 2005). Amongst the various nanostructures that have recently been developed for use in practical applications, nanofibres produced from synthetic and natural polymershave received increased attention due to their ease of fabrication and the ability to control their compositional, structural and functional properties (Burger*etal*.,2006). The nanofibres can be produced from a wide range of polymers and these fibres have extremely high specific surface area due to their small diameters. Nanofibre mats can be highly porous with excellent pore interconnection. These unique characteristics plus the functionalities from the polymers themselves impart nanofibres with many desirable properties for advanced applications. There are a number of different processing techniques that can be used for the fabrication of nanofibres, such as drawing, template synthesis, phase separation, self-assembly and electrospinning.

Electrospinning is currently the most promising technique to produce continuous nanofibres on a large scale and the fibre diameter can be adjusted from few nm to microns. Also, electrospinning is a relatively easy and fast process to produce nanofibres. Although the first patent on electrospinning technique was published as early as in 1934, this technique has not been well established until recent times (Subbiahet al., 2005). In electrospinning, polymer nanofibres are obtained by the application of a strong electrical field between a grounded target and a polymer solution that is pumped from a storage chamber through a small capillary orifice. Fibres are collected as a non-woven mesh or membrane on a collector plate that acts as the counter electrode. The fibres range in diameter between 10 and 1000 nm depending on the solution and the process con-ditions, and may be produced in a range of different lengths. The key advantage of producing fibres with extremely small di-ameters is their large surface-to-mass ratio, high porosity, and superior mechanical performance (Frenot and Chronakis, 2003; Kim and Reneker, 1999). Moreover, the fibre functionalities may be dominated by the molecules located on the surface of the fibre thus allowing the engineering of fibre properties by tailoring the fibre surface compositions and morphologies. This has led to diverse applications of nanofibres in tissue engineering, wound healing, drug delivery, medical implants, dental applications, biosensors (Li and Xia, 2004; Zhang et al., 2006), military protective clothing, filtration media, and other industrial applications (Doshi and Reneker, 1995; Huang et al., 2003).

Table 1: Comparison of processing techniques in obtaining nanofibres

Process	*Production*	*Scaling*	*Repeatability*	*Convenience*	*Control on fibre dimensions*
Drawing	Laboratory	x	√	√	x
Template Synthesis	Laboratory	x	√	√	√
Separation	Laboratory	x	√	√	x
Self-Assembly	Laboratory	x	√	x	x
Electrospinning	Laboratory & industrial	√	√	√	√

Ramakrishna, 2005

The properties of electrospun polymer fibres can be further modified by spinning polymer blends to create composite nanofibres that may be able to better fulfil specific industrial requirements in terms of their material properties, thereby increasing the potential array of applications.

Electrospinning setup and process

A schematic setup of a typical electrospinning apparatus (Kriegelet al., 2008)is shown in Fig. 1. In this setup, a polymer precursor is placed in aninjector equipped with a blunt ended stainless steel capillary. The injector is placed in a syringe pump which permits adjustment and precise control of the solution flow rates. The polymer solution is pumped through a metallic capillary needle, which is connected to a high voltage power supply that is typically capable of producing voltages between 1 and 30 kV and is operated in positive DC mode. The setup also contains a grounded target collector which completes the circuit and allows for an electric field to be established between the capillary tip and the target. The involved amperages are typically very low and usually do not exceed a few Amperes. Instead of us-ing a grounded metal plate, rotating drums or disks have been used as the target to allow a collection of single fibres rather than the production of non-woven, randomly-deposited fibre mats. In electrospinning, the application of the electrical field between the electrodes introduce additional forces and as a result, the hemispherical surface of the solution at the tip of the capillary is distorted to form a conical shape known as a Taylor cone (Taylor, 1969). Once the electrical field reaches a critical value so that the shape maintaining the surface forces are exceeded, a charged polymer solution jet is ejected from the tip of the Taylor cone. This jet is accelerated by the electrical field towards the grounded collector. As the jet travels through the electrical field, charges accumulate on the surface of the jet. These charges may be unevenly distributed resulting in a whipping or bending motion of the jet. As a result solvent evaporation may rapidly occur, while the polymer chains within the jet tend to stretch and orient. The thin jet containing the polymer molecules is then deposited on the collector plate (Huang et al., 2003).

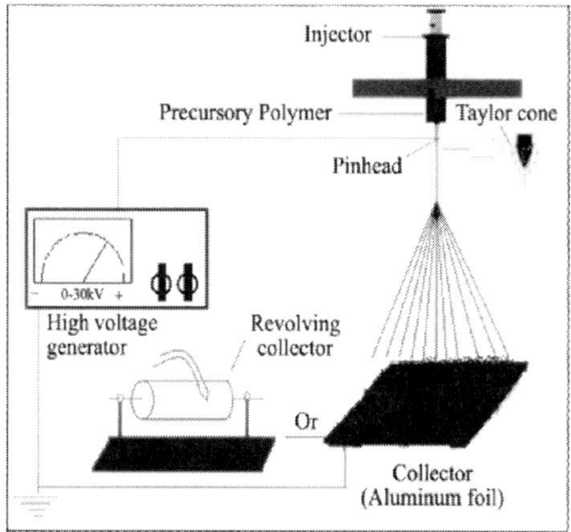

Fig. 1: Electrospinning Assembly

Until now, more than 100 polymers and many inorganic materials have been electrospun into nanofibres (Ramakrishna, 2005). Electrospinning of biopolymers using polysaccharides, proteins and phospholipids are of great interest in various industries because they are (a) nontoxic, edible, and digestible, (b) biocompatible and biodegradable, and (c) renewable and sustainable giving rise to a broader utilization especially in fields such as biomedical sciences, pharmaceuticals, cosmetics, and other related fields (Kreigel et al., 2008). The extremely fine electrospun nanofibres have been used in a wide range of advanced applications, and many new applications have been explored. The potential applications of electrospun nanofibres in pest management include smart delivery of pesticides and agrochemicals, bio sensors in detection of pests and diseases and filtration to remove pathogens.

Nanofibres in smart delivery of pesticides through seed treatment

Seed treatment is an important component in integrated pest management in which dry or slurry treatment of seeds with chemical pesticides has been routinely carried out for protection against various soil borne pests and diseases. The efficacy of conventional seed treatment methods is limited due to loss of treated pesticides by leaching, evaporation, and degradation. Hence, over the past two decades, consistent efforts have been made to devise a novel delivery system for seed-protectant chemicals that can place a specified quantity of pesticide onto the seed and further ensure its prolonged and effective usage in pest control.

Controlled-release formulations can ameliorate pesticide losses by conventional seed treatment methods (Wilkins, 1990) with additional benefits like lower application rates, improved specificity, ease and safety in handling, and minimum biological or ecological side effects (Bahadir and Pfister, 1990). Several methods of controlled-release formulations including biologics (Dutt and Khuller, 2001), polymers (Chorny et al., 2002) silicon-based materials (Liu et al., 2006), or metals (Hussain et al., 2003) are extensively being tested for improving the pesticide delivery. The polymerization technique for controlled-release has got major advantages like ability to control the molecular weight of the polymer, the active agent/polymer weight ratio, the hydrophobic-hydrophilic balance of the polymer (via appropriate comonomers), and possibly the distribution of groups along the backbone (Kenawy et al., 1992).

Rameash et al. (2012) developed a novel seed coating protocol using electrospun nano-fibres for controlled release of pesticides. Systemic pesticides were incorporated into a biodegradable polymer system and in turn coated on the surface of seeds using electrospinning technology for sustained releasein cotton and wheat. A systemic insecticide, imidacloprid and a systemic fungicide, tebuconazole were used for the seed coating study. Imidacloprid [1-(6-chloro-3- pyridinylmethyl) -N-nitroimidazolidin-2-ylideneamine] is a widely used systemic insecticide. It has been used as seed dressing, soil treatment, and foliar treatment in different crops (Ping et al. 2010) for controlling aphids, whiteflies, thrips, scales, psyllids, plant bugs, and other insect pests. Tebuconazole, [(RS)-1-p-chlorophenyl-4,4-dimethyl-3-(1H-1,2,4-triazol-1-ylmethyl)pentan-3-ol] is a systemic fungicide used for disease control on fruit, nut, cereal and vegetable crops world-wide (Ahmed et al., 2001). Biopolymers polylactic acid [PLA] and poly (lactic-co-glycolic acid) [PLGA] were selected as carrier for the pesticides. These polymers have excellent biocompatibility and biodegradability, and are useful materials as base material for sustained-release formulation and have received much attention recently on the controlled release of drugs, delivery of proteins, cell encapsulation, and tissue regeneration, gene therapy etc. The biopolymers incorporated with the pesticides were coated on the seed surface as nanofibres thorough electrospinning process. Persistent toxicity studies conducted under poly house conditions confirmed the slow release of pesticides for up to 54 days.

Nanofibres are being widely used in the area of drug delivery, be-cause the dimensions and the surface properties of the fibres can be specifically tailored to adjust and control the release of active ingredients making them a highly efficient carrier system. So far, applications have mainly focused on rate control as well as the chemical integrity of drugs. Sensitive compounds such as human *j*-nerve growth factor has been incorporated into fibres by mixing them with bovine serum albumin (BSA), a known enzyme stabilizer, and electrospinning the mixture together (Chew et al., 2005) and it was found that the growth factor had a sustained release over the course of several months and remained bioactive after the spinning process. Using BSA as a model protein-analogous drug, Qi et al. (2006) produced fibres containing BSA in poly (L-lactic acid), while Zhang et al., (2006) coaxially electrospun nanofibres with a BSA and poly (ethylene glycol) loaded core and a shell of poly (caprolactone). Using the latter approach, the surface and release behaviour of BSA was tuned by varying the composition of the core-shell fibres. Buschle-Diller (2006) studied the encapsulation of antibiotics in nanofibres composed of poly (L-lactic acid) and poly *(e*-caprolactone) and found good retention of activity after pro-cessing. The controlled release of a model drug from gelatin-PVA nanofibres was demonstrated with the efficiency of encapsulation and the excellent control over delivery (Yang et al., 2007).

Similar approaches have been tried for the controlled release of pheromones in management of insect pests. Pheromones, dispensed in a quantifiable way, are being used for disrupting the mating communication between male and female pest insects. The advantage is that pheromones are very species specific, so have no adverse effect on surrounding organisms, but only to the desired species. However the efficacy of a pheromone trap mainly depends on the dispensing system. Hummelet al.,(2011) incorporated the pheromone compound (E,Z)-7,9-dodecadien-l-yl acetate into an electrospun nanofibre dispenser system based on biodegradable polyester. The resultant smart delivery system increased the matis disruption of European grapevine moth *Lobesiabotrana* (Lepidoptera: Tortricidae) for upto 7 weeks.

By virtue of their nano-scale diameter and very large surface area, electrospun fibres may offer a number of additional advantages like enhanced bioavailability of the pesticides, improving the timed release of pesticide molecules, and increased responsiveness to changes in the environment (e.g., relative humidity and temperature changes). Furthermore, electrospun fibers/ particles are more suitable for encapsulating thermallylabile active pesticides like imidacloprid, as the electrospinning process takes place at ambient conditions unlike conventional melt spinning process. The major advantages of using nanofibre biopolymer system in pesticide applications include,

- Ability to control the release rate of pesticide by altering the molecular weight of the polymer, the pesticide/polymer weight ratio
- Very high surface area of the nano structures requires lower chemical pesticides which ensures minimum biological or ecological side effects
- The nano polymer coating provides ample permeability to water and oxygen molecules, which ensures a 100% germination
- Improved specificity, ease and safety in handling and
- Enhanced safety to non-target organisms in soil and plant ecosystem

Nanofibres have enormous potential in the food and agriculture industry and the applications are emerging continuingly. In the food industry, the fibers may find uses as ingredients if they are composed solely of edible polymers and GRAS ingredients, (e.g., fibers could contain functional ingredients such as nutraceuticals, antioxidants, antimicrobials, and favors), as active packaging materials or as processing aids (e.g., catalytic reactors, membranes, filters) (Lala et al.,2007). While the potential for nanofibres is huge, there are also considerable challenges ahead. Most of the studies in this area have been conducted on fibres produced on a very small scale, using a needle based system to electrospin nanofibres from a polymer solution. While large scale production of nanofibres using pilot scale equipment is now possible, more work is needed to evaluate the characteristics and performance of the fibres produced on the large scale equipment to ensure consistent fibre quality and address the environmental issues associated with solvent/solution based electrospinning technology.

References

Ahmed, N.E., Kanan, H.O., Inanaga, S., Ma, Y.Q., Sugimoto, Y. 2001. Impact of pesticide seed treatments on aphid control and yield of wheat in the Sudan. *Crop Protection*, **20(10):** 929-934

Bahadir, M., Pfister, G. 1990. Controlled release formulations of pesticides. In *Controlled Release, Biochemical Effects of Pesticides and Inhibition of Plant Pathogenic Fungi*; Bowers, W. S., Ebing, W., Martin, D., Wegler, R., Eds.; Chemistry of Plant Protection 6; Springer-Verlag:. New York, pp 1"60.

Burger, C., Hsiao, B.S. and Andchu, B. 2006.Nanoûbrous materials and their applications.*Annual Review of Materials Research,* **36**: 333.

Buschle-Diller, G., Hawkins, A. and Cooper, J. 2006.Electrospun nanofibres from biopolymers and their biomedical applications. In: *Modified Fibres with Medical and Speciality Applications*, Edwards, J. V., Ed., Springer.

Chew, S. Y., Wen, J., Yim, E. K. F., and Leong, K. W. 2005. Sustained release of proteins from electrospun biodegradable fibres.*Biomacromolecules.* **6:** 2017

Chorny, M., Fishbein, I., Danenberg, H.D. and Golomb, G. 2002. Lipophilic drug loaded nanospheres prepared by nanoprecipitation: effect of formulation variables on size, drug recovery and release kinetics.*J Control Release* **83:**389-400.

Doshi, J. and Reneker, D. H. 1995. Electrospinning process and applications of electrospun fibres. *Journal of Electrostatics* **35:** 151.

Dutt, M. , and Khuller, G.K. 2001. Liposomes and PLG microparticles as sustained release antitubercular drug carriers—an in vitro–in vivo study. *International Journal of Antimicrobial Agents*, **18:** 245-252.

Frenot, A. and Chronakis, I. S. 2003. Polymer nanofibres assembled by electrospinning.*Current Opinion in Colloid & Interface Science.***8:** 64.

Hegde, R,R. Dahiya, A. and Kamath, M. G. 2005. Nanofibre nonwovens. http://web.utk.edu/~mse/Textiles/Nanofibre%20Nonwovens.htm accessed on 13.12.2011

Huang, Z. M., Zhang, Y. Z., Kotaki, M. and Ramakrishna, S. 2003.A review on polymer nanofibres by electrospinning and their applications in nanocom- posites. *Composites Science and Technology* **63**: 2223.

Hummel HE, Hein DF, Breuer M, Lindner I, Greiner A, Wendorff JH, Hellmann C, Dersch R, Kratt A, Kleeberg H, Leithold G. 2011. Organic nanofibres containing insect pheromone disruptants: a novel technical approach to controlled release dispensers with potential for process mechanization. CommunAgricApplBiol Sci. **76(4)**:809-17

Hussain, N., Singh, B., Sakthivel, T. and Florence, A.T. 2003.Formulation and stability of surface-tethered DNA-gold-dendron nanoparticles.*Int. J Pham.*, **254**:27-31.

Kenawy, E.R., Sherrington, D.C. and Akelah, A. 1992.Controlled release of agrochemical molecules chemically bound to polymers *Eur. Polym. J.*, **28:** 841-862

Kim, J.S. and Reneker, D. H. 1999. Mechanical properties of com-posites using ultrafine electrospun fibres.*Polymer Composites.***20:** 124.

Kriegel, C., Arrechib, A., Kevin ,K., McClementsa, D. J. and Weissa, J. 2008. Fabrication, Functionalization, and Application of Electrospun Biopolymer Nanofibres. Critical Reviews in Food Science and Nutrition **48:** 775–797

Lala, N. L., Ramaseshan, R., Li, B. J., Sundarrajan, S., Barhate, R. S., Liu, Y. J., and Ramakrishna, S. 2007. Fabrication of nanofibres with antimicro-bial functionality used as filters: Protection against bacterial contaminants. *Biotechnology and Bioengineering.***97:** 1357.

Li, D. and Xia, Y. N. 2004. Electrospinning of nanofibres: Reinventing the wheel?*Advanced Materials.***16:** 1151.

Liu, F., Wen, L., Li, Z., Yu, W., Sun, H.Y. and Chen, J.F. 2006. Porous hollow silica nanoparticles as controlled delivery system for water-soluble pesticide. *Materials Research Bulletin.***41:** 2268-2275.

Ping Lifeng, Chunrong Zhang, Yahong Zhu, Min Wu, Fen Dai, Xiuqing Hu, Hua Zhao and Zhen Li. 2010. Imidacloprid adsorption by soils treated with humic substances under different pH and temperature conditions. African Journal of Biotechnology. **9 (13):** 1935-1940

Qi, H. X., Hu, P., Xu, J., and Wang, A. J. 2006. Encapsulation of drug reservoirs in fibres by emulsion electrospinning: Morphology char-acterization and preliminary release assessment. *Biomacromolecules.***7:** 2327.

Ramakrishna, S. 2005. An Introduction to Electrospinning and Nanoûbers. World Scientiûc, Singapore.396 p.

Rameash, K., Juan P Hinestroza, Fredrick O Ochanda and Priscilla P Luz. 2012. Biodegradable Nanofibres as Carrier for the Controlled Release of Pesticides, In: International Conference on Plant Health Management for Food Security, 28-30 November, 2012, Hyderabad. Pp. 44-45

Subbiah,T., Bhat, G. S., Tock, R.W., Pararneswaran, S. and Ramkumar, S.S. 2005. Electrospinning of nanoûbers. *Journal of Applied Polymer Science.***96**:557.

Taylor, G. 1969. Electrically Driven Jets, Proceedings of the Royal Society of London Series a-Mathematical and Physical Sciences. **313(1515):** 453.

Wilkins, R. M. 1990. Biodegradable polymer methods.Controlled Delivery of Crop-Protection Agents; Taylor and Francis: Bristol, PA, pp 149"165.

Yang, D. Z., Li, Y. N., and Nie, J. 2007. Preparation of gelatin/PVA nanofibres and their potential application in controlled release of drugs.*Carbohydrate Polymers.***69:** 538. 112.

Zhang, Y. Z., Feng, Y., Huang, Z. M., Ramakrishna, S. and Lim, C. T. 2006. Fabrication of porous electrospun nanofibres.*Nanotechnology.***17:**901.

20

Nano-Pheromones – Frontier Areas in Nanotechnology

K. Subaharan

Nanotechnology offers unique opportunities for the development of novel materials and applications. It is set to offer a platform to revolutionize agriculture sector from production, protection, processing and storage (Kuzma and Verhage, 2006). Bulk of nanomaterials produced is used in other sectors; only ten per cent is of use in food, beverages and packaging sector. Studies have projected the opportunities that nanotechnologies might bring to developing countries in the fields of water, energy, food and agriculture and health (Salamanca- Buentello et al., 2005) The key contributions due to application of nanotechnology in agriculture are delivery systems that aid in slow release and efficient delivery of agroinputs.

The need to produce an inexpensive and abundant food supply for a growing population warrants higher use of fertilizers and pesticides. The increasing use of pest icides has a negative impact on farmers, consumers, non-target organism and the environment. Around the globe, scientists are searching for alternative methods of pest control. Continuous use of increasing amounts of toxic chemicals seeps into our soil, leech into our aquifers and vaporize into the atmosphere, thereby causing harm to the environment. A challenge to balance between crop production and environmental protection can be achieved by adopting nanotechnology. Nanotechnology can increase the productivity through genetic improvement of crop plants (Kuzma, 2007), by site specific delivery of gene and drugs, delivery systems for agro-chemicals (Maysinger, 2007).

Modern pest monitoring and pest control systems across the globe depend on chemo-ecological approach. Over the past decade a widely accepted picture of insect olfaction has emerged. Olfactory sensory cells housed in sensilla on the insect antennae express cell - type specific odorant receptor proteins. Sensory cells expressing the same odorant receptor conveys their signals into specific and distinct areas, so-called glomeruli, in the antennal lobes, the first centres of processing of olfactory information in the brain. The molecular mechanisms of odorant detection, signal transduction, as well as physiological processing of olfactory information have been investigated in detail (Benton 2006; Larsson et al. 2005; Wicher et al. 2008). Many advances in the field have led to identification of pheromone and kairamones for management of insect pests.

Pheromones are used by insects to transmit information from one individual to another. Among the various types of pheromones released by insects the pheromones play important role in the field of plant protection. Release of pheromone by female signals their readiness for mating, thus allowing the males to track them down. Pheromones are species specific and the males are attracted to the compounds that are released only by their own species. This allows protecting the plants from herbivory with highly selectivity.

In India, pheromone lures are used for trapping fruit flies, stem borers in sugarcane and rice, boll worms *Heliothis armigera* and *Spodoptera litura*, coconut red palm weevil, *Rhychophorus ferrugineus* and *rhinoceros* beetle *Oryctes rhinoceros*. The mode of application of pheromone in the field ranges from spraying of a functionalized fluid containing pheromones, the mechanical distribution of solid particles incorporating the pheromones, or the evaporation of pheromones from dispensers. Many of these established distribution methods for pheromones including the dispenser methods have limitations. The pheromones applied via spraying or mechanical distribution are sensitive to wind turbulence and heavy rain. Loss of concentration of the compound in the locale warrants increased frequency of application that adds up to cost. In the case of dispensers the concentration of pheromones distributed in the neighborhood of the dispenser has to be exceedingly high.

The success of the pheromone/ kairamone technology depends of development of an effective delivery system. Nano materials are a novel carrier/ dispenser for the volatile signaling molecules. The nano materials have highly controlled spatiotemporal release rates of pheromones / kairomones and improved climatic stability.

While selecting the nano matrix for the release of pheromone the following points are to be considered. The ideal dispenser should have a constant release rate during the whole fight period of the pest, independent of weather conditions.

Previous work has shown that zeolites and nano fibres are suitable matrices for manufacturing dispensers because of their ability for retaining substances depending on their polarities and molecular sizes. Moreover, chemical structure and properties of these materials can be adapted to release substances at an adequate emission rate over a long period of time. It is essential to identify the threshold window that can provide optimum emission interval, as the catches decrease below and above this interval.

The series of steps involved in developing a nanomatrix for loading pheromone and kairamone are:

Identification of electro-physiologically – active compounds

The first step in identifying the volatile components involved in host location and recognition is to determine which of the tens to hundreds of compounds being released by plants are perceived by the olfactory system. The volatiles emitted by the preferred host plants will be collected by dynamic head space sampling. The volatiles so collected in an odor collection cartridge will be purged with solvent for use in gas chromatography. In case of pheromones the compounds released by insects will be collected and identified.

The combined gas chromatography – electroantennographic detection (GC-EAD) will be used to simultaneously separate volatile compounds present in the extracted plant headspace mixture of from insects to screen insect antennae for olfactory responses toeluted compounds. Compounds that elicit electro-physiological response are identified using gas chromatography – mass spectrometry (GC-MS) and verified using purified standards.

Electrophysiological studies provide useful information on sensitivity and selectivity of olfactory receptors. However they do not indicate the specific behavioral responses that such a perception might elicit, or weather a stimulus might act as an attractant or repellent. Moreover, a compound that stimulates chemosensory receptors may not elicit behavioural activity in an organism until the component is blended with other compounds.

Based on information gained from GC –EAD studies described above, the behavioral bioassay is carried out using a wind tunnel. The composition of odor blends to be tested will be based on the relative proportions of the EAG active components emitted; with a systematic subtraction of individual components until changes in insects behaviour towards a blend is observed.

Development of nanomatrix

On identifying the physiologically active compounds using the biological and chemical detectors, suitable nanomatrix has to be identified for loading and releasing the semiochemicals. Nanofibres and zeolites offer an ideal platform for release of pheromones/ kairamones. The nanofiber can be used as pheromone carriers similar to the case of drug carriers in nanomedicine. The nanofiber web is produced by electrospinning. In addition to nanofibers the zeolites can also be used as a matrix for the release of pheromones as they have a network of channels.

Electrospinning of solution containing the matrix polymer polyamide 6 (PA6) and pheromone will yield smooth nanofibers with diameters from 150 to 600 nm. The fiber diameters are rather narrowly distributed within the 200–400 nm range whereas the ribbons display a very broad distribution extending up to 1 μm and above. Characterization of the nano fiber along with pheromone is done in Cryo scanning electron microscopy and transmission electron microscopy.

Zeolites can be procured form vendors and their crystalline structures were verified by X-ray diffraction, by comparing the patterns of diffraction of the zeolites with those of published standards (van Ballmoos, 1984). The crystallinity is determined by comparing the areas of the most intense peaks with those of a standard whose crystallinity is 100% (ASTM 1985).

Thermal Gravimetric techniques are used to assess the release rate of the pheromone from nanofibres and zeolites.These nano dispensers can be evaluated in the field in comparison with the available commercial lure.

The nanofibers loaded with pheromone when distributed across the field resemble the spider webs. Electrospun nanofibers offer a set of unique advantages during such application. As they possess high stiffness and strength due to strong chain and crystal orientations, it enhances the fiber stability when applied in the field. The small diameter of the nanofibers adds to this effect since the strength of fibers is known to increase with decrease in the diameter. Furthermore, small fiber diameters within the nanometer scale reduce the resistivity with respect to airflow. Hence they are excellent choice as carrier systems for pheromones in the field as they are less disturbed by wind flow and air turbulence. The pores in the zeolites alter the emission kinetics and this indicates that the microporous materials are suitable semiochemical dispensers.

The expected advantages of the nanomatrix are highly controlled spatiotemporal release rates of pheromones/kairomones and improved climatic stability. Nanofibres and zeolites have been used for the controlled drug release in biomedical area. When nanomatrix is used for the release of pheromones could be extended for a longer period spanning several weeks to months which would help to scale down the cost in the chemistry used and the labour and logistics involved in using the semiochemicals for pest management.

References

ASTM. 1985. Standards for Catalysis. D-3942-80; ASTM:West Conshohocken, PA.

Benton, R., Sachse, S., Michnik, S.W. and Vosshall, L.B. 2006. Atypical membrane topology and heteromeric function of Drosophila odorant receptors in vivo. *PLoS Biology*, **4:**240-257.

Chau, C.F., Wu, Shiauan-Huei and Gow-Chin Yen 2007. The development of regulations for food nanotechnology. *Trends in Food Science and Technology*, **18:**269-280. Maysinger, Dusica. 2007. Nanoparticles and cells good companions and doomed partnerships. *Org. Biomol. Chem.*, **5:**2335-2342.

Kuzma, J. 2007. Moving forward responsibly: Oversight fot the nanotechnology-biology interface. *Journal of Nanoparticle Research*, **9:**165-182.

Kuzma, J. and VerHage, Peter 2006. Nanotechnology in Agriculture and Food Production: Anticipated Applications. Project on Emerging Nanotechnologies and The Consortium on Law, Values and Health and Life Sciences. Centre for Science, Technology and Public Policy (CSTPP). September 2006.

Larsson, M.C., Domingos, A.I., Jones, W.D., Chiappe, M.E., Amrein, H., Vosshall, L.B. 2005. Or83b encodes a broadly expressed odorant receptor essential for Drosophila olfaction. Neuron, **43:**703-714.

Salamanca-Buentello, F.S., Persad, D.L., Court, E.B., Martin, D.K. and Daar, A.S. 2005. Nanotechnology and the developing world. *PLoS Medicine*, **2(4)e97:**300-303.

von Ballmoos, R. Collection of Simulated XRD PowderPatterns for Zeolites; Butterworths Scientific Limited, Guilford, U.K., 1984.

Wicher, D., Schäfer, R., Bauernfeind, R., Stensmyr, M.C., Heller, R., Heinemann, S.H., Hansson B.S. 2008. Drosophila odorant receptors are both ligand-gated and cyclicnucleotide-activated cation channels. *Nature*, **452:**1007-1011.

21

Parapheromone for Fruitfly Management

S. Sithanantham

Background

The demand for development of Integrated Pest Management (IPM) strategies is increasing, since many problems have appeared with the use of synthetic pesticides and semiochemicals – informative molecules used in insect-insect or plant-insect interaction – are being considered as an alternative or complementary approach within IPM strategies. (Heuskin et al., 2011). Indeed, such communication molecules like para-pheromones do not present any related adverse effects on beneficial organisms in the ecosystem.

For fruitflies management, several eco-friendly control options have been developed and recommended across the globe which include trapping, safe disposal of infested fruits, repulsion with botanical products and physical covering of the developing fruits, besides area-based control through mass trapping and sterile-insect technique. Nevertheless, the use of para-pheromones-based lures in trap systems appears to have greater potential to be among the more widely adopted practices (Verghese, 2009).

Definition

Insect para-pheromones are chemical compounds of anthropogenic origin, not known to exist in insect systems in nature, but structurally related to natural pheromone constituents who in some way affect physiologically of behaviorally the insect pheromone communication system eliciting similar responses as that of a true pheromone (Renou and Guerrero, 2000). Modifications at the chain and/or at the polar group, isosteric replacements, halogenation or introduction of labeled atoms have been the most common modifications of the pheromone structure. Such pheromone analogues can replace pheromones when these are costly to prepare or unstable under field conditions. Males of many Tephritid fruit fly species are strongly attracted to such compounds which either occur naturally in plants (eg. Methyl eugenol) or are synthetic analogues of plant-borne substances (Cue lure).

Chemistry & Properties

Name: **Methyl eugenol**

Synonyms: 1, 2-Dimethoxy-4-(2-propenyl) benzene; 4-Allyl-1, 2-dimethoxybenzene; 4-Allylveratrole

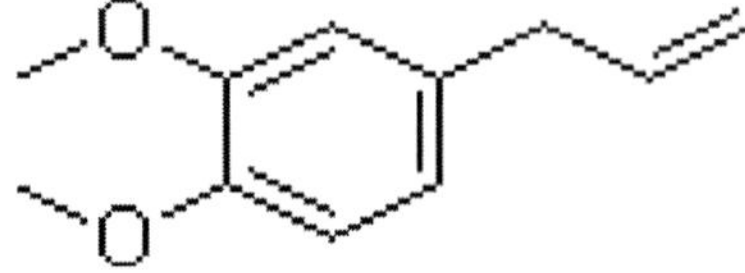

Molecular Structure

Name: **Cue lure**

Synonyms: 4-[4-(Acetyloxy) phenyl]-2-butanone,4-(p-hydroxyphenyl) -2-butanone acetate

Molecular Structure

Properties of Methyl eugenol and Cue lure

Properties	Methyl eugenol	Cue lure
Density	1.035	1.099
Melting point	-4 °C	-
Boiling point	248 °C	bp0.2
Refractive index	1.533-1.535	D25 1.5061
Flash point	117 °C	>230°C
Water solubility	insoluble	insoluble

Target fruit fly species

The tephritid fruit flies are a globally important production constraint to several horticultural crops in both tropical and temperate ecosystems. Their important target crops in India and elsewhere in Australasia are tree fruits such as mango, guava, citrus, apple, plums, papaya and banana, besides cucurbit vegetables such as gherkins, water melon, musk melon, snake gourd, ribbed gourd, ash gourd, bottle gourd, bitter gourd and pumpkin (Verghese et al., 2002).

The Dacinae species of fruit flies made up of the two major genera- *Bactrocera* Macquart and *Dacus* F. - are strongly attracted to either methyl eugenol (ME) or cue-lure (CL). Owing to their powerful male-specific attractiveness, both ME and C-L lures have been used successfully in male annihilation systems against *B. dorsalis* and *B. cucurbitae,* (Chambers et al., 1974).

Methyl eugenol (ME) has been used for detection of oriental fruit fly, *Bactrocera dorsalis* (Hendel), and the Pacific fruit fly, *B. xanthodes* (Broun), and cue-lure is used for *B. kirki* (Froggatt), melon fly, *B. cucurbitae* (Coquillett), and the Queensland fruit fly, *B tryoni* (Froggatt). (Metcalf & Metcalf., 1992).

While Ramani (1998) and Madhura and Verghese et al. (2004) have provided identification guidance of the species in South India, more recently, David and Ramani (2011) have provided a more comprehensive key for identification of the fruit fly species occurring in the region.

Para pheromones in fruit fly trap system

Major purposes for fruit fly traps

The emerging interest in India to deploy effective methyl eugenol-based trap systems for monitoring and/ or mass trapping of fruit flies as means of improving the marketable yields in general and to qualify for export through establishing fruit fly-free zones in the country is noteworthy. This scenario has prompted commercial-focus research and development initiatives to improve the efficiency of methyl eugenol based trap-lure systems by more efficient combinations of trap designs and dispenser systems and the recent progress made, are illustrated in this paper. One of the prime objectives of such R&D has been to identify trap design-dispenser combinations which can enhance their attractiveness and/or field life, to catch not only the commonly known Oriental fruit fly, *Bactrocera dorsalis* (Hendel) but also the other associated fruit fly species such as *Bactrocera correcta (*Bezzi*), Bactrocera zonata (*Saunders*),* *Bactrocera caryae* (Kapoor) and *Bactrocera* sp. that also commonly occur in mango and guava ecosystems in South India.

A fly-free area is an area with no detectable populations of a fruit fly species (Malavasi et al. 1994). There are two levels of fly-free areas:

1. A fly-free zone, where an entire geographical or political entity is recognized to be free from target flies, and
2. A fly-free production fields, where flies are managed in a way that ensures absence of flies from specific production areas but not necessarily from large geographic or political entities. Several factors come into play in

characterizing and establishing a fly-free area, including physical, ecological, biological, political, economic, and social factors (Malavasi et al. 1994).

Males of most *Bactrocera* species respond to either raspberry ketone (RK, or the man-made analogue cue lure, CL) or methyl eugenol (ME). Traps containing these chemicals are widely used in detection programs (Verghese et al., 2004, Jessup et al., 2007).

Trapping requirements for fruit crops

Among the practical requirements for slow release lures dispenser are the following:

Coverage for full fruiting duration

For fruit crops like mango, the fruiting duration may extend to 10-12 weeks. For several cucurbit crops, also there is such long fruiting period. Presently, available commercial lures (mainly wooden blocks) may need replacement in about 3-6 weeks. Hence, the need for longer field life lures dispensers for both ME and CL trapping system.

Extended/multiple fruiting duration

In crops like guava, the fruiting period may extend to 8-10 months. In special agro-ecozones like Kannyakumari, mango can fruit twice a year (main season: March-June; special season: Nov-Jan). In these cases, the farmers may require several replacements of the presently available lures in this a year. They would welcome longer-life lures so to sender continuous tapping of fruit flies less laborious.

Year round trapping

It has been shown beneficial to keep fruit fly traps with ME lures throughout the year to bring about cumulative reduction in their populations so reducing the fruit infestations in subsequent fruiting seasons. In Tamilnadu, such positive impact of ME trapping has long back been demonstrated for mango, mandarin orange and plum (Balasubramanian et al., 1972; Lakshmanan et al., 1973). Such year-round trapping system will also greatly benefit from slow-release and long-life dispenser.

Maximizing catch efficiency & attractancy duration

Because of their complex biological activity, the dispersion of parapheromone in the environment may need to be monitored the elaboration of slow-release devices ensuring a controlled release of these biologically active volatile compounds. These sensitive molecules also need to be protected from degradation by UV light and oxygen. Many studies were conducted on estimation of release-rate from commercialized or experimental slow-release devices. The influence of climatic parameters and dispenser type were estimated by several researchers in order to provide indications about the on-field longevity of lures. This review outlines slow-release studies conducted recently.

Research elsewhere in alternative release systems

The scope for extending the attractancy duration of these insect lures through use of nano-technology has also been hypothesized (Bhattacharyya et al., 2010).

A variety of carbon-based, metal and metal oxide-based dendrimers(nano-sized polymers) and biocomposite nanomaterials are being developed which include single-walled and multi-walled carbon nanotubes (SWCNT/MWCNT), magnetized iron (Fe) nanoparticles, aluminium (Al), copper (Cu), gold (Au), silver (Ag), silica (Si), zinc (Zn) nanoparticles, Zinc Oxide (ZnO), titanium oxide (TiO) and Cerium Oxide (CeO) etc. -Nanomaterials and biocomposites exhibit useful properties like stiffness, permeability, crystalinity, thermal stability, solubility and biodegradability needed for formulating nanopesticides, which can also offer large specific surface area and hence increased affinity to the target, Nanoemulsions, nanoencapsulates, nanocontainers and nanocages are some of the nanopestcide delivery techniques being discussed recently for plant protection. Nanoemulsions may be potentially better pesticide delivery medium due to better kinetic stability, smaller size, low viscosity and optical transparency. Development of nanomaterials that can be used as a coating or protective layer to enable slow release may also be pursued based on the possibility of using chitosan nano particles, a highly degradable anti-bacterial material for slow release fertilizer.

Resins and gels

Numerous reservoir-type controlled release devices (CRDs) have been developed to overcome the high volatility of pheromones and achieve a sustained release over a period of several weeks, essential for effective pest control. However, the current devices are typically polymeric, involve multi-step preparation protocols, exhibit low pheromone-holding capacities and are not readily biodegradable. Moreover, most of the devices leak if they are broken or compressed. Hence the scope for a bio-based molecular gelator as an efficient alternate material for developing reservoir-type releases devices.

Min-U-Gel formulations with ME were developed for spot applications in male annihilation programs in California for eradication of *B. dorsalis* (Chambers et al., 1974, Cunningham and Suda 1985). Min-U-Gel is a high grade of attapulgite clay (anhydrous magnesium aluminum silicate) mixed with naled or malathion and ME to form a gel male annihilation formulation. Acti-Gel is a refined grade of Min-U-Gel. However, Min-U-Gel and similar thickened formulations are not long-lived when weathered in areas with warm temperatures and high rainfall (Cunningham et al., 1975a, b; Cunningham and Suda, 1985, Vargas et al., 2000).

Jadhav et al. (2011) have pointed out that molecular gels find applications in the agricultural industry by demonstrating their capability in developing efficient controlled release devices for pheromones, which are potential biopesticides. Such new devices are readily biodegradable, exhibit high pheromone-loading capacity and on deliver the compounds uniformly at high concentration for a prolonged time.

Molecular gelators (MGs) are low molecular weight amphiphilic molecules that self-assemble through non-covalent forces to form a volume-filling 3-D network. Within the network, solvent molecules are immobilized by physical interactions such as surface tension, thereby converting the liquid into a coherent gel. In some cases, extensive solvent–network cohesion causes the gel to be stable even at the boiling point of a solvent. The interactive nature of gels enables them to show a sustained-release property. Such gelators are of particular interest due to their non-polymeric nature, higher biocompatibility and more importantly, they exhibit extremely low minimum gelation concentration (MGC, usually #5% wt/v).

SPLAT strategy

Vargas et al. (2008) have indicated the scope for Specialized Pheromone and Lure Application Technology (SPLAT) methyl eugenol (ME) and cue-lure (CL) as "attract-and-kill" sprayable formulations containing spinosad compared with other formulations under Hawaiian weather conditions against, *Bactrocera dorsalis* and *Bactrocera cucurbitae* respectively. Field tests were conducted with three different dispensers (Min-U-Gel, Acti-Gel, and SPLAT) and two different insecticides (naled and spinosad). SPLAT ME with spinosad was equal in performance to the standard Min-U-Gel ME with naled formulation up to 12 wk. SPLAT C-L with spinosad was found to be equal in performance to the standard Min-U-Gel C-L with naled formulation during weeks 7 to12, but not during weeks 1Ð 6. In subsequent comparative trials, SPLAT ME spinosad compared favorably with the current standard of Min-U-Gel ME naled for up to 6 wk, and it was superior from weeks 7 to 12 in two separate tests conducted in a papaya (*Carica papaya* L.) orchard and a guava (*Psidium guajava* L.) orchard, respectively. In outdoor paired weathering tests (fresh versus weathered), C-L dispensers (SPLAT spinosad, SPLAT naled, and Min-U-Gel naled) were effective up to 70 d. Weathered ME dispensers with SPLAT spinosad compared favorably with SPLAT naled and Min-U-Gel naled, and they were equal to fresh dispensers for 21Ð28 d, depending on location. These studies have indicated that SPLAT ME and SPLAT C-L sprayable attract-and-kill dispensers containing spinosad are a promising substitute for current liquid organophosphate insecticide formulations used for area wide suppression of *B. dorsalis* and *B. cucurbitae* in Hawaii.

Leblanc et al. (2011) have evaluated the performance of solid male lure (cuelure (C-L)/raspberry ketone (RK) against *Bactrocera tryoni* and methyl eugenol (ME) against oriental fruit fly, *B. dorsalis* both formulated with insecticide, in Tahiti Island (French Polynesia). Captures of *B. tryoni* in traps with BactroMAT CL stations, Mallet C-L, Mallet MC wafers (containing both ME and RK), and Specialized Pheromone and Lure Application Technology (SPLAT) C-L were as high as with the standard liquid C-L formulation until 8 weeks, but thereafter the effectiveness of Mallet C-L baited traps declined. Captures of *B. dorsalis* with Mallet ME wafers outperformed any other ME formulation. For control applications, the weathered SPLAT-MAT-ME-spinosad lure and kill formulation was equal to fresh material for up to 4 weeks. SPLAT C-L was more persistent than weathered SPLAT-MAT-ME under Tahitian climatic conditions, which suggested that SPLAT-MAT-ME may need to be reapplied at shorter intervals and in greater amounts for suppression of *B. dorsalis* than is required to suppress *B. tryoni* with SPLAT-MAT-C-L. Mallet ME and MC wafers and SPLAT-MAT-ME/C-L were more convenient and safer to handle than standard liquid insecticide formulations, and should be considered for monitoring and control programs in Pacific island nations.

SPLATTM is a waxy formulation of biologically inert materials used to control the release of semiochemicals with or without pesticides. Previous research in Hawaii has shown that the SPLAT matrix emits ME or C-L at effective pest suppression levels for a time interval ranging from 4-8 wk (Vargas et al., 2009a, 2010b). Similarly, solid lure insecticide dispensers have been developed and successfully evaluated for monitoring and male annihilation traps (Vargas et al. 2009b, 2010a) allowing for the elimination of liquid lures and insecticides.

Other approaches

Fibrous (adsorbent) boards, saturated with the appropriate lure and a toxicant and distributed widely, have also been used successfully to eradicate invasive *Bactrocera* populations (Hancock et al., 2000; Seewooruthun et al., 2000).

Todd (2010) has indicated that *Bactrocera cucurbitae* and *B. dorsalis* are important agricultural pests of the Pacific region. Detection and control of these species rely largely on traps baited with male-specific attractants (parapheromones), namely cue lure for *B. cucurbitae* and methyl eugenol for *B. dorsalis*. Recently, a solid formulation (termed a wafer) has been developed that contains both chemicals (plus DDVP, an insecticide), and data from field tests are available that compare the effectiveness of liquid versus solid formulations of the lures in attracting males of these two *Bactrocera* species. Trapping in two areas of Oahu, Hawaii, showed that Jackson traps baited with wafers captured (1) similar numbers of *B. cucurbitae* males and (2) significantly greater numbers of *B. dorsalis* males than Jackson traps baited with standard liquid lures. Additional tests examined the effect of ageing on the attractiveness of the wafers and revealed that Jackson traps baited with wafers aged for 6 or 8 weeks captured similar or significantly more males of both *Bactrocera* species than Jackson traps baited with fresh liquid lures.

At present, CL and ME are applied as liquids to cotton wicks positioned inside traps. Moreover, the insecticide naled is often added (also in liquid form) to the lures before application to the cotton wicks. Thus, the current procedure involves considerable handling time for measuring and applying the liquids as well as potential health risks resulting from accidental contact or ingestion of the insecticide.

Research on field life duration release rate-India

Early workers demonstrated the utility of deploying traps for monitoring and population suppression in several tree fruit crops, especially mango and guava ecosystems (Balasubramanian et al.,1972; Belavadi,1979; Lakshmanan et al.,1973). Research to refine fruit fly trap design and dispenser systems for fruit flies was attempted by several workers in India also, mainly targeting the fruit fly groups infesting fruit tree crops (Jalaluddin et al., 1998; Madhura and Viraktamath., 2003).Some aspects of the R&D on field life extension of lures in India are summarized in Table.1.

Leading research by ICAR-led R & D network

Under the Indian Council of Agricultural Research (ICAR)- Department for Industrial Development (DFID)- joint R & D network project on fruit flies management led by two eminent entomologists-Dr.Stonehouse and Dr. Abraham Verghese, provided major inputs (Table 1).

Table.1: Recent R&D on extending parapheromone lure field life in India- examples

Methyl eugenol ME/cue lure CL/others-OT	Lab. study -LS /Field study -FS	Aspect of release system studied	Important findings	Country/crops applicable	Reference
CL	FS	Soaking time and wood source	Soaking time- limited role; but wood source was more important for greater catches and field life	All tree fruit crops-India	Sing et al. (2005)
ME	FS	Solvent for blocks	Hexane was found to be safe alternative to ethanol, ether or benzene in extending the field life	All tree fruit crops-India	Patel et al. (2005b)
ME	FS	Liquid Dispenser type	Cotton wick in vial could give greater catches and longer field life	All tree fruit crops-India	Sithanantham et al. (2006)
ME	FS	Solid Dispenser type	Plywood discs - hexane as solvent	All tree fruit crops-India	Sithanantham et al. (2010)
ME		Dispenser role	Extending field life possible with increased disc volume; also trap design can help in maximizing catches	All tree fruit crops-India	Sithanantham (2011)
ME	LS/FS	Nanogel	Slow release rate and extended field life of lure	Major tree fruit crops-India	Deepa Bhagat et al. (2012)

Patel et al. (2005a) found that wooden substrates which absorbed more of the attractant soakate (lure mix) were relatively more efficient and longer-life dispensers for methyl eugenol and cue lure, whereas Patel et al. (2005b) identified ether as a be a potential substitute for ethanol as the dispensing liquid for methyl eugenol, while for cue lure hexane or ether could be effective alternatives.

Recent and ongoing R & D at SABRC

The Sun Agro Biotech Research Centre, (SABRC) Chennai,a non-profit Scientific and Industrial Research Organization-recognized by Government of India-DSIR (in close collaboration linkage for commercialization with Sun Agro Biosystems Private Limited) launched commercial-focus R&D to develope more efficient combinations of trap sign and dispenser systems, with holistic attention to species-specific responses among the fruit fly complex (Sithanantham et al., 2006; 2009; 2010, 2011; Sithanantham, 2011). The prime objective of the R&D

at SABRC has been to identify alternative of trap design-dispenser combinations with enhanced attractiveness and/or field life, so to maximize catch both the main species, *B. dorsalis,* and also the associated species such as *Bactrocera correcta (*Bezzi*), Bactrocera zonata (*Saunders*), Bactrocera caryae* (Kapoor) and *Bactrocera* sp., which are also commonly occur in mango and guava ecosystems in South India (Sithanantham et al., 2006).This was reckoned as very important since these trap-lure combinations are to be used either for mass trapping or for monitoring the occurrence for population dynamics studies, besides for quarantine detection of their presence in sensitive or benchmark locations (Sithanantham et al., 2006, 2010).

To start with, a combination of cylindrical white plastic jar as trap and methyl eugenol dispensed as liquid in wick system as lure were initially selected for improving the trapping efficiency, including the field attractancy life (Sithanantham et al., 2006). This improved trap-lure combination has been utilized for characterizing the seasonal abundance of fruit flies in mango ecosystems in five benchmark locations in Tamil Nadu.

Another cost-effective combination more recently found promising was a collapsible plastic sheet trap along with methyl eugenol impregnated in plywood or wooden discs; also the relative utility of wick-based and vial-based dispenser systems in reciprocal combination of the two promising trap designs was also determined (Sithanantham, 2011).

Cotton wick for liquid dispenser

Comparison of two substrates-cotton wick and chalk stick- to identify more effective and adsorbent dispensers for delivering the liquid attractant in the cylindrical jar traps, showed that cotton wick was more efficient (Fig.1)

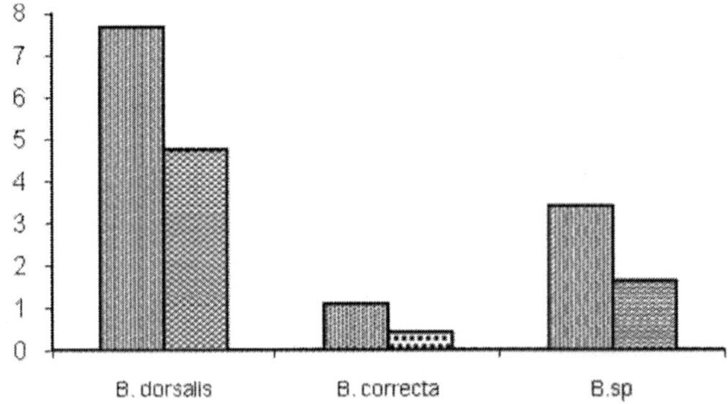

Fig. 1: Comparison of species wise catches for cotton wick as dispenser

Assessment of the relative catches of the constituent species of fruit flies showed similar pattern of species response by *B. dorsalis, B. correcta and B.sp*, with cotton wick being more efficient for each of the species (Fig.2). Such comparison between substrates as dispensers has not been done earlier at species level in mango ecosystem in India and elsewhere. Further, this information is of significant commercial value towards standardizing a promising trap design-dispenser combination for fruit tree crop ecosystems in India (Sithanantham and Boopathi, 2010).

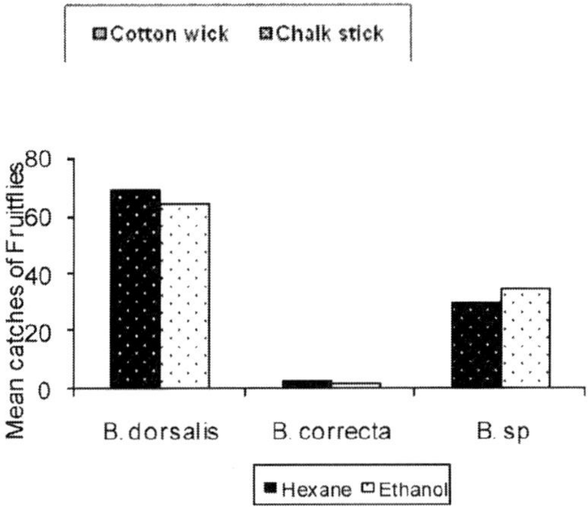

Fig. 2: Species-wise catches for the two solvents in Guava ecosystem (*Source*: Sithanantham et al., 2010)

Dispensing liquid- hexane as alternative choice

Among the dispensing (solvent) liquids used, while ethanol has been recommended, hexane was identified as cheaper and more readily available alternative.

The relative effects of hexane and ethanol and catches of the constituent fruit fly species have also been compared (Fig.2)

Recent results on Nanotechnological applications

Deepa Bhagat et al. (2012) have conducted detailed laboratory and field studies in guava ecosystem and reported positive results with nanogel prepared with methyl eugenol (ME) using a low-molecular mass gelator. This formulation was found to be very stable at open field conditions and slowed down the evaporation of pheromone significantly, which could enable its easy handling and transportation without refrigeration, besides reduction in the frequency of pheromone lure replacement/recharging in the orchard.They have been able to achieve an increased field-life of the pheromone (ME) when immobilized in a nanogel and demonstrated that such pheromone nanogels exhibit high residual activity, excellent efficacy in open orchard even during adverse seasons and are also environment friendly. Thus the scope for simple, practical and low cost green chemical approach has been demonstrated.

Future needs and R & D threats

No doubt the goal is extended field life of para-pheromone lures and slow release systems can play a major role towards this goal, while clearly recognizing that choice of the technology potions should depend on being cost-effective and user-friendly.The following may well be the key components of our future R&D initiatives:

Critical need for extending field life of lures with slow release systems

Both solid and liquid dispensing systems may be considered, as they may have their own niches across the different purposes and ecosystems where the trapping of fruitflies with parapheromones as key attractants are required

Trap design combinations to maximize the impact of slow release lures

Recent and ongoing research has clarified that while dispenser system improvement to extend the field life of lures would be the main focus, it will be beneficial to link up the promising dispenser types with compatible/beneficial trap design so to maximize the catches for the same amount of attractant dispensed.

Collaboration in R & D and commercialization of the technologies

The scope for inter-disciplinary team effort is indeed critical, since the technology options may well include organic chemists, nanotechnology experts, fruitfly taxonomists, laboratory behavioral experts, and field ecologists among others. With due compliance with IPR regimes, it is possible that public-private-partnership (PPP) mode of R&D could culminate in sustainable commercialization of the promising technology options developed.

Acknowledgement

Grateful appreciations of the author is extended to colleagues at SABRC and to also the collaborating R&D institutions for extending their expertise and cooperation.

References

Balasubramanian, M., Abraham, E.V., Vijayaraghavan, S., Subramanian, S., Santharaman T.R.G. and Gunasekaran. C.R. 1972. Use of male annihilation technique in the control of Oriental fruit fly, *Dacus dorsalis* Hendel. *Indian Journal of Agricultural Sciences,* **42**: 975-977.

Belavadi, V.V. 1979. Bionomics of the oriental fruit fly, *Dacus dorsalis* Hendel. (Diptera: Tephritidae) on guava (*Psidium gujava* L.) and its control by male annihilation. *M.Sc (Agri.) Thesis, University of Agricultural Sciences, Bangalore, India*, pp. 95.

Bhattacharyya A, Bhaumik A, Usha Rani P, Suvra Mandal, Epidi TT, 2010. Nano-particles –a recent approach to insect pest control. *Afr. J. Biotechnol.* **9**: 3489-3493.

Chambers, D. L., R. T. Cunningham, R. W. Lichty, and R. B. Thraikill. 1974. Pest control by attractants: a case study demonstrating economy, specificity, and environmental acceptability. *Bioscience* **24**: 150 Ð152.

Cunningham, R. T., D.L. Chambers, and A. G. Forbes. 1975a. Oriental fruit ßy: thickened formulations of methyl eugenol in spot applications for male annihilation. *J. Econ. Entomol.* **68**: 861-863.

Cunningham, R.T., D.L. Chambers, L.F. Steiner, and K. Ohinata. 1975b. Thixcin-thickened sprays of cue-lure naled: investigation of rates of application for use in male annihilation of melon ßy. *J. Econ. Entomol.* **68**: 857-860.

Cunningham, R. T., and D. Y. Suda. 1985. Male annihilation of the oriental fruit fly (Diptera: Tephritidae): a new thickener and extender for methyl eugenol formulations. *J. Econ. Entomol.* **78**: 503-504.

David., K.J. and Ramani, S. 2011. An illustrated key to fruit flies (Diptera: Tephritidae) from Peninsular India and the Andaman and Nicobar Islands. *Zootaxa,* **3021**: 1-31.

Deepa Bhagat., S.K. Smanta and S.Bhattacharya. 2012. Efficient Management of Fruit Pests by Pheromone Nanogels. Scientific reports. p.1-8.

Hancock, D.L., R. Osborne, S. Broughton, and P. Gleeson. 2000. Eradication of *Bactrocera papayae* (Dipteral: Tephiritidae) by male annihilation and protein baiting in Queensland, Australia. pp.381-388. In Tan, K.H., ed., Area Wide Control of Fruit flies and Other Insect Pests. Penerbit University Sains Malaysia, Pulau Pinang, Malaysia.

Heuskin, S., Francois. J.V., Eric. H., Jean. W.P and Georges. L. 2011. The use of semiochemical slow-release devices in integrated pest management strategies. *Biotechnol. Agron. Soc. Environ.* **15(3)**:459-470.

Jalaluddin, S.M., Natarajan, K., Sadakathulla, S. and Rajukkannu, K. 1998. Effect of height and dispensers on catches of guava fruitflies. In: *Proceedings of Symposium of Pest Management in Horticulture Crops,* Bangalore, India, pp. 34-39.

Jadhav, S.R., B.S. Chiou., F.W. Delilah., Gloria D.H., Gregory. M.G and George.J. 2011. Molecular gels-based controlled release devices for pheromones. *Soft matter*. **7**:864-867.

Jessup, A.J., B. Dominiak, B. Woods, C.P.F. De Lima, A. Tomkins, and C.J. Smallridge. 2007. Area-wide management of fruit flies in Australia, pp. 685–697. *In* Vreysen, M.J.B., Robinson, A.S., and Hendrichs, J., eds., Area-Wide Control of Insect Pests. Springer, Dordretch, The Netherlands.

Lakshmanan, P.L., Balasubramaniam, G. and Subramaniam, T. R. 1973. Effect of methyl eugenol in the control of the Oriental fruit fly *Dacus dorsalis* Hendel on mango. *Madras Agricultural Journal,* **60(7):** 628-629.

Leblanc, L., Roger.I.V., Bruce.M., Rudolph.P and Jaime.C. P. 2011. Evaluation of cue-lure and methyl eugenol solid lure and insecticide dispensers for fruit fly (Diptera: Tephritidae) monitoring and control in Tahiti. *Florida Entomologist* **94(3):** 510-516.

Madhura, H.S. and Viraktamath, C.A. 2003. Efficacy of different traps in attracting fruit flies (Diptera: Tephritidae). *Pest Management in Horticultural Ecosystem,* **9(2):** 53-154.

Malavasi, A., G. G. Rohwer, and D. S. Campbell. 1994. Fruit fly-free areas: strategies to develop them, pp. 165-180. *In* C. O. Calkins, W. Klassen & P. Liedo [eds.], Fruit Flies and the Sterile Insect Technique. CRC Press, Boca Raton, FL.

Metcalf, R. L., and E. R. Metcalf. 1992. Fruit flies of the family Tephritidae. *In* R. L. Metcalf and E. R. Metcalf [eds.], Plant kairomones in insect ecology and control. Routledge, Chapman & Hall, New York.

Patel, R.K., Jhala, R.C., Joshi, B.K., Sisodiya, D.B., Verghese, A., Mumford, J.D. and Stonehouse, J.M. 2005a. Effectiveness of solvents for soaked-block annihilation of male fruit flies in Gujarat. *Pest Management in Horticultural Ecosystem,* **11(2)**: 123-125.

Patel, R.K., R.C.Jhala, B.K.Joshi, D.B.Sisodya, A.Verghese, J.D. Mumford, and J.M.Stonehouse (2005). Effectiveness of Solvents for Soaked –Block Annihilation of Male Fruit Flies in Gujarat. *Pest management in horticulture Ecosystems.* **11:**123-125.

Patel, Z.P., Jhala, R.C., Jagadale, V.S., Sisodiya, D.B., Mumford, J.D, Verghese, A. and Stonehouse, J.M. 2005b. Effectiveness of woods for soaked block annihilation of male fruit flies in Gujarat. *Pest Management in Horticultural Ecosystem,* **11(2):** 117-120.

Ramani, S. 1998. Biosystematic studies on fruit flies (Diptera: Tephritidae) with special reference to the fauna of the Karnataka and Andaman and Nicobar islands. Ph.D. Thesis, University of Agricultural Sciences, Bangalore, India, pp.230.

Renou,M and Guerrero,A.2000.Insect Parapheromones in Olfaction Research and Semio-chemical -Based Pest Control Strategies.Annual Review of Entomology. **45:** 605-630.

Seewooruthun, S.I., S. Permalloo, B. Gungah, A.R. Soonnoo, and M. Alleck. 2000. Eradication of an exotic fruit fly from Mauritius, pp. 389–394. *In* Tan, K.H., ed., Area-wide Control of Fruit Flies and Other Insect Pests. Penerbit Universiti Sains Malaysia, Pulau Pinang, Malaysia.

Sithanantham S, 2011. Recent progress in improving para-pheromone-based trap-lure systems for fruit flies *Bactrocera spp*. (Diptera: Tepheritidae) in India: overview Proceeding international symposium on Insect Pest Management. Ed. by Sahayaraj K, St. Xavier's college, Palayamkottai, TN, India, November (28-30), 56.

Sithanantham, S. 2011. Insect traps and lures. In*: Proceedings of National Training Workshop on Appropriate Pest Management Technologies,* Tamil Nadu Agricultural University.

Sithanantham, S. and Boopathi, T. 2010. Fruit flies trap system improvement 2. Methyl eugenol source and dispenser comparison. *Hexapoda,* **17(1):** 66-68, 2010.

Sithanantham, S., Boopathi, T., Selvaraj, P., Gajalakshmi, S., Vasumathi, S., Kannan, M., Kalyanasundaram, S., Manisegaran, S., Ganapathy, N., Revathi, K., and Jesudasan, R.W.A. 2006. Consortium approach towards monitoring and management of fruit flies in Tamil Nadu, India. In: *Proceedings of Organic Crop Protection Technologies for Promoting Export Agri-Horticulture*, Chennai, India.

Sithanantham, S., Gajalakshmi, S., Ranjith, W.A.C. and Richard Kennedy, R. 2009. Pest management in mango with special reference to Off-Season fruiting in Kanyakumari. In: *Proceedings of National Level Training Cum Seminar on Off-Season Mango Production*, (eds.) Kennedy, R., Rajan, S., Joshua, P. and Kumar, N., Kanyakumari, Tamil Nadu Agricultural University, Coimbatore, India.

Sithanantham, S., Ranjith, W.A.C. and Gajalakshmi, S. 2011. Improved pest management in off-season mango, with emphasis on fruit fly trapping systems. In: *Proceedings of National Consultation Workshop on Off-season Mango Production in India*, (eds.) Kennedy, R. and Kumar, N., Tamil Nadu Agricultural University, Kanyakumari.

Sithanantham, S., Ranjith, W.A.C., Gajalakshmi, S. and Kannaiyan, J.J. 2010. Fruit fly trap systems improvement 1. Testing of two dispensing liquids for methyl eugenol. *Hexapoda,* **17(1):** 64-65, 2010.

Todd, E.S. 2010. Capture of *Bactrocera* Males (Diptera: Tepheritidae) in Parapheromone-Baited Traps: A comparison of liquid versus solid Formulations. *Proceedings of the Hawaiian Entomological Society,* **42:**1-8.

Vargas, R. I., Burns, R. E., Mau R. F. L., Stark, J. D., Cook, P., and Piñero, J. C. 2009b. Captures in methyl eugenol and cue-lure detection traps with and without insecticides and with a Farma Tech solid lure and insecticide dispenser. *J. Econ. Entomol.* **102:** 552-557.

Vargas, R. I., J. D. Stark, M. H. Kido, H. M. Ketter, and L. C. Whitehand. 2000. Methyl eugenol and cue-lure traps for suppression of male oriental fruit flies and melon flies (Diptera: Tephritidae) in Hawaii: effects of lure mixtures and weathering. *J. Econ. Entomol.* **93:** 81-87.

Vargas, R. I., Mau, R.F.L., Stark, J.D., Piñero, J.C., Leblanc, L., and Souder, S.K. 2010a. Evaluation of methyl eugenol and cue-lure traps with solid lure and insecticide dispensers for monitoring and male annihilation in the Hawaii area wide pest management program. *J. Econ. Entomol.* **103:** 409-415.

Vargas, R.I., Piñero, J.C., Jang, E.B., Mau, R.F.L., Stark, J.D., Gomez, L., Stoltman, L., and Mafra-Neto, A. 2010b. Response of melon fly (Diptera: Tephritidae) to weathered SPLAT-spinosad-cue-lure. *J. Econ. Entomol.* **103**: 1594-1602.

Vargas, R.I., Piñero, J.C., Mau, R.F.L., Stark, J.D., Hertlein, M., Mafra-Neto, A., Coler, R., and Getchell, A. 2009a. Attraction and mortality of oriental fruit flies to SPLAT-MAT-methyl eugenol with spinosad. Entomol. Exper. Appl. **131**: 286-293.

Vargas, R.I., John D. S., Mark H., Agenor M.N., Reginald.C and Jaime C.P. 2008. Evaluation of SPLAT with Spinosad and Methyl Eugenol or Cue-Lure for "Attract-and-Kill" of Oriental and Melon Fruit Flies (Diptera: Tephritidae) in Hawaii. *J.Econ.Entomol.* **101(3)**:759-768.

Verghese, A. 2009. Integrated pest management in mango.In: *Proceedings of National Symposium on Production, Post harvest Technology and Marketing in Mango,* Horticultural College and Research Institute, Tamil Nadu Agricultural University, Periyakulam, Tamil Nadu, India, pp.157-162.

Verghese, A., Madhura, H.S., Jayanthi, P.D.K and Stonehouse, J.M. 2002. Fruit flies of economic significance in India with special reference to Bactrocera dorsalis (Hendel). In: *Proceedings at 6th International Symposium on Fruit Flies of Economic Importance*, Stellenbosch, South Africa.

Verghese, A., H.S. Madhura, P.D. Kamala Jayanthi, and J.M. Stonehouse. 2004. Fruit flies of economic significance in India, with special reference to *Bactrocera dorsalis* (Hendel), pp. 317–324. *In* Barnes, B.N., ed., Proceedings of the 6th International Symposium on Fruit Flies of Economic Importance. Isteg Scientific Publications, Irene, South Africa.

22

Nanotechnological Concepts in Plant Disease Management

V. Jayakumar and K.S. Subramanian

In recent years application of nanotechnology in various fields including agriculture has received increased attention. Nanotechnology offers an important role in improving the existing crop management techniques especially in plant disease management by controlled and targeted release of agrochemicals to enhance their efficiency. The nano-based delivery systems are beneficial because of improvement in efficacy due to their higher surface area, higher solubility, induction of systemic activity, higher mobility and lower toxicity (Sasson et al., 2007). The pharmacokinetic parameters of nano particles may be altered according to size, shape and surface functionalization. They can also be used to alter the kinetic profiles of drug release leading to more sustained release of drug there by reducing the frequent application (Sharon et al., 2010). The successful use of metal nanoparticles in medical streams as antimicrobial agents has also led to their applications in controlling phyto pathogens (Nair et al., 2010). Nanoscale materials with novel properties can make the agricultural systems "smart". These smart systems deliver chemicals in a controlled and targeted manner as similar to the proposed use of nanodrug delivery in humans (Roco, 2003). "Smart Delivery Systems" in agriculture should possess combinations of time controlled, specifically targeted, preprogrammed, self-regulated and multifunctional characteristics to avoid biological barriers for successful targeting. The other area of application of nanotechnology in crop protection is developing nano-formulations. The nano-formulations may overcome the problems of traditional fungicide application *viz*., leaching of chemicals, degradation by photolysis, hydrolysis, microbial degradation *etc* (Nair and Sakthikumar, 2013).The development of nano devices and nanomaterials could open up novel application in plant biotechnology especially in pathogen detection and plant-pathogen interaction (Mahendra et al., 2012). In addition, the nanoparticles mediated plant transformation can be applied for genetic modification of plants to address the specific problems in plant protection.Application of nanotechnology in plant disease management is broadly categorized and discussed below.

Nanoparticles as fungicides

Materials of both organic and inorganic origin are used for NP synthesis and application in plant disease management. Among inorganic materials NPs of Ag, Zn, Mg and their oxides are reported to possess antimicrobial property. Silver NPs with broad spectrum of antimicrobial activity reduce various plant diseases caused by spore producing fungal pathogens. The small size of the active ingredient [diameter of 1–5 nm] effectively controls fungal diseases like powdery mildew (Park et al., 2006). Silver NPs were also reported to control the development of plant pathogenic fungus *viz*., *Bipolarissorokiniana* and *Magnaporthegrisea* (Jo et al., 2009), *Colletotrichum gloeosporioides* (Aguilar-Méndez, 2011), *Fusarium culmorum* and *F. oxysporum* (Kasprowicz et al. 2010) by strikingly decreasing the pathogen in a short period of time. The use of Ag NPs as an alternative to pesticides for the control of sclerotium forming phytopathogenic fungi was also investigated. Apart from silver ZnO NPs were also reported to control the post-harvest pathogenic fungi, *Botrytis cinerea* and *Penicillium expansum* (He et al., 2011).Bioassay of elemental and nanosulphur against *Aspergillus niger* showed that nano sulfur was more efficient (Choudhury et al., 2010). The high stability, chemical versatility and biocompatibility of silica NPs are utilized for pesticide delivery (Li et al., 2007).

Among the organic materials chitosan has been widely used in pharmaceutical and medical areas due to its favourable biological properties such as biodegradability, low toxicity, biocompatibility, bacteriostatic and fungistatic properties. The antimicrobial property of chitosan is attributed to its poly-cationic nature (Fujimoto et al., 2006; Liu et al., 2006) and the length of the polymer chain also enhances its antifungal property. Chitosan amended culture medium inhibits the radial growth of many fungal plant pathogens, *viz*., *Alternaria alternata, B. cinerea, C. gloeosporioides, Rhizopus stolonifer* (El Ghaouth, 1992), *Scelerotinia sclerotiorum* (Cheah et al., 1997) and *Fusarium* spp (Kendra and Hadwiger, 1984). In addition to fungistatic effect, chitosan is effective at eliciting plant immunity against diseases of plenty of crops. The inhibitory effect of chitosan has also been reported on plant diseases caused by viruses and viroids (Pospienzny and Atabekov, 1989). Considering the broad spectrum host defense induction, Yin et al., (2010) termed Chistosan as "plant disease vaccine". The NPs of Chitosan can be utilized well in plant disease management both directly or as carrier material.

Nanoparticle based delivery systems

In traditional pesticide application only a very low concentration of active ingredient reaches the target site of crops due to problems such as leaching of chemicals, degradation by photolysis, hydrolysis and by microbial degradation. For successful crop protection the chemical should remain active in the spray environment (heat, rain etc), penetrate to the target, penetrate the organism (fungi etc), resist the defense of pathogen, cost effective to formulate and manufacture and provide economic returns. This can be possible by applying the concept of pesticide delivery systems (Tsuji, 2001). The nano particle based delivery systems can combine all these properties to obtain the fullest biological efficacy.

Smart delivery systems

Smart delivery systems in agriculture should possess combinations of time controlled, specifically targeted, preprogrammed, self-regulated and multifunctional characteristics to avoid biological barriers for successful targeting. The nano-formulated material should be designed with unique property so that the active ingredient is released in required location or time in response to certain stimuli e.g., heat, moisture, pH, magnetic field *etc*. A pesticide Karate® Zeon is a product of Syngenta containing a synthetic insecticide lambda-cyhalothrin that normally breaks open in contact with leaves. The encapsulated product "gutbuster" made it as smart delivery system, i.e., breaks open to release only when it comes in contact with alkaline environment (in the stomach of insects). These kind of smart delivery systems not only overcome the problem of traditional pesticide application but also increase the efficacy to several folds. By smart delivery system it is possible to have a distribution of properly functionalized nanoparticles throughout the plant vascular system and guide them to targeted sites. These targeted nanoparticles can be successfully used to unload chemicals (fungicides, insecticides, etc.), or other substances (plant hormones, elicitors, nucleic acids) of interest into localized areas of plant tissues.

Nanoformulations of chemical pesticides

Agrochemicals are conventionally applied to crops by spraying and/or broadcasting. Hence repeated application is necessary to have an effective control which might cause some unfavorable effects such as soil and water pollution. Nano-encapsulated agrochemicals should be designed in such a way that they possess all necessary properties such as effective concentration (with high solubility, stability and effectiveness), time controlled release, enhanced targeted activity and less ecotoxicity with safe and easy mode of delivery thus avoiding repeated application.

A pesticide avermectin faced problems of UV inactivation during the field application, which was overcome by developing NPs based formulation. The use of porous hollow silica nanoparticles, with ashell thickness of nearly 15nm and a pore diameter of 4–5 nm was prepared to provide shielding protection to pesticides from degradation by UV light. Porous hollow silica nanoparticles carrying pesticide, i.e., avermectin loaded into the inner core showed a sustained-release pattern from the carrier. The slow release of encapsulated avermectin by the NPs carrier was reported for about 30 days (Li et al., 2007).Surface modified hydrophobic nanosilica has been successfully used to control a range of agricultural insect pests. Properly functionalised lipophilic nanosilica gets absorbed into the cuticular lipids of insects by physiosorption and damages the protective wax layer and inducesdeath by desiccation (Mewis, 2001). The use of such nanobiopesticideis more acceptable since they are safe for plants and cause less environmental pollution in comparison to conventional chemical pesticides.

Other materials such as nano-clays possess good biocompatibility, low toxicity and potential for controlled release (Bin Hussein et al., 2002). In addition to the use of NPs for pesticide delivery, the use of nanosized aqueous dispersion formulations was reported to enhance the bioavailability of pesticides. A nano-sized aqueous dispersion formulation (Banner MAXX) was released by Syngenta with broad spectrum systemic fungicidal action for the control of leaf spots, blights, rusts and powdery mildew diseases on ornamentals, turf, and other landscape plantings (Latin, 2006; Wong and Midland, 2004).

Delivery of microbes and microbial products

Microbial products such as enzymes, antibiotics and toxins that are promising as bio control agents against plant pathogens need stabilization and directed delivery mechanism towards identified targets. The specificity of fungal interactions *viz.* mycoparasitism of *Trichoderma* with plant pathogenic fungi are sources of effective biocontrol agents such as mycolytic and cuticle-degrading enzymes, antibiotics (viridian, gliovirin) and toxins (destruxins, bassianin), which need a delivery mechanism for its application in field. Experiments were conducted to explore the potential of biocompatible and biodegradable nanomaterials like chitosan or clayas enzyme stabilizing and delivery agents.

Chitosan and montmorillonite clay NPs were used to prepare nanoformulations that stabilized the *Myrothecium* enzyme complex. In the bioassay, biological activity against the plant pathogen *Fusarium* and cotton mealy bug, *Phenacoccus gossypiphilous* was observed due to controlled slow release of the enzymes (Ghormade et al., 2011).

Nanotechnology for Delivering Genetic Materials into Plants

Delivery of genetic material such as DNA and alteration of gene expression is important for the development of disease resistant crop plants. Gene delivery systems for plant transformation face obstacles such as targeting of delivery system, transportation through the cell membrane, uptake and degradation in endo lysosomes and intracellular trafficking of DNA to the nucleus. To circumvent the above mentioned obstacles, NPs were employed to develop efficient gene transformation vehicles. It was reported that Au NP SEM bedded carbon matrices were employed for the delivery of DNA during transformation of plant cells that carried higher amount of genetic material as compared to the microparticles (Vijayakumar et al., 2010). The DNA coated NPs also gained easy access to the plant cell due to its size.

Nanofibre arrays were reported to deliver genetic material to cells quickly and efficiently. The application of fluorescent labeled starch-nanoparticles as plant transgenic vehicle was reported, in which the nanoparticle biomaterial was designed in such a way that it bind and transport genes across the cell wall of plant cells by inducing instant aneouspore channels in cell wall, cell membrane and nuclear membrane with the help of ultrasound (Jun et al., 2008). It is possible to integrate different genes on the nanoparticle at the same time and the imaging of fluorescent nanoparticle is possible with fluorescence micro scopethus understanding the movement of exterior genes along withthe expression of transferred genes. Hence successful generation of pores on cell wall and cell membrane by suitable agents help in nanoparticle mediated DNA transfer that might be more successful in regenerative calli and soft tissues.

The ability of surface functionalized mesoporous silica nanoparticles (MSNs) to penetrate plant cell walls also opens up new ways to precisely manipulate gene expression at single cell level by delivering DNA and its activators in a controlled fashion (Torney et al., 2007).It was reported that a honeycomb MSN system with 3-nm pores transport DNA and chemicals into isolated plant cells and intact leaves. MSNs were loaded with gene and its chemical inducer and the ends were capped with gold nanoparticles to protect the molecules from leaching out. Uncapping of gold nanoparticles enabled release of chemicals and triggered gene expression in plants under controlled-release conditions. In this method the minimum amount of DNA that is required to detect marker expression was 1000-foldless than that required for the conventional delivery method.

Changes in agricultural technology have been a major factor shaping modern agriculture. Among the latest line of technological innovations, nanotechnology occupies a prominent position in transforming plant disease management strategies, agriculture and food production.

References

Aguilar-Méndez M.A, Martín-Martínez E.S, Ortega-Arroyo L, Cobián-Portillo G, Sánchez-spíndola E 2011. Synthesis and characterization of silver nanoparticles: effect on phytopathogen *Colletotrichum gloesporioides*. *Journal of Nanoparticle Research*, **13:** 2525-2532.

He, L., Y. Liu, A. Mustapha and M. Lin 2011. Antifungal activity of zinc oxide nanoparticles against *Botrytis cinerea* and *Penicilliumexpansum*. *Microbiological Research*, **166**: 207-215.

Latin, R. 2006. Residual efficacy of fungicides for control of dollar spot on creeping bentgrass. *Plant Disease*, **50**:571-5.

Li Z.Z, Chen J.F, Liu F, Liu A.Q, Wang Q, Sun H.Y. 2007. Study of UV-shielding properties ofnovel porous hollow silica nanoparticle carriers for avermectin. *Pest Management Science*, **63:**241-246.

Mahendra R, Shivaji D, Aniket G, Kamel-Abd-Elsalam 2012. Strategic nanoparticle-mediated gene transfer in plants and animals - a novel approach. *Current Nanoscience*, **8:**170-179

Park, H.J., S.H. Kim, H.J. Kim, S.-H.Choi 2006. A new composition of nanosizedsilica–silver for control of various plant diseases. *Plant Pathol. J.*, **22**:295-302.

Roco, M.C. 2003. Nanotechnology convergence with modern biology and medicine, *Current Opinion in Biotechnology*, **14**: 337-346.

Sharon M, Choudhary AK, Kumar R 2010. Nanotechnology in agricultural diseases and food safety. *Journal of Phytology,* **2:** 83-92.

Torney, F., B.G. Trewyn, S.-Y. Lin, K. Wang 2007. Mesoporous silica nanoparticlesdeliver DNA and chemicals into plants, *Nat. Nanotechnol*., **2:** 295-300.

Tsuji K. 2001. Microencapsulation of pesticides and their improved handling safety. *Journal of Microencapsulation,* **18**:137-147.

Vijayakumar P.S, Abhilash O.U, Khan B.M, Prasad B.L.V. 2010. Nanogold-loaded sharp-edged carbon bullets as plant-gene carriers. *AdvFunct Mater.*, **20:**2416-2423

Yin, H., X. Zhao and Y. Du 2010. Oligochitosan: A plant disease vaccine- A review. *Carbohydrate Polymers*, **82**: 1-8.

23

Nano-biotechnology for Plant Disease Management

R. Selvakumar

Introduction

Since past decades, the application of nanotechnology in the areas of medicine, materials science and electronics have been widely accepted and appreciated due to its potential and benefits. However, only during the recent years, researchers from other disciplines have started to see the potential applications of nanoscience in the field of agriculture (Robinson and Morrison, 2009). Recently, a generalized and process based framework to enable identification and characterization of the outputs (publications, patents etc.), and map them to the different agricultural research theme areas through the filter of links in the agri-value chain were developed by Kalpana Shastri et al., (2010) from National Academy of Agricultural Research Management (NAARM), Hyderabad, India. The frame work also permits assessing the implications for technology transfer, and impacts on society and environment. This frame work mainly relies on the identification of nano research and relating them agri-food thematic areas (Fig.1).

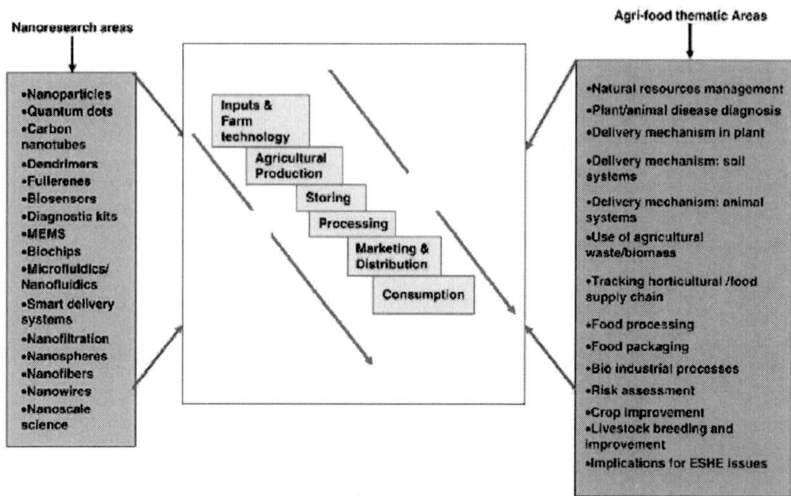

Fig. 1: Frame work for integrating nano research and agri-food thematic areas

The above mentioned interaction and integration of the nanotechnology with agriculture be best understood only when biologist read nanotechnology in terms of biological aspects. Nanobiotechnology is one such key area where scientists are focusing to understand the relationship between nanotechnology and biotechnology for the benefit for mankind through health, environment, agriculture, nutrition and crop improvement.

Biological aspects of nanobiotechnology

The term 'Nanobiotechnology' was coined by a Lynn W. Jelinski, a biophysicist at Cornell University, USA. Nanobiotechnology is a thriving new area of research at the interface between the life sciences and nanotechnology, which deals with structures of dimensions ranging from 1nm to 100nm, below the range of lithographic fabrication techniques (Lowe, 2000). Nanobiotechnology aims to exploit biomolecules and the processes carried out by them for the development of novel functional materials and nano-bio based devices. Oded Shoseyov and Ilan Levy (2008) defined nanobiotechnology as the engineering, construction and manipulation of entities in the 1 to 100 nm range using biologically based approaches for the benefit of biological systems. In simple terms nanobiotechnology can be stated as "Convergence of technology with biology at the nano level".

In this presentation, I have made a small attempt to classify nanobiotechnology into two main areas namely;

1. Nanoscale materials from biology
2. Nanoscale materials for biology (with importance to agriculture)

1. Nanoscale materials from biology through molecular self assembly

Recent advances in molecular biology have lead to the development of nanoscale materials using molecules derived from biological origin. The best examples are proteins, DNA and lipids. Molecular self-assembly is emerging as a new route to produce novel materials and to complement other materials. These biomolecules directly or indirectly involve naturally in molecular self assembly resulting in nanostructures with defined property.

Micelles formation

One of the well-known examples is the silk assembly. The monomeric silk fibroin protein is approximately 1 μm but a single silk worm can spin fibroins into silk materials over 2 km in length, two billion times longer! (Fig. 2) (Feltwell, 1990; Winkler et al., 1999). Likewise, spiders are grand master materials engineers who can produce many types of spider silks through self-assembly of the building blocks in a variety of ways, thus, producing spider silk fiber with tremendous strength and flexibility. These building blocks are often at the nanometer scale. These biomolecules undergo spontaneous organization/molecular self assembly into a well-defined and stable macroscopic structure using non-covalent bonds like hydrogen bonds, ionic bonds, water-mediated hydrogen bonds, hydrophobic and van der Waals interactions. Although each of these forces is rather weak, their collective interactions can produce very stable structures. The key elements in molecular self-assembly are chemical complementarities and structural compatibility. Like hands and gloves, both the size and the correct orientation, i.e. chirality's are important in order to have a complementary and compatible fitting (Zhang, 2002).

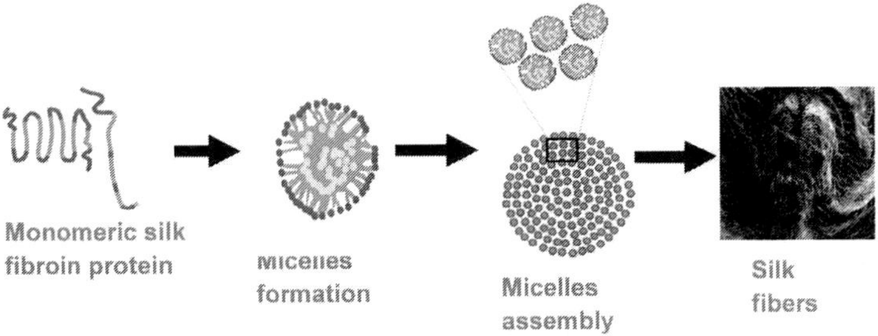

Fig. 2: Molecular self assembly of monomeric fibroin protein to silk fiber

Self-assembling peptide systems as nanostructures

Based on the previous research on peptide self assembly, Zhang (2002) classified self assembling peptides into four main types namely, Type I, II, III and type IV based on the assembling mechanism. Type I peptides were called "molecular Lego", form beta sheet structures in aqueous solution because they contain two distinct surfaces, one hydrophilic, the other hydrophobic. These peptides form complementary ionic bonds with regular repeats on the hydrophilic surface. The hydrophilic surface has alternating positive and negatively charged amino acids forming several moduli as mentioned below (Fig. 3).

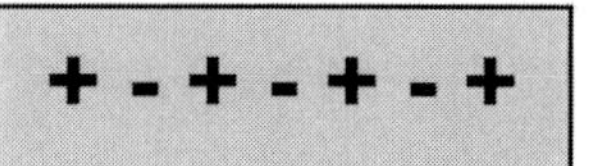

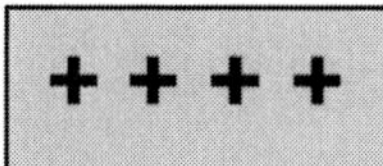

Fig. 3: Complementary ionic possible modules

Type II peptides were called as "Molecular Switches" in which the peptides could drastically change its molecular structure. Type III peptides, like "Molecular Paint" and "Molecular Velcro", undergo self-assembly onto surface rather with among themselves.

They form monolayers on surfaces for specific cell pattern formations or to interact with other molecules. Type IV peptides are surfactant- like peptides which undergo self-assembly to form nanotubes and nano vesicles having an average diameter of 30–50 nm.

Based on the above mentioned types various types of protein nanostructures like microbial S-layers, nanopores, protein based hydrogels, nanoropes, nanotubes, nanospheres, nanofibrils and filaments and protein arrays have been reported.

Enzyme based synthesis of metal nanoparticles

Enzymes are proteins that catalyze chemical reactions. The organism uses its proteins (especially oxido-reductase enzymes) for either extracellular production or intracellular production of metallic nanoparticle. Incase of biosynthesis of nanoparticle microorganism and plants have been well studies for its potential to synthesis nanoparticles. In yeast, the membrane bound (as well as cytosolic) oxido-reductases and quinines play an important role in the process of CdS nanoparticle formation. The oxido-reductases are pH-sensitive and work in alternative manner. At a lower value of pH, oxidase gets activated while a higher pH value activates the reductase (Nelson and Cox, 2005; Botham and Mayes, 2006).

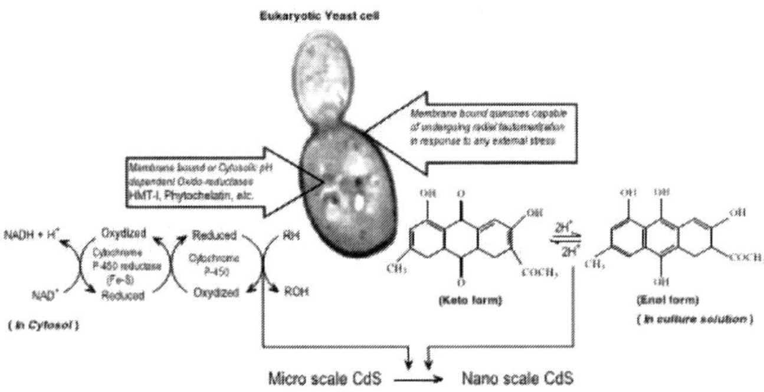

Fig. 4: Involvement of oxido-reductase enzymes in formation of CdS nanoparticles (Prasad and Jha, 2010)

The transformation seems to be negotiated at two distinct levels, at the cell membrane level immediately after addition of the CdS solution which triggers auto-merization of quinines and low pH- sensitive oxidases and makes molecular oxygen available for the transformation. This transformation causes the conversion of microscale CdS to nanoscale (Fig. 4). Similar to microorganisms, plants also uses its enzymes used in metabolic pathways to synthesis nanoparticles. Unlike microorganisms, the enzymes of metabolic pathways used to synthesis nanoparticles are different from one type of plant to other. For example, the mechanism of synthesis of silver nanoparticles was reported to vary between xerophyte, mesophytes and hydrophytes. A detailed discussion on the metabolic pathways and enzymes involved in the silver nanoparticles with suitable example has been covered in my presentation.

DNA nanostructures based on self assembly

DNA nanoengineering is dreamy, but difficult. Researchers have been putting together carefully chosen segments of DNA to form sheets, tubes, even simple machines such as tweezers since the early 1980s. Chemists looking to create complex self-assembling nanostructures are turning to DNA. Paul Rothemund, a computational bioengineer at the California Institute of Technology in Pasadena, made DNA orgami. The DNA orgami looked like viewed using smiley face and several other shapes, when observed through atomic-force microscopy.

Knowing the sequence of the virus twist and turn, he was able to write complementary DNA sequences, about 16-base-pairs long that would essentially staple the folds in place. He ordered the 'staples' from a DNA-synthesis company, mixed them with his virus in a buffer that stabilized the DNA and then heated and cooled the mixture, allowing the single stranded viral DNA to bind with the staples. Using similar approach lot of DNA nanostructures of different shapes have been reported (Fig. 5).

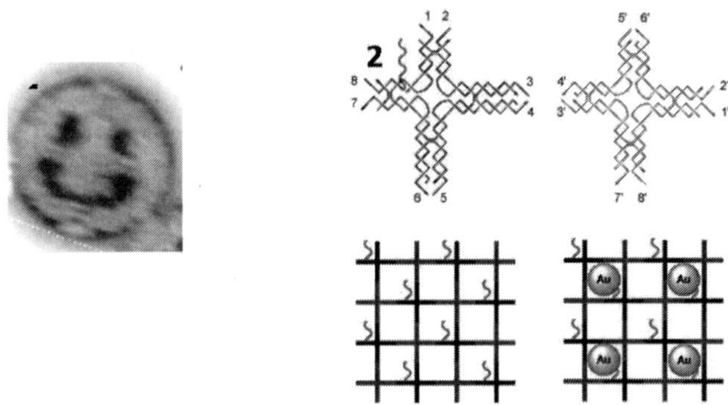

Fig. 5: (1) Smiley made of virus and DNA **(2)** DNA tiles with and without gold nanoparticles

Lipid based nanostructures

Lipids suitable for development of colloidal carriers have the generally regarded as safe (GRAS) status which can be used as drug delivery vehicle. Lipids firmly incorporate the drug during storage and have the great advantage of being identical or very similar to physiologically occurring compounds. They are also advantageous due to its biocompatible toxicity (degradation products including), and scaling up the manufacture process. It has the potential for controlled release and protects the drug until it reaches its site of action and hence can act as an efficient nanovector. In my presentation I have described the various methods to make the lipid nanostructures such as liposomes, neosomes and solid lipid nanoparticles (SLP) (Fig. 6)

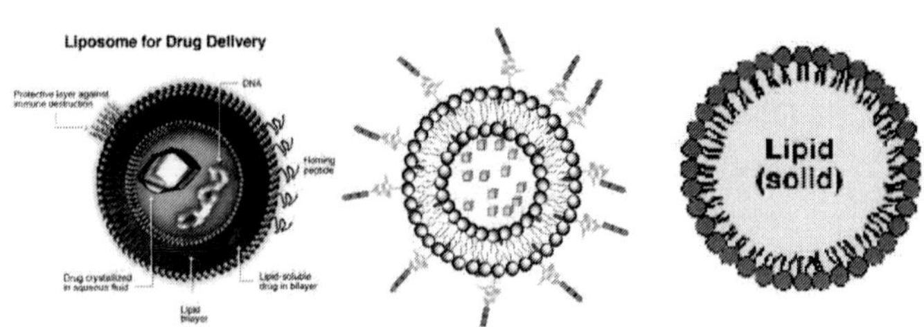

Fig. 6: Lipid based nanostructures: liposome (left), Neosomes (middle) and SLP (right)

2. Nanoscale materials for biology (with importance to agriculture)

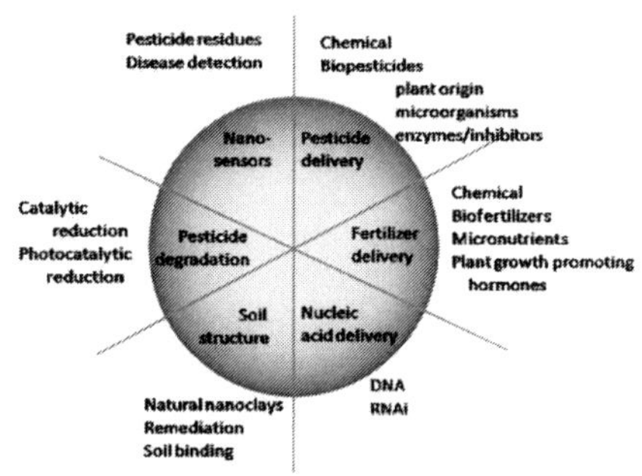

Fig. 7: Applications of nano-biotechnology in plant protection and nutrition (Ghormade *et al.*, 2011)

The nanotechnology and nanomaterials have been found to have profound applications in various aspects of biology and biology related fields. Biomedical application of nanotechnology (Nanomedicine), biomimetics, biotemplating and De Novo-designed structures, DNA-based nanotechnology and nanoelectronics, molecular motors and devices have been considered to be the potential areas of research which will improve the human health and living. Although many researchers have discussed about the above mentioned areas, I am restricting my lecture to the applications of nanoscale materials for agricultural applications especially to plant protection and nutrition. Potential applications of nanotechnology in agriculture are: delivery of nanocides–pesticides encapsulated in nanomaterials for controlled release; stabilization of biopesticides with nanomaterials; slow release of nanomaterial assisted fertilizers, biofertilizers and micronutrients for efficient use; and field applications of agrochemicals, nanomaterials assisted delivery of genetic material for crop improvement (Fig.7). Nanosensors for plant pathogen and pesticide detection, and nanoparticles for soil conservation or remediation are other areas in agriculture that can benefit from nanotechnology. Enzyme immobilization for nanobiosensor using nanomaterials involves the high value low volume application of enzymes (Kim et al., 2006). Usually costly, large enzyme volumes are required for biocontrol in agricultural fields that would be practical if spray applications combined high volume with low value. Cost-effectiveness of such biocontrol preparations can be achieved by immobilization of enzyme /inhibitors on nanostructures, providing large surface areas, to increase the effective concentration of the preparation (Ghormade et al., 2011).

Conclusion

Nano-biotechnology is one of the emerging fields of science that is being exploited in drug designing, smart delivery of inputs and early detection of plant and human diseases. Since, biotechnlogy had been extensively researched for the past twenty five years which provides ample opportunities for integration of biotech with nanotechnology to gain insights into the biological process at the molecular scale. The application of nano-biotechnology is very appropriate for early detection of plant diseases well before symptoms visually recognized. This is one of the hottest fields in biological sciences going to benefit mankind in the years to come.

References

Botham, K.M. and P.A. Mayes 2006. Biologic Oxidation–Harper's Illustrated Biochemistry, Lange-McGrawHill, London, p.47.

Ghormade, V., M.V. Deshpande and K.M. Paknikar. 2011. Perspectives for nano-biotechnology enabled protection and nutrition of plants, *Biotechnology Advances* **29**:792–803.

Kalpana, R. Sastry, H.B. Rashmi , N.H. Rao and S.M. Ilyas, 2010. Integrating nanotechnology into agri-food systems research in India: A conceptual framework. *Technological Forecasting & Social Change* **77**:639–648.

Kim, J. and J.W. Grate, P. 2006. Wang Nanostructures for enzyme stabilization. *Chem Engineer Sci* **61**:1017–26.

Lowe, C.R. 2000.Nanobiotechnology: the fabrication and applications of chemical and biological nanostructure. *Curr. Opin Struct. Biol.,* **10**:428-434.

Nelson, D.N., M.M. Cox, Lehninger Principles of Biochemistry, Freeman Publn, NY, 2005. 798.

Prasad, K. and A.K.Jha. 2010. Biosynthesis of CdS nanoparticles: An improved green and rapid procedure. *Journal of Colloid and Interface Science,* **342**:68-72.

Robinson, D.K.R. and M. Morrison: Nanotechnology Developments for the Agrifood sector-Report of the Observatory NANO. Institute of Nanotechnology, UK; 2009 [http://observatorynano.eu/].

Zhang, S. 2002. Emerging biological materials through molecular self-assembly, *Biotechnology Advances*, **20**:321–339.

24

Biosensors in Agriculture

K. Ilamurugu

Introduction

Biosensor technology is a powerful alternative to conventional analytical techniques, harnessing the specificity and sensitivity of biological systems in small, low cost devices. Despite the promising biosensors developed in research laboratories, there are not many reports of applications in agricultural monitoring. The authors review biosensor technology and discuss the different bio-receptor systems and methods of transduction. The difference between a biosensor and a truly integrated biosensor system are defined and the main reasons for the slow technology transfer of biosensors to the market place are reported. Biosensor research and development has been directed mainly towards health care, environmental applications and the food industry. The most commercially important application is the hand-held glucose meter used by diabetics. The agricultural / veterinary testing market has seen a number of diagnostic tests but no true biosensor systems have made an impact. The need for fast, on- line and accurate sensing opens up opportunities for biosensors in many different agricultural areas *in situ* analysis of pollutants in crops and soils, detection and identification of infectious diseases in crops and livestock, on-line measurements of important food processing parameters, monitoring animal fertility and screening therapeutic drugs in veterinary testing. Future challenges in the commercial development of biosensor are also addressed.

Definition

A self-contained integrated device which is capable of providing specific quantitative or semi-quantitative analytical information using a biological recognition element which is in direct spatial contact with a transducer element (IUPAC,1996).

Parts of biosensor

A biosensor has two components: a receptor and a detector. The receptor is responsible for the selectivity of the sensor. Examples include enzymes, antibodies, and lipid layers. The detector, which plays the role of the transducer, translates the physical or chemical change by recognizing the analyte and relaying it through an electrical signal. The detector is not selective. For example, it can be a pH- electrode, an oxygen electrode or a piezoelectric crystal.

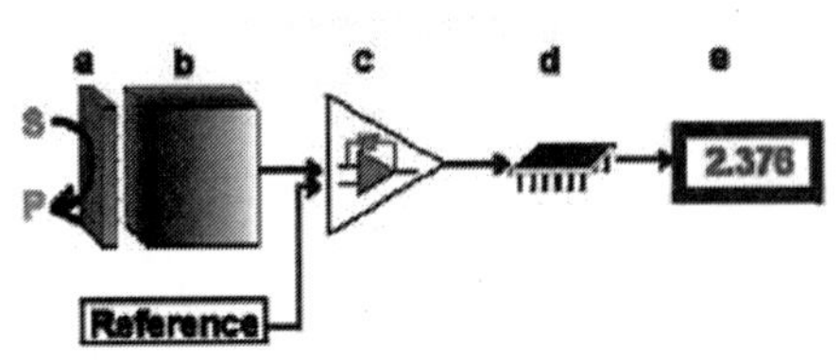

The biocatalyst (a) converts the substrate to product. This reaction is determined by the transducer (b) which converts it to an electrical signal. The output from the transducer is amplified (c), processed (d) and displayed (e).

Principles of biosensors

The principle of detection is the specific binding of the analyte of interest to the complementary biorecognition element immobilised on a suitable support medium. The specific interaction results in a change in one or more physico-chemical properties (pH change, electron transfer, mass change, heat transfer, uptake or release of gases or specific ions) which are detected and may be measured by the transducer. The usual aim is to produce an electronic signal which is proportional in magnitude or frequency to the concentration of a specific analyte or group of analytes, to which the biosensing element binds (Turner et al., 1986; Powner & Yalcinkaya, 1997).

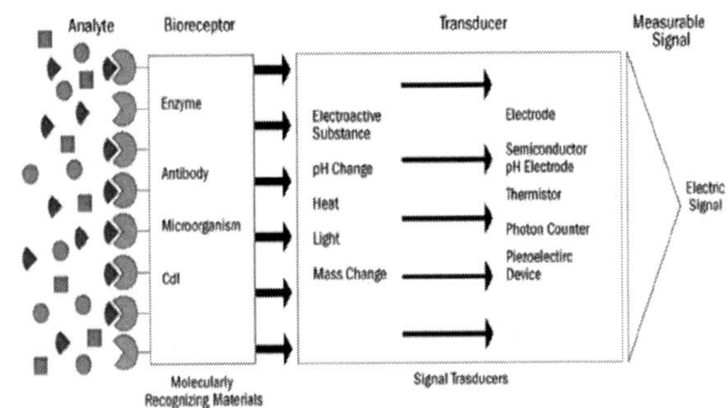

History of biosensors

Professor Leland C Clark Jnr = the father of the biosensor (1918–2005) In 1956, Clark published his definitive paper on the oxygen electrode. In 1962, he described "how to make electrochemical sensors more intelligent" by adding "enzyme transducers as membrane enclosed sandwiches". In 1975, YSI (Yellow Spring Instruments Co., Ohio, USA) produced the first of many biosensor- based laboratory analysers to be built by companies around the world.

Guilbault and Montalvo were the first to detail a potentiometric enzyme electrode, a urea sensor based on urease immobilised at an ammonium selective liquid membrane electrode. Lubbers and Opitz in 1975 to describe a fibre-optic sensor with immobilised indicator to measure carbon dioxide or oxygen. They extended the concept to make an optical biosensor for alcohol by immobilising alcohol oxidase on the end of a fibre-optic oxygen sensor.

Route in 1975, bacteria could be harnessed as the biological element in microbial electrodes for the measurement of alcohol. In 1976, Clemens et al. incorporated an electrochemical glucose biosensor in a bedside artificial pancreas and this was later marketed by Miles (Elkhart) as the Biostator. In the same year, La Roche (Switzerland) introduced the Lactate Analyser LA 640 in which the soluble mediator, hexacyanoferrate, was used to shuttle electrons f rom lactate dehydrogenase to an electrode. In 1982, in vivo application of glucose biosensors was reported by Shichiri et al. who described the first needle-type enzyme electrode for subcutaneous implantation. The idea of building direct immunosensors by fixing antibodies to a piezoelectric or potentiometric transducer had been explored since the early seventies, but it was a paper by Liedberg et al. The BIAcore (Pharmacia, Sweden) produced commercial model of this technology in 1990. In 1987, a pen-sized meter for home blood-glucose monitoring formed the basis for the screen-printed enzyme electrodes launched by MediSense (Cambridge, USA). The electronics have since been redesigned into popular credit-card and computer-mouse style formats, and MediSense's sales showed exponential growth, reaching $175 million per annum by 1996, when they were purchased.

Boehringer Mannheim, Bayer and Life Scan now have competing mediated biosensors and the combined sales of the four companies dominate the world market for biosensors and are rapidly displacing conventional reflectance photometry technology for home diagnostics.

Types of biosensors

Depending on the method of signal transduction, biosensors can also be divided into different groups: electrochemical, optical, thermometric, piezoelectric or magnetic. Amperometric devices are the most commonly reported class of biosensors. Amperometric detection typically relies on an enzyme system that catalytically converts analytes into products that can be oxidised or reduced at a working electrode, maintained at a specific potential. The main advantage of this transducer is the low cost and the use of disposable electrodes. The high reproducibility of these single use electrodes eliminates the requirement for repeated calibration. Limitations of amperometric transducers include interference from electroactive compounds.

Potentiometric biosensors

Potentiometric biosensors make use of ion-selective electrodes in order to transduce the biological reaction into an electrical signal. In the simplest terms this consists of an immobilised enzyme membrane surrounding the probe from a pH-meter (Figure 1 a,b,c), where the catalysed reaction generates or absorbs hydrogen ions. The reaction occurring next to the thin sensing glass membrane causes a change in pH which may be read directly from the pH-meter's display. Typical of the use of such electrodes is that the electrical potential is determined at very high impedance allowing effectively zero current flow and causing no interference with the reaction.

There are three types of ion-selective electrodes which are of use in biosensors:

1. Glass electrodes for cations (e.g. normal pH electrodes) in which the sensing element is a very thin hydrated glass membrane which generates a transverse electrical potential due to the concentration-dependent competition between the cations for specific binding sites. The selectivity of this membrane is determined by the composition of the glass.

   The sensitivity to H^+ is greater than that achievable for NH_4

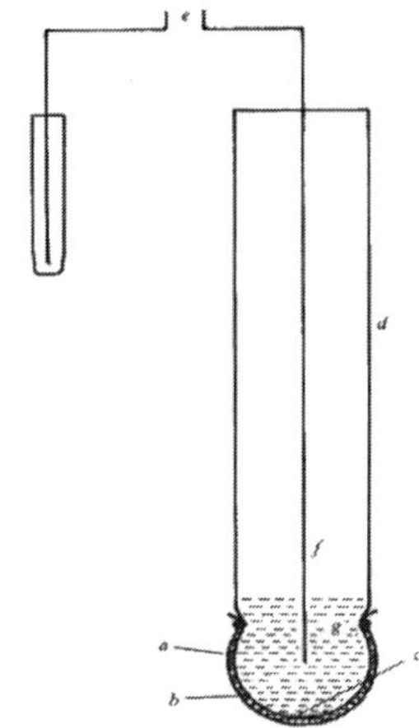

Fig. 1(a): A simple potentiometric biosensor. A semi- permeable membrane **(a)** surrounds the biocatalyst **(b)** entrapped next to the active glass membrane **(c)** of a pH probe **(d)**. The electrical potential **(e)** is generated between the internal Ag/AgCl electrode **(f)** bathed in dilute HCl **(g)** and an external reference electrode

2. Glass pH electrodes coated with a gas-permeable membrane selective for CO_2, NH_3or H_2S. The diffusion of the gas through this membrane causes a change in pH of a sensing solution between the membrane and the electrode which is then determined.

3. Solid-state electrodes where the glass membrane is replaced by a thin membrane of a specific ion conductor made from a mixture of silver sulphide and a silver halide. The iodide electrode is useful for the determination of I- in the peroxidase reaction and also responds to cyanide ions.

Biosensors which involve H release or utilization necessitate the use of very weakly buffered solutions (i.e. < 5 mM) if a significant change in potential is to be determined. The relationship between pH change and substrate concentration is complex, including other such non-linear effects as pH-activity variation and protein buffering. However, conditions can often be found where there is a linear relationship between the apparent change in pH and the substrate concentration. A recent development from ion-selective electrodes is the production of ion-selective field effect transistors (ISFETs) and their biosensor use as enzyme-linked field effect transistors (ENFETs, Figure 3). Enzyme membranes are coated on the ion-selective gates of these electronic devices, the biosensor responding to the electrical potential change via the current output. Thus, these are potentiometric devices although they directly produce changes in the electric current. The main advantage of such devices is their extremely small size (<< 0.1 mm2) which allows cheap mass-produced fabrication using integrated circuit technology. As an example, a urea-sensitive FET (ENFET containing bound urease with a reference electrode containing bound glycine) has been shown to show only a 15% variation in response to urea (0.05-10.0 mg ml^{-1}) during its active lifetime of a month. Several analytes may be determined by miniaturised biosensors containing arrays of ISFETs and ENFETs. The sensitivity of FETs, however, may be affected by the composition, ionic strength and concentrations of the solutions analysed.

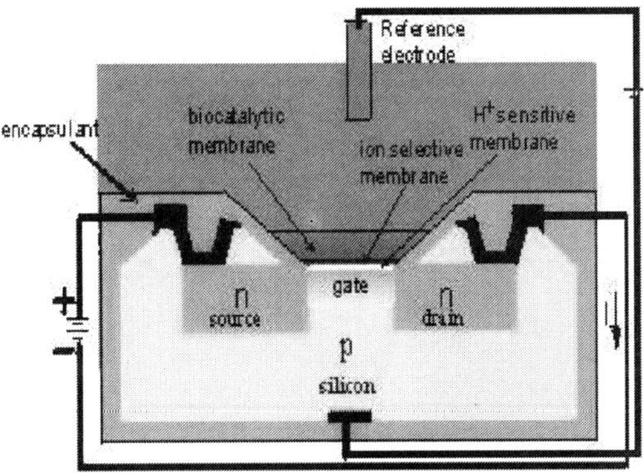

Fig. 1 (b): Schematic diagram of the section across the width of an ENFET

The actual dimensions of the active area are about 500 mm long by 50 mm wide by 300 mm thick. The main body of the biosensor is a p-type silicon chip with two n-type silicon areas; the negative source and the positive drain. The chip is insulated by a thin layer (0.1 m thick) of silica (S_iO_2) which forms the gate of the FET. Above this gate is an equally thin layer of H^+-sensitive material (e.g. tantalum oxide), a protective ion selective membrane, the biocatalyst and the analyte solution, which is separated from sensitive parts of the FET by an inert encapsulating polyimide photopolymer. When a potential is applied between the electrodes, a current flows through the FET dependent upon the positive potential detected at the ion-selective gate and its consequent attraction of electrons into the depletion layer. This current (I) is compared with that from a similar, but non-catalytic ISFET immersed in the same solution.

Amperometric biosensors

Amperometric biosensors are quite sensitive and more suited for mass production than the potentiometric ones (Ghindilis et al., 1998). The working electrode of the amperometric biosensor is usually either a noble metal or a screen-printed layer covered by the biorecognition D component (Wang 1999). Carbon paste with an embedded enzyme is another economic option (Cui et al. 2005). At the applied potential, conversion of electroactive species generated in the enzyme layer occurs at the electrode and the resulting current (typically nA to mA range) is measured (Mehrvar and Abdi 2004). The principle of the previously mentioned YSI 23A (Magner 1998) can serve as an example.

Glucose + GOD (FAD) gluconolactone + GOD (FADH2) (1)

GOD ($FADH_2$) + O2 GOD (FAD) + H_2O_2 (2) $H_2O_2O_2$+2H++2e-(3)

The reactions (1) and (2) are catalyzed by glucose oxidase (GOD) containing FAD as a cofactor. The last reaction is the electrochemical oxidation of hydrogen peroxide at the potential of around +600 mV.

Amperometric biosensors can work in two- or three-electrode configurations. The former case consists of reference and working (containing immobilized biorecognition component) electrodes. The main disadvantage of the two-electrode configuration is limited control of the potential on the working electrode surface with higher currents, and because of this, the linear range could be shortened. To solve this problem, a third auxiliary electrode is employed. Now voltage is applied between the reference and the working electrodes, and current flows between the working and the auxiliary electrodes. A common screen-printed three electrode sensor is shown in (Fig. 1 a,b,c)

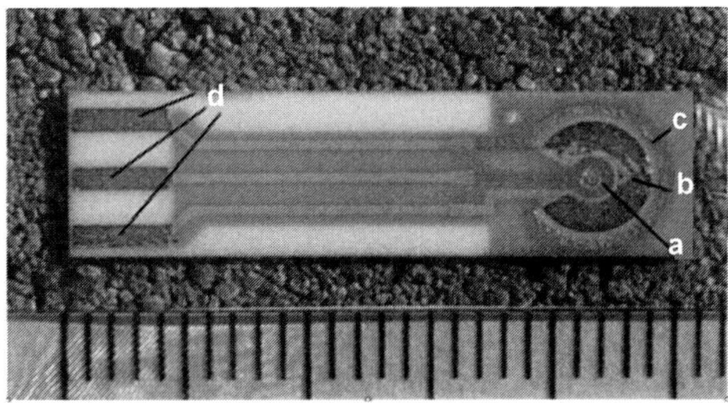

Fig. 1 (c): Example of the three-electrode screen-printed sensor produced by BVT (Brno, Czech Rep.). The sensor body is made from ceramics. A gold working electrode **(a)** is surrounded by an Ag/AgCl reference electrode **(b)** and gold auxiliary electrode **(c)**. Letter d means silver output contacts. The ruler in the bottom is in millimeter scale.

The amperometric biosensors are often used on a large scale for analytes such as glucose, lactate (Ohnuki et al. 2007), and sialic acid (Marzouk et al. 2007). Biological agents such as model Bacillus cereus and Mycobacterium smegmatis (Yemini et al. 2007), the serological diagnosis of Francisella tularensis (Pohanka and Skl dal 2007), a pharmacology study (Pohanka et al. 2007c) and the detection of pesticides and nerve agents (Liu et al. 2006) have also been described. A metabolism apparatus of whole cells can be used for certain analytes such as the measurement of phenol with immobilized Pseudomonas sp. cells (Skl dal et al. 2002). Biosensors based on AChE and butyrylcholinesterase (BChE) can be employed for rapid detection of organophosphates and carbamates (Skl dal 1996) due to strong enzyme inhibition (Krejzov et al. 2005). The AChE amperometric biosensor based on a nanoporous carbon matrix was used for the dichlorvos assay (Sotiropoulou and Chaniotakis 2005) and a similar device based on the screen-printed carbon electrode modified with Prussian blue was tested for aldicarb, paraoxon and parathion-methyl (Suprun et al. 2005). Amperometric biosensors were evaluated also for assays with nucleic acid acting as a marker and/or biorecognition component; uropathogens were assayed using their 16S rRNA (Liao et al. 2006). Opposite direction to the flow of electrons).

Optical biosensors

In recent years, a key stimulus for the development of optical biosensors has been the availability of high- quality fibres and optoelectronic components. The optical biosensor format may involve direct detection of the analyte of interest or indirect detection through optically labelled probes and the optical transducer may detect changes in the absorbance, luminiscence, polarisation or refractive index. The advantages of optical biosensors are their speed, the immunity of the signal to electrical or magnetic interference and the potential for higher information content (spectrum of information available) but the main drawback can be the high cost of some instrumentation.

Piezo-Electric Biosensors

The piezoelectric biosensor is based on measuring frequency changes directly related to mass change on the sensor surface. One of the main advantages of this technique is the real-time binding reaction detection, allowing kinetic evaluation of affinity interactions (feature similar to the surface plasmon resonance biosensors) and in addition, the low cost of the instrumentation required. Limitations of this transduction method are the need for a calibration of each crystal and the possible variability when coating the surface with the antigen or antibody.

Calorimetric biosensors

If the enzyme catalyzed reaction is exothermic, two thermistors may be used to measure the difference in resistance between reactant and product and, hence, the analyte concentration.

Biosensors offer great advantages over conventional analytical techniques. The selectivity of the biological sensing element offers the opportunity for development of highly specific devices for real-time analysis in complex mixtures, without the need for extensive sample pre-treatment or large sample volumes. Biosensors also promise highly sensitive, rapid, reproducible and simple to operate analytical tools.

Despite optimism for the potential of biosensors, their emergence from the research laboratory to the marketplace has been slow. The obstacles to exploitation have been fundamentally related to the presence of biomaterial in the biosensor (immobilisation of biomolecules on transducers, stability of enzymes and antibodies), the development of the sensor device (sensitivity and reproducibility issues) and the integration of biosensors into complete systems. Another major problem for the realistic mass production of biosensors has been the cost factor.

A biosensor system can be defined as the combination of elements such as a method of sampling automatically or manually, a biosensor, a system for replenishing or replacing the biosensor and a data analysis system to implement a biological model which provides information to a human or automated controller. The integration of fluidics, electronics, separation technology and biological subsystems is crucial for the development of biosensor systems. The sensor sampling system biointerface is a key target for the construction of an integrated system. From the literature, there are many examples of biosensors successfully tested in a laboratory or at prototype level but few examples, even in research, of integrated biosensor systems that offer automatic monitoring in complex matrices.

Biosensors on the nanoscale

- Molecular sheaths around the nanotube are developed that respond to a particular chemical and modulate the nanotube's optical properties.
- A layer of olfactory proteins on a nanoelectrode react with low-concentration odorants (SPOT- NOSED Project). Doctors can use to diagnose diseases at earlier stages.
- Nanosphere lithography (NSL) derived triangular Ag nanoparticles are used to detect streptavidin down to one picomolar concentrations.
- The School of Biomedical Engineering has developed an anti-body based piezoelectric nanobiosensor to be used for anthrax, HIV hepatitis detection.

Applications

The potential applications of biosensors in agriculture and food processing are numerous and each has its own requirements in terms of the concentration of analyte to be measured, the required precision of output, the sample volume needed, the time taken to complete the assay, the time necessary for the biosensor to be ready to be used again and the cleaning requirements of the system. The size of the possible market may also have an impact on the type of biosensor specified, as some are more amenable to mass production than others. Some examples are given below:

- Glucose monitoring in diabetes patients ← historical market driver
- Other medical health related targets
- Environmental applications e.g. the detection of pesticides and river water contaminants
- Remote sensing of airborne bacteria e.g. in counter-bioterrorist activities
- Detection of pathogens
- Determining levels of toxic substances before and after bioremediation
- Detection and determining of organophosphate
- Routine analytical measurement of folic acid, biotin, vitamin B12 and pantothenic acid as an alternative to microbiological assay

- Determination of drug residues in food, such as antibiotics and growth promoters, particularly meat and honey.
- Drug discovery and evaluation of biological activity of new compounds.
- Protein engineering in biosensors
- Detection of toxic metabolites

Biosensors for environmental monitor

Various Biosensors can be developed and used for monitoring the quality of air, water and soil.

Some of them include :

- B.O.D Biosensors
- Pesticide Biosensors
- Phenol Biosensors
- Biosensors for detecting Heavy metal

BOD Biosensors

BOD is used as a measure of water quality. The greater the BOD of a water sample the more polluted it is. BOD is a measure of the amount of metabolizable material in a sample. Conventionally BOD-5 technique is used to measure BOD and it takes 5 days to estimate BOD using this technique.

Two types of BOD Biosensors are in use:

- Film type BOD Biosensor
- Respirometer type BOD Biosensor

In a film type BOD Biosensor a microbial film is sandwiched between a porus membrane and oxygen permeable membrane of the Clarke oxygen electrode. The microbial cells utilize the oxygen while metabolizing any material in the sample. The amount of oygen utilized is proportional to the amount of metabolizble material in the sample. The Clarke oxygen electrode measures the amount of oxygen consumed and thus BOD can be estimated.

Pesticide residues in crop and soil samples

The analysis of pesticide residues is an important concern due to their high toxicity and the serious risk that they represent for the environment and human health. Analysis for pesticides is usually carried out by gas chromatography

(GC) or high performance liquid chromatography (HPLC). However, these methods require laborious extraction and clean up steps that increase analysis time and the risk of errors. The development of biosensors is a growing area, in response to the demand for rapid, simple, selective and low cost techniques for pesticides. The main principle of the biosensors developed is based on the correlation between toxicity of a pesticide and a decrease in the activity of a biomarker suchas an enzyme. This activity can be registered by employing different transducers, e.g. amperometry, potentiometry, spectrometry, fluorimetry or thermometry for detection of different substrates or products of enzymatic reaction. Organophosphorus and carbamate insecticides selectively inhibit cholinesterases. The enzyme acetylcholinesterase (AChE) catalyses the hydrolysis of acetylcholine to acetic acid and choline:

AchE

$(CH_3)_3N(OH)CH_2CH_2OCOCH_3+H_2O$

$CH_3COOH+\ (CH_3)_3N(OH)CH_2CH_2OH$

Several authors have used a pH-sensitive transducer in the development of AchE-based biosensors (Tran-Minh et al., 1990; Andres & Narayanaswamy, 1997), measuring the pH change generated by the release of acetic acid during the enzymatic reaction. Xavier et al. (2000) described an optical fibre biosensor for the determination of the pesticides propoxur and carbaryl, widespread insecticides in vegetable crops. The optical transducer was a pH indicator (chlorophenol red) covalently bound to controlled pore glass beads. In the presence of a constant acetylcholine concentration, the colour of the pH-sensitive layer changed proportionally to the carbamatenconcentration in the sample solution. This colour change varied the reflectance signal (at 602 nm) measured by the fibre optic device. An incubation time of 6 min was chosen and the linear dynamic range for the determination of propoxur and carbaryl was 0.03 – 3.00 & 0.8–6.0mgl-1, respectively. The optode produced reproducible (relative standard deviation, RSD 5%) and stable responses for over 4 months. The biosensor was successfully applied to the quantification of propoxur in vegetable samples (onion and lettuce) below the maximum concentrations allowed by the Spanish law (3.0mgkg-1 in fresh or frozen fruits and vegetables). Another example is the multi-enzymatic electrochemical sensor developed by Starodub et al. (1999), that was based on a capacitative pH-sensitive electrolyte– insulator–semiconductor (EIS)- structure withsil icon nitride ion-sensitive layers as transducers. Determination of heavy metal ions and phosphororganic pesticides in contaminated potatoes and cabbage sap was performed by the sensor array. The multi-enzyme analysis followed by mathematical processing is an effective approach to develop computer-controlled sensor arrays for analysis of toxic substrates.

The amount of choline generated in the enzymatic reaction could also be directly related to the enzyme activity. Nunes et al. (1999) used as substrate for the cholisterases enzymes, a salt of acetylthiocholine or butyrylthiocholine (ATCh, BTCh), and the thiocholine produced during the enzymatic reaction was anodically oxidised on a screen-printed electrode. The Amperometric biosensor procedure consisted of the deposition of a small drop of substrate or sample (50 ml) on a horizontally positioned biosensor strip representing the microelectrochemical cell. An incubation time of 10 min for enzyme inhibition was needed before addition of the substrate. The biosensors were suitable for single use and no complicated and time-consuming procedures for regeneration of enzyme activity after inhibition were necessary. The linear range of the biosensor for N methylcarbamates (aldicarb, carbaryl, carbofuran, methomyl and propoxur) varied from 5x10-5 to 50 mg kg-1. The cost of carbamate analysis with a biosensor was much lower compared to chromatographic methods, and the analysis was conducted on crop extracts and on vegetable juices at ppb concentration levels without any pre-treatment. By combining native and recombinant variants of acetylcholinesterase with data processing by artificial neural networks, Bachmann et al. (2000) reported multi- analyte detection for organophosphates and carbamates. The assay duration was 40 min and the limit of detection for paraoxon and carbofuran in drinking water in separate analyses was of 0.5μgml-1. For inhibition analysis of binary mixtures the sensor displayed a resolution error of 0.4μgml-1 for paraoxon and 0.5μgml- 1 for carbofuran. The multisensor approach needs improvement as the legal required limits are 0.5 mgml- 1 for total and 0.1 mgml-1 for individual pesticide concentration in drinking water (EC Council Directive, 30.08.1980). The hydrophobicity of organophosphorus and carbamic pesticides in water miscible organic solvents was also evaluated for the development of new biosensors. Palchetti et al. (1997) described a choline Amperometric biosensor for carbofuran in real samples (fruits and vegetables), using buffers containing 1% v/ v acetonitrile.

Biosensors have been reported for pesticides based on the inhibition of acetylcholisterase but using different signal transduction methods. Roda et al. (1994) developed a chemiluminiscence-based flow method for the determination of organophorus and carbamate pesticides. The choline formed by the acetylcholisterase was oxidised by choline oxidase and the H_2O_2 produced was via the luminol/peroxidase luminiscent reaction. The detection limits for paraoxon were 125 μg l and the results obtained were in good agreement with those obtained by a commonly used colorimetric test. Lui et al. (1997) developed a biosensor for methamidophos (one of the most commonly used organophosphate insecticides in South East Asia), based on acetylcholinesterase immobilised onto magnetic particles in a photometric flow injection system. The biosensor could detect methamidophos in lettuce and cabbage at 12 and 3 mgkg-1 vegetable material, respectively. The determination of organophosphate and carbamate pesticides in spiked drinking water and fruit juices was also carried out using a photothermal biosensor (Pogacnik & Francko, 1999). With this approach, 0.2 mgml-1 of paraoxon was detected in less than 15 min, without any pretreatment step.

In the literature, there are also other biosensors described based on the property of some toxic compounds to reduce the intensity of certain natural processes such as photosynthesis and bioluminescence. Koblizek et al. (1998) developed a biosensor for traces of herbicide residues based on a chlorophyll–protein reaction center complex, which measures oxygen levels. If the soil contained a herbicide, the chemical reacted with the biosensor protein and inhibited oxygen production. The biosensor measured herbicides that inhibit photosynthesis, 50% of all herbicides used in agriculture. The test was ultra-sensitive, with detection limits similar to the more complex, highly sensitive enzyme-linked immunosorbent assay (ELISA).

Researchers are starting to address analytical problems as the determination of pesticides using molecular imprinted polymers based biomimetic sensors. Sergeyera et al. (1999) reported molecularly imprinted polymer membranes for atrazine as a recognition element of a conductometric sensor. The biosensor developed demonstrated high selectivity and sensitivity (detection limit 5 nM), rapid response (few minutes) and long stability (over 6 months). Also there is reported a rapid, cheap and disposable sensor for 2,4- dichlorophenoxyacetic acid, based on molecular imprinted polymers as recognition elements and electrochemical detection using screen printed electrodes (Kroger et al., 1999).

Process control

Bacteriological food safety

Food borne pathogens cause economic loss, human suffering and even death. Pathogen detection using culture techniques and bioassays such as ELISA for determining and enumerating pathogens in food is well established. However, these methods are very elaborate, very time consuming and cannot be used as on-site monitoring techniques. In recent years, various types of biosensor have been developed which could help in overall quality control in food processing plants by detecting pathogens within minutes of sampling. If pathogens are found with on- or near-line biosensors, then food processors can make decisions more quickly about applying treatments, minimising the chance of a contaminated final product. The general approach for the biosensors described in the literature is an immune affinity step to capture and concentrate bacteria on beads, a membrane or a fiber optic probe tip, followed by detection of bound bacteria by laser excitation of bound fluorescent antibodies, acousto gravimetric wave transduction, surface plasmon resonance or electrochemical methods. The main problem facing the production of biosensors for direct detection of bacteria is the sensitivity of the assay in real samples, an issue that still requires improvement. The infectious dosage of pathogens such as Salmonella or Escherichia coli O157: H7 is 10 cells and the existing coliform standard for E. coli in water is 4 cells 100 ml-1.

Poultry products are presumed to be a major cause of human food borne illness due to the relatively high frequency of contamination with pathogens Salmonella spp. and Campylobacter spp. The Threshold Immunoassay System, a biopharmaceutical- contaminant detection system based upon the light-addressable potentiometric sensor (LAPS) has been used to detect Salmonella rapidly (in less than 15 min) and reliably, to levels slightly above 100 colony forming units (CFU) (Dill et al., 1999). The immunoassay system utilized solution phase binding of antibodies to Salmonella, and the immune complex formed is then captured on biotin coated nitro-cellulose membrane. Finally, a signal generator (an anti-fluorescein urease conjugate) was bound to the immunocomplex. Detection and quantitation of the immunocomplex was made via changes in pH at the silicon chip surface as a result of the conversion of urea to carbon dioxide and ammonia. The limit of detection of the silicon chip-based biosensor is substantially better than the values obtained using enzyme-linked methods and comparable with a polymerase chain reaction (PCR)-chemiluminescent method. Experiments involving the detection of salmonella from chicken carcass washing (showing a recovery of 90%) indicated that this technology could be placed into onsite facilities and used to evaluate the extent of salmonella contamination in the poultry industry.

Ye et al. (1997) described a piezoelectric biosensor for detection of Salmonella typhimurium. The device consists of a quartz crystal wafer sandwiched between two metal electrodes. These electrodes provide a means of connecting the device to an external oscillator circuit that drives the quartz crystal at its resonant frequency. A change in mass on the surface of the electrode thus changes the resonant frequency of the quartz crystal microbalance (QCM) device (Fig. 2). The antibody against Salmonella was immobilised onto the gold electrode-coated quartz crystal surface through a polyethylenimine–glutaraldehyde technique. The Salmonella cells reacted specifically and bound to the crystal surface resulting in an increase in mass that was directly related to the concentration in the solution. The biosensor responded to concentrations of S. *typhimurium* in the range of 5.3x 105 to 1.2x109 CFUml- 1 in 25 min. Another biosensor technique for Salmonella detection developed by Seo et al.

(1999) was a direct method, in which Salmonell a binding to specific antibodies attached to a waveguide surface were detected in minutes by measuring interferometrically the alteration in phase velocity of a guided optical wave. The biosensor was able to detect S. typhimurium in chicken carcass wash fluid inoculated at a level of 20 CFU/ml after 12 h of nonselective enrichment. The planar optic biosensor showed promise as a fast, sensitive and economical means of detecting food pathogens.

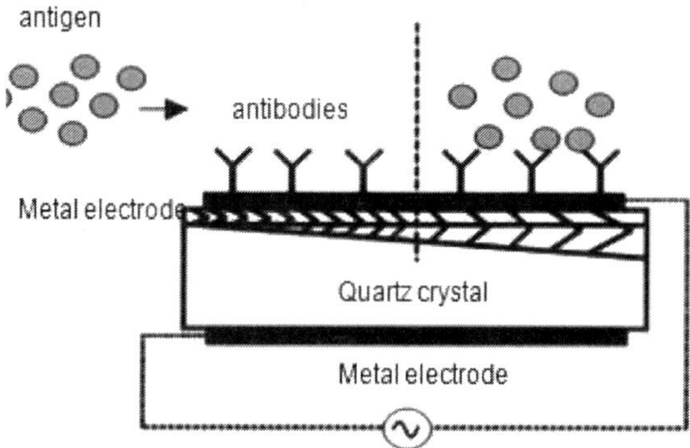

Fig. 2: Piezoelectric crystal biosensor; mass deposition at the surface alters the frequency of the resonant of the crystal.

One of the many applications of surface Plasmon resonance (SPR) technology is the detection of *E. coli* O157:H7. Surface plasmon resonance is a quantum optical-electrical phenomenon based on the fact that energy carried by photons of light can be transferred to electrons in a metal. The wavelength of light at which coupling occurs is characteristic of the particular metal and the environment of the metal surface illuminated. This transfer can be observed by measuring the amount of light reflected by the metal surface. Figure 3 shows how a change in the chemical environment 100 nm above the thin metal layer results in a shift in the wavelength of light, which is absorbed rather than reflected. The most common practical implementation of SPR is to use a metal-coated optical prism, but other practical implementations have been demonstrated including metal-coated diffraction gratings, optical fibres and planar waveguides. The SPR biosensor has potential for use in rapid, real-time detection and identification of bacteria, and to study the interaction of organisms with different antisera or other molecular species (Fratamico et al., 1998). The lower detection limit of the BIAcore system (a commercial example of an SPR system) is approximately 10 pg of analyte mm^{-2}.

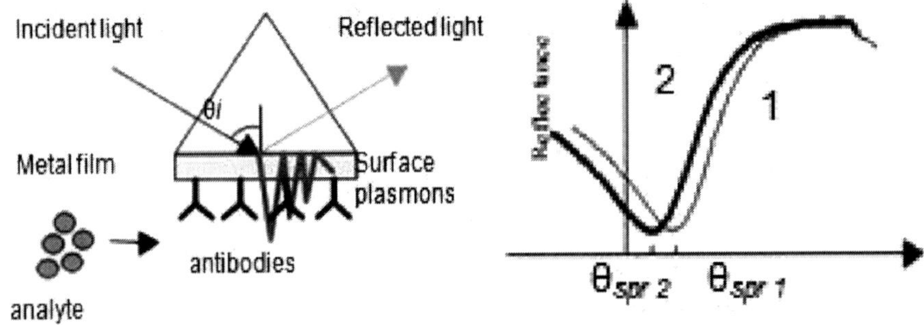

Fig. 3: Surface plasmon resonance (SPR) biosensor; surface plasmons are excited at a specific incident angle (yspr), resulting in a massive reduction in reflectivity at that angle; biomolecular.

Interactions at the metal surface are monitored as a change of the optimal angle required for surface plasmons excitation on the metal layer; SPR curves before (1) and after (2) analyte binding.

An evanescent wave biosensor for detection of Staphylococcal enterotoxin A in potato salad, milk and mushrooms has also been reported (Rasooly & Rasooly, 1999).

Electrochemical methods coupled with magnetic separation were used to detect Salmonella (Che et al.,1999) and E. coli (Perez et al., 1998). The methods were completed in less than 2 h and the detection limit was 5-103 cell ml-1 for the Salmonella and 105 cell ml-1 for the *E. coli.* Recently, there have been important developments in the use of nucleic acid-based assays for detection of foodborne pathogens (Smith et al., 2000; Olsen, 2000). In the future biosensor/ deoxyribonucleic acid (DNA)-chip technology could be an alternative to detect microbial diseases.

Quality control

Improper package design or temperature abuse during handling may cause fruits and vegetables in modified-atmosphere packages to be exposed to low, injurious O_2 levels associated withth e production of fermentation volatiles, quality loss and eventually product breakdown. Excessively low package O_2 also may promote growthof dangerous pathogens (e.g. Clostridium botulinum). The detection of ethanol would provide a sensitive technique for low-O_2 injury identification.

A commercial ethanol biosensor composed of a chromagen and immobilised enzymes: alcohol oxidase and peroxidase has been tested. Alcohol oxidase catalyses oxidation of ethanol into acetaldehyde and H_2O_2 in the presence of O_2, and peroxidase (an H_2O_2 decomposing enzyme) catalyses oxidation of the chromagen causing a colour change. The biosensor detected ethanol to levels below the human olfactory threshold [10 ml l_1 (_1 Pa) ethanol in gas phase at 58C with a 15 s exposure]. The onset of low O_2 injury was detected in lightly processed lettuce, cauliflower, broccoli and cabbage modified-atmosphere packages as measured by accumulation of headspace ethanol (Smyth et al., 1999). The response of the biosensor was very similar to the one measured by gas chromatography, which is expensive and requires technical expertise. The biosensor could also be useful to monitor ethanol during controlled-atmosphere storage of apples, rot development in stored potato tubers or any application where ethanol accumulation can be associated with a loss of quality.

Natural amines in plants are involved in various physiological processes such as the fruit development and senescence. Their levels can vary depending on variety, ripening and storage conditions. Moreover, microbial contamination results in an increase of the biogenic amine content of fruits and vegetables.

Electrochemical biosensors for the determination of the amine content in fruits have been assembled using diamine oxidase (DAO) and polyamine oxidase (PAO) covalently immobilised onto polymeric membranes. Both enzymes in the presence of their substrates produced H_2O_2 that was detected at a platinum electrode. The detection limit of DAO and PAO biosensor for the polyamines spermidine and spermine is 10_6 mol/l witha good repeatability of 3 RSD%. The linear dynamic range for the analysis of polyamines in real samples was 2x10-6 to 5x10 -5 the PAO biosensor lifetime was 45 days. The biosensor was applied for determination of amine variations during apricot and sweet cherry ripening under different storage conditions (Esti et al., 1998). Apricots and cherries were stoned and ground to a puree, the homogenate was filtered and the resulting juice was used for the analysis. Biogenic amines (e.g. histamine, putrescine, cadaverine, tyramine, cystamine, agmatine, spermide) may also be formed during storage of foodstuffs. Recently, Carsol and Mascini (1999) and Niculescu et al . (2000) developed biosensors for monitoring freshness in fish samples, based on electrochemical oxidation of enzymatically produced H_2O_2 in the presence of these chemical indicators.

Biosensing principles, such as the Amperometric glucose sensor, have been applied in process control and the food industry. Maines et al. (1996) reviewed the requirements for sensing in the food industry and discussed the potential of enzyme electrodes to fulfil the need and the challenges presented by technology transfer into food applications. Examples of successful commercialised sensing instruments are the meatcheck and biocheck sensors. The meatcheck is a four-electrode array attached to a knife, which can be inserted into meat to measure the glucose gradient immediately below the surface. The size of the gradient is related to microbial activity on the surface of the meat and is regarded as a sound indicator of meat quality. The device provides in seconds what laboratory-based microbiology takes days to test. The biocheck method transformed the glucose sensor into a device capable of detecting and quantifying microorganisms in aqueous solutions. The system transferred electrons from the respiratory pathways of micro-organisms, and detected bacteria in under two minutes. Concentration of lactic acid is an important parameter for the meat industry as it characterises the state of freshmeat. Lactic acid is caused by anaerobic glycolysis from glycogen post mortem in muscles. The lactic acid concentration leads to conclusions concerning the pre-mortem metabolic situation, physical stress and deficiency in the meat quality. Bergann et al. (1999) reported an enzymatic biosensor based on immobilized lactatoxidase as bioreceptor and an amperometric transducer. The biosensor estimates lactic acid without special sample preparation, very quickly and at low cost. The increasing demand for on-line evaluation of milk quality directs the industry to look for practical solutions, and biosensors are a promising possibility. Eshkenazi et al. (2000) developed a multi-enzymatic amperometric biosensor f or lact ose in freshraw milk. The characteristics of the biosensor (easy operation, rapid response, long stability) suggested that this method could be used as an economical on-line lactose measurement technique in the milking parlour. Also in the literature there are biosensors described for fat in milk. Schmidt et al. (1996) reported a microbial biosensor based on thick film technology for free fatty acids. The biosensor measured the oxygen up taken by respiratory activity of the immobilized microorganisms. Oxygen was determined by electrochemical reduction. The sensor could be applied to milk samples without previous pretreatment, having a short response, high sensitivity and easy handling. However, biological research is needed to determine how sensor derived information can be used to improve the product quality other than by separating the milk into sources of high and low quality.

Biosensors for detection and identification of infectious disease in crops

Some microorganisms, particularly certain bacteria and fungi, are pathogens that attack crops and cause disease, sometimes in epidemic proportions. Fungal infection and aflatoxin production can occur at any stage of plant growth, harvesting, drying, processing and storage. Exposure by ingestion or inhalation of aflatoxins may lead to the development of serious medical conditions (structural and functional damage of t he liver, hepatic encephalopathy, immune suppression, lower respiratory infections, gastrointestinal haemorrhage, anorexia, malaise, fever). US government agencies monitor and tightly regulate aflatoxin levels innanimal feed and various food products. Action levels above which human food products are removed from commerce are typically 20 ppb. Recently an immuno-affinity fluorimetric biosensor was developed for detecting and quantifying aflatoxins, a group of chemically related mycotoxins formed by common fungi (*Aspergillus flavus*, *A. parasiticus* and *A. nomius*) found in maize, cottonseed, peanuts, and other nuts, grains and spices. The sample was filtered through a column containing sepharose beads to which the polyclonal aflatoxin-specific antibodies were attached.

The beads with the attached bound aflatoxin were subsequently rinsed to remove any unbound or nonspecifically bound impurities and interferents. After the rinse, an eluant solution was passed through the beads causing the antibodies to release the bound aflatoxin. The analyte was then collected and placed in a fluorimeter. There were essentially two subsystems within this biosensor: a fluidics subsystem, which performed mechanical sample-handling and processing, and an electro-optical system incorporating a miniature fluorometer that measured and reported the toxin level to the user. The two systems were controlled by a microprocessor that direct ed the fluidics and electrooptical system to perform the analysis. The aflatoxin biosensor provided a portable multi sample, rapid sampling and measurement capability with minimal sample handling and consumables. It provided very high sensitivity from 0-1 to 50 ppb, in less than 2 min with a 1ml sample and made over 100 measurements before refurbishment was required (Carlson et al., 2000). Schutz et al. (2000) developed a biosensor to detect marker volatiles released by potato tubers infested with Phytophtora infestans, offering a possible solution to the problem of screening large numbers of seed potatoes for fungal infestation. The biosensor, based on the intact antennae of Colorado potato beetle (*Leptinotarsa decemlineata*), was able to detect one single diseased potato tuber within up to 100 kg potato tubers, offering a promising early warning technique. Insect antennae are very suitable for the construction of highly sensitive biosensor, because of their remarkable sensory abilities.

Animal production

Oestrus detection

In the cattle breeding industry, where artificial insemination techniques are employed, the successful prediction of oestrus onset leads to considerable cost saving in herd management. Visual observation is the most commonly used way of detecting oestrus, however, oestrus does not always accompany ovulation and can occur in pregnant animals. Monitoring progesterone levels in milk is a more effective method not only of predicting ovulation but also for detecting pregnancy and fertility problems. The optimum fertility rates are achieved when insemination is performed 3 days after the progesterone level falls to below 5 ngml-1 of whole milk. ELISA test kits have allowed oestrus detection with 98% specificity but these methods require time and skills not abundant on a typical farm. Development of a real-time milk progesterone biosensor would provide a very useful tool for fertility monitoring.

Several approaches have been developed by researchers to determine progesterone concentration in milk. Pemberton et al. (1998) reported a disposable screen printed amperometric progesterone biosensor, operated in a competitive immunoassay format. The biosensor relies upon a reduction in the binding to the sensor surface of alkaline phosphatase-labelled progesterone in the presence of endogenous milk progesterone. The enzyme substrate is napthyl phosphate. The 1-napthol generated in the reaction is electrochemically oxidised, producing a signal inversely proportional to the concentration of unlabelled progesterone in milk. By using screen-printing technology, it is possible to mass produce the transducer element and fabricate sensors at low cost, enabling the screen-printed electrodes to be disposable devices. Mottram et al. (2000a) are transferring this technology, developing an automated ovulation system for progesterone in whole fresh milk, linked to a herd management database. The prototype biosensing system detects concentrations of progesterone between 3 and 30 mgml-1 in whole fresh milk (normal physiological levels), with a good regression (coefficient of determination R2=0.965). Another biosensor for oest rus det ection, monitoring on-line bovine progesterone during milking, employed an enzyme immunoassay format for molecular recognition but in this case the transducer was optical (Claycomb et al., 1998). The endogenous progesterone present in milk competed with progesterone labelled with horseradish peroxidase (HRP) for binding the active biosensing surface (a nitrocellulose membrane). The substrate was a solution of 3, 30, 5, 50-tetramethylbenzidine (TMB) and H2O2 and the transduction was based on the oxidation of TMB, generating a blue product. A spectrophotometer read a signal at 650nm that is inversely proportional to the concentration of progesterone in milk.

Veterinary drug residue screening

The use of antibiotics and chemotherapeutics in animal husbandry has led to the occurrence of veterinary drug residues in foods of animal origin. Traditional microbial screening methods have insufficient sensitivities to meet new regulations and classical physiochemical techniques, such as chromatographic methods and mass spectrometry are often precluded due to the level of experience, skills and cost involved. Immunological techniques have become increasingly popular for monitoring levels of therapeutic substances. Sulphonamides are a family of chemotherapeutics that are widely used for therapeutic and prophylactic purposes in animal diseases. In the treatment of mastitis, sulphonamides are usually administrated in the case of infections caused by Gram-negative pathogens, e.g. *E. coli.* Toxicological data show that sulphonamides have anti thyroid effects in both animals and humans. In Europe, a maximum residue limit of 0.1m g kg^{-1} has been established for total sulphonamides in milk. A surface plasmon resonance biosensor was compared with existing methods (microbial inhibitor assays, microbial receptor assays, ELISA, HPLC) for detection of sulfamethazine (SMZ) residues in milk (Mellgren et al., 1996). The Pharmacia BIAcore (a commercial surface plasmon resonance system) indicated the occurrence of less than 0.9 mg of SMZ per kg of milk (concentration below the detection limit of HPLC) and offered sufficient advantages (no sample preparation, high sensitivity rapid and fully analysis in real time) to be an alternative for the control of residues and contaminants in food. Baxter et al. (1999) reported the first study conducted to determine the feasibility of performing on-site drug screening at an abattoir, using an immune biosensor. The biosensor was used for screening for SMZ in pig bile samples, using a predetermined threshold limit of 0.4 $mgml^{-1}$. This method gave an accurate indication that the corresponding tissue sample contained SMZ residues in excess of the maximum residue limit. All positive control pigs were correctly identified during testing. A false positive rate of 0.3% was obtained but no false negatives were generated. Enrofloxacin is a synthetic antimicrobial agent of the fluoro quinolone family, especially designed for use in veterinary medicine (to treat mastitis). In the cow, enrofloxacin has a long elimination time in milk. In general, microbiological inhibitor assays for routine control of inhibitory substances in milk fail to detect fluoro quinolone residues at low levels. A rapid, sensitive surface plasmon resonance biosensor for enrofloxacin and its main metabolite, ciprofloxacin, in milk was developed (Mellgren & Sternesjo, 1998). The drug salbutamol (SBL) is a beta-agonist that may be used illegally as an animal growth promoter. Elliot et al. in 1998 using the commercially available surface plasmon resonance biosensor instrument, (Biacore, Uppsala, Sweden) analysed salbutamol (SBL) in urine samples of calves treated withSBL orally for 3 days. This work showed that biosensor-based veterinary drug residue testing procedures can generate results in real time without the need for time-consuming sample preparation.

Setford et al. (1999) developed a screen-printed device to measure penicillin G levels in milk. The biosensor was based on an ELISA affinity assay coupled to Amperometric determination of bound enzyme label activity. The system is ideal as a field-based screening tool for beta-lactam quantification in milk. More recently Delwiche et al. (2000) reported an enzyme immunoassay for beta-lactam penicillins, but in this case the system used a photometric sensor as transducer.

Veterinary diagnosis

Biosensors are a promising tool to diagnose and thereby aid in controlling animal diseases, but there are very few examples of biosensors applied to veterinary diagnosis and they are mainly piezoelectric immunosensors. Wu et al. (1997) reported a liquid piezoelectric immunoassay to determine the infection of rabbit serum by adult worm antigen of S. japonicum. In 1998 a direct piezoelectric flow injection analysis (FIA) immunoassay was developed for the detection of African swine fever virus (Uttenthaler et al ., 1998). The calibration curves in buffer and in serum could be determined and proved the suitability of the quartz crystal biosensor for the classification of positive and negative pig serum samples. More recently, Su et al. (2000) reported a piezoelectric immunosensor for porcine reproductive and respiratory syndrome virus (PRRSV). The proposed biosensor was used to screen pigs suspected to have been infected with the virus and to provide positive or negative results in a few minutes. Kumar (2000) developed a method for diagnosis of tuberculosis and other infections caused by mycobacteria. The preliminary results presented with the piezoelectric immunosensor for mycobacterial antigen (in gas and liquid phase), offer an enormous potential. For instance, detection of the antigen in Saliva

could constitute a non-invasive method of screening high-risk population. The potential could then be demonstrated for animal disease exposures to be detected without the need for blood sampling and offline laboratory analysis. Biosensor systems to detect infectious diseases at ports and in field situations without the need for expensive veterinary support would be a major asset for monitoring and controlling animal diseases. Biosensor technology could also be applied to detect mastitis infection by sensing markers such as enzyme Nacetylglucosaminidase (NAGase) in milk. This enzyme is released into milk as a result of tissue damage when the cow is resisting a clinical intra-mammary infection. Mottram et al. (2000b) showed the potential for a sensor based on the ability to convert 1-naphthyl N-acetyl-b-D glucosaminidine to 1-naphthol, which can be detected electrochemically.

Future challenges

Application and commercialisation of biosensor technology has lagged behind the output of research laboratories. Although many biosensor-related patents are filed eachyear, very few play a prominent role in clinical diagnostic, food industry, environmental agricultural or veterinary applications. There have been many reasons for the slow technology transfer from the research laboratories to the marketplace: cost considerations, stability and sensitivity issues, quality assurance and instrumentation design. Many of the main barriers are technical, methods of sensor calibration, methods of producing inexpensive and reliable sensors, stabilization and storage of biosensors, and total integration of the sensor system.

The behaviour of biomolecules adsorbed and immobilized on surfaces during the fabrication and use of biosensors needs to be better understood. In recent years, there has been remarkable progress in surface chemical analysis and scanning probe microscope techniques such as atomic force microscopy (AFM) and scanning tunneling microscopy (STM) but unfortunately our knowledge of manipulation and improvement of protein stability is far from complete. The limited stability of proteins (enzymes, antibodies) acts as a brake on the development of biosensors.

One of the future challenges is to develop cost effective methods for sequencing, interpreting and storing deoxyribonucleic acid (DNA) sequences. The development of DNA probes is a promising area of research in biosensors. Deoxyribonucleic acid sensors could be used to detect polymorphism or mutations in genes within plants, animals and microorganisms. Biomimetic systems will also accelerate biosensor development and applications. Some researchers are trying to overcome the poor stability of biological molecules by developing artificial molecular recognition systems with predetermined selectivity for various substances. So far, molecular imprinted polymers have been prepared with affinities for proteins, amino acid derivatives, sugars, vitamins, pesticides, and pharmaceuticals. Some advantages of molecular imprinting versus biological receptors are the low cost, ease of synthesis and relative long-term stability.

The development of automated manufacturing technologies is extremely important in the commercial mass production of biosensors. Deposition techniques such as screen-printing and ink-jet printing allow printing of materials at very high precision and speed, producing large numbers of inexpensive and reproducible biosensors. Thin-film deposition techniques such as Langmuir- Blodgett technology are able to create layers less than 1 mm thick, suitable for the development of micro sensors. Another area of intense research act ivity is silicon fabrication technology. The idea of lab-on-a chip is very attractive. The fabrication of credit-card sized micro laboratories will rely on advanced micro fabrication and micromachining technologies. It is now technically feasible to miniaturize and integrate com plex components such as pumps, valves, mixers and flow cells in a single chip along with the biosensor. Biosensor advancement in the commercial world could also be accelerated by the use of intelligent instrumentation, electronics, and multivariate signal processing methods such as chemo metrics and artificial neural networks. Increasing attention will have to be paid to the engineering of both the basic components and the device as a whole. It is in this area where agricultural engineers will have a key role in applying their knowledge of systems to improve sampling, calibration and data analysis to provide instructions for a farmer or processor rather than raw data. A biosensor array strategy, adaptable to multiple analytes detection, will allow spreading development costs over several products. These future improvements will produce devices more competitive with the presently available instruments, and be able to operate under field conditions.

Conclusions

The application of biotechnology would seem to be vital in maintaining the productivity and health of crops and livestock in the face of environmental concerns, limited resources and population increases. Biosensors could play an important role in providing powerful analytical tools to the agricultural diagnosis sector, particularly where rapid, low cost, high sensitivity and specificity measurements in field situations are required. This review summarised on-going developments in this field.

A range of molecules with biorecognition properties can be used as the sensing element in biosensors. A wide range of transducers is also available to engineer new biosensing devices. There are many different ways to combine biology, chemistry, physiscs, mathematics and engineering in order to develop new biosensors with applications in agriculture. The promise shown by biosensor technology is very real, however there are some technological obstacles that need to be overcome. Advances in areas such as surface chemical analysis, protein stabilisation, and autom ated manufacturing technologies would widen the market and allow biosensors to be more competitive in the agricultural market.

References

Andres R T; Narayanaswamy R 1997. Fibre-optic pesticide biosensor based on covalently immobilized acetylcholinesterase and thymol blue. *Talanta*, **44:** 1335–1352.

Bachmann, T., Leca B., Vilatte F., Marty, J.L., Fournier D., Schmid, R.D. 2000. Improved multianalyte detection of organophosphates and carbamates with disposable multielectrode biosensors using recombinant mutants of Drosophila acetylcholinesterase and artificial neural networks. *Biosensors & Bioelectronics*, **15:**193–201.

Baxter G A., Oconnor M C., Haughey S A., Crooks S R H; Elliott, CT. 1999. Evaluation of an immunobiosensor for the on-site testing of veterinary drug residues at an abatt oir. Screening for sulfamethazine in pigs. *Analyst*, **124 (9):**1315–1318.

Bergann T., Gifley K., Abel P, 1999. Concentration of lactic acid in carcasses and freshmeat-estimation withan enzymatic-biosensor measuring system. *Fleischwirtschaff*, **79(1):**84-87.

Carlson, M A., Bargeron C B., Benson D C., Fraser A B., Phillips T E., Velky J T., Groopman J. D., Strickland, PT., Ko HW. 2000. An automated, handheld biosensor for aflatoxin. *Biosensors & Bioelectronics*, **14:**841-848.

Carsol M A., Mascini M., 1999. Diamine oxidase andputrescine oxidase immobilized reactors in flow injection analysis: a comparison in substrate specificity. *Talanta*, **50 (1):**141-148.

Claycomb R W., Delwiche M J., Munro C J., Bon Durant R H. 1998. Rapid enzyme immunoassay of bovine progesterone. *Biosensors & Bioelectronics*, **13(11):**1165–1171.

Che Y H., Yang Z P., Li Y B., Paul D., Slavik M. 1999. Rapid detection of Salmonella typhimurium using an immunoelect rochem ical method coupled with immunomagnetic separation. Journal of Rapid Methods and Automation in Microbiology, **7(1):**47–59.

Delwiche, M., Cox E., Goddeeris B., Van Dorpe C., De Baerdemaeker J., Decuypere E., Sansen W 2000. A biosensor to detect penicillin residues in food. Transactions of the ASAE, **43 (1)**:153-159.

Dill, K., Stanker LH. and Young CR. 1999. Detection of salmonella in poultry using a silicon chip-based biosensor. *Journal of Biochemical and Biophysical Methods*, **41**:61-67.

Eshkenazi, I., Maltz, E., Zion B., Rishpon J. 2000. A threecascaded- enzymes biosensor to determine lactose concentration in raw milk. *Journal of Dairy Science,* **83(9):**1939-1945

Koblizek M; Masojidek J; Komenda J; Kucera T; Pilloton R; Mattoo A K; Giardi, M.T. 1998. A sensitive photosystem IIbased biosensor for detection of a class of herbicides.

Biotechnology and Bioengineering, 60 (6), 664– 669 Kroger S; Turner A P F; Mosbach K; Haupt K. 1999. Imprinted polymer based sensor system for herbicides using differential pulse voltammetry on screen-printed electrodes. *Analytical Chemistry,* **71(17):**3698–3702.

Kumar A. 2000. Biosensors based on piezoelectric crystal detectors. http://www.tms.org/ pubs/ journals/JOM/0010/Kumar/ Kumar-0010.htmc.

Lui J; Gunther A; Bilitewski U. 1997. Detection of methamidophos in vegetables using a photometric flow injection system. *Environmental Monitoring and Assessment,* **44(1–3):**375-382.

Maines A., Ashworth D., Vadgama P. 1996. Enzyme electrodes for food analysis. *Food Technology and Biotechnology*, **34(1):**31-42.

Mellgren C., Sternesjo A., Hammer P., Suhren G., Bjorck L., Heeschen W 1996. Comparison of biosensor, microbiological, immunochemical and physical methods for detection of sulfamethazine residues in raw milk. *Journal of Food Protection,* **59(11):**1223–1226.

Mellgren C., Sternesjo A. 1998. Opt ical immunobiosensor assay for determining enrofloxacin and ciprofloxacin in bovine milk. *Journal of AOAC International*, **81(2)**:394–397.

Mottram T., Hart J. and Pemberton R. 2000a. A sensor based automatic ovulation prediction systemfor dairy cows. Proceedings of 5thAISEM Conference. Lecce, Italy

Mottram T., Hart J. andPemberton, R. 2000b. Biosensing techniques for detecting abnormal and contaminated milk. Robotic Milking. Proceedings of the International Symposium.

Niculescu M., Nistor C., Frebort I., Pec P., Mattiasson B.C. and Soregi E. 2000. Redox hidrogel-based amperometric bienzyme electrodes for fish freshness monitoring. *Analytical Chemistry,* **72(7):**1591-1597.

Nunes G S., Barcelo D., Grabaric B S. ,Diaz Cruz J M., Ribeiro M L 1999. Evaluation of a highly sensitive Amperometric biosensor with low cholinesterase charge immobilized on a chemically modified carbon paste electrode for trace determination of carbamates in fruit, vegetable and water samples. *Analytica Chimica Acta,* **399(1–2)**: 37–49.

Olsen, J.E. 2000. DNA-based methods for detection of foodborne bacterial pathogens. *Food Research International,* **33(3–4):** 257–266.

Palchetti, I., Cagnini, A., DelCarlo, M., Coppi, C., Mascini M., Turner, APF. 1997. Determination of anticholinesterase pesticides in real samples using a disposable biosensor. *Analytica Chimica Acta,* **337(3)**:315–321.

Olsen, J.E., 2000. DNA-based methods for detection of foodborne bacterial pathogens. *Food Research International,* **33(3–4)**: 257–266.

Palchetti I., Cagnini, A., DelCarlo M., Coppi C., Mascini M., Turner APF. 1997. Determination of anticholinesterase pesticides in real samples using a disposable biosensor. *Analytica Chimica Acta,* **337(3)**:315–321.

Pemberton R M., Hart J P., Foulkes J A. 1998. Development of a sensitive, selective, lectrochemical immunoassay for progesterone in cow's milk based on a disposable screenprinted amperometric biosensor. *Electrochimica Acta*, **43(23):**3567–3574.

Perez FG., Mascini M., Tothill I E., Turner APF. 1998. Immunomagnetic separation with mediated flow. injection analysis amperometric detection of viable Escherichia coli O157. *Analytical Chemistry*, **70(11):**2380–2386.

Elliot C T., Baxter G A., Hewitt S A., Arts C J M., Van Baak M., Hellenas K E. and Johannson A 1998. Use of biosensors for rapid drug residue analysis without sample deconjugation or clean up: a possible way forward. *Analyst*, **123(12):**2469–2473.

Pogacnik L., Franko M. 1999. Determination of organophosphate and carbamate pesticides in spiked samples of tap water and fruit juices by a biosensor with photot hermal detection. *Bi osensors & Bioelectronics*, **14(6):**569–578.

Powner E T., Yalcinkaya F. 1997. Intelligent biosensors. Sensor Review, 17(2):107–116. Rasooly L; Rasooly A 1999. Real time biosensor analysis of Staphylococcal enterotoxin A in food. *International Journal of Food Microbiology,* **49(3):**119–127.

Roda A., Rauch P., Ferri E., Girotti S., Ghini S., Carrea G. and Bovara R. 1994. Chemiluminiscent flow sensor for the determination of paraoxon and aldicarb pesticides. *Analytica* **294 (1):** 35-42.

Seo, K.H., Brackett, R.G., Hartman N F. and Campbell DP. 1999. Development of a rapid response biosensor for detection of Salmonella Typhimurium. *Journal of Food Protection*, **62(5):**431–43.

Sergeyera T A., Piletsky S A., Brovko A A., Slinchenko E A., Sergeeva L M., El'skaya, A.V. 1999. Selective recognition of atrazine by molecularly imprinted polymer membranes. Development of conductimetric sensor for herbicides detection. *Analytica Chimica Acta,* **392 (23):**105–111.

Setford S J., Van Es R M., Blankwater Y J. and Kroger, S. 1999. Receptor binding protein amperometric affinity sensor for rapid beta-lactam quantification in milk. *Analytica Chimica Acta,* **398(1)**:13–22.

Schmidt A; Standfub-Gabisch C; Bilitewski U (1996). Microbial biosensor for free fatty acids using an oxygen electrode based on thick film technology. *Biosensors & Bioelectronics*, **11 (11):**1139–1145

Smith T J., O'Connor L., Glennon M., Maher M. 2000. Molecular diagnostics in food safety: rapid detection of food-borne pathogens. *Irish Journal of Agricultural and Food Research*, **39 (2):**309–319.

25

Development of Diagnostics Kits for Plant Diseases

R. Manimekalai

Introduction

Crops diseases are caused by bacteria, virus, fungi and phytoplasmas. The detection of pathogenic bacteria and viruses in plants, planting material, vectors or natural reservoirs is essential to ensure safe and sustainable agriculture. The molecular based techniques, have evolved significantly during recent past that allows the rapid and reliable detection of pathogens. The detection and diagnosis of plant pathogens is essential to take control measures and subsequently for eradication of the disease that causes. Molecular detection is largely based on PCR or RT-PCR amplification following purification of nucleic acids from the samples, with the extraction of the target DNA. Variants of PCR, such as simple or multiplex nested-PCR in a single closed tube, co-operative-PCR, and real-time monitoring of amplicons or quantitative PCR, allow high sensitivity in the detection of one or several pathogens in a single assay. Recently, genomics approaches help in accurate detection of pathogens. Even though the advancements had taken place in molecular detection techniques such PCR, RT- PCR and micro arrays, there is still problem of detection exists for many plant pathogens because of their low titers and presence of inhibitory substance in nucleic acid preparations. In these conditions, nanotechnology offers potential to diagnose low titre pathogens and increase the sensitivity of the diagnosis. Nanotechnology based detection technique takes less time and that can give results within a few hours, that is simple, portable and accurate and does not require any complicated technique for operation so that even a simple farmer can use the portable system.

Nanotechnology for crop diagnosis

Nanoparticles are typically in size range of 1–100 nm (1 nm= 10^{-9} m) and can have different shapes and compositions. Their very small size imparts physical and chemical properties that are very different from those of the same material in bulk form. These properties include a large surface to volume ratio, enhanced or hindered particle aggregation depending on type of surface modification, enhanced photoemission, high electrical and heat conductivity and improved surface catalytic activity. Nanoparticles are also structurally robust and their physical properties can be tailored by variation of particle size, shape and composition. Their nano-size is within the typical size range of biomolecules and cellular organelles. This would allow a nearly one-on-one interaction between nanoparticle and biomolecule of interest. The properties of nanoparticles make them of high potential for use in *in vitro* diagnostics.

Types of nanoparticles employed in diagnostics

Gold NPs, silver particles and quantum dots (semiconductors) are most widely used, nanoparticles but new materials are becoming available as more molecular entities are discovered as amenable to nanoscale design and fabrication. Crystal materials like those of gallium, phosphate, quartz, and ceramic are chosen for their durability and piezoelectric properties of developing and retaining an electric potential (charge) when subjected to mechanical stress.

Gold nano particles (AuNP)

The properties of gold nanoparticles make them to use in diagnostics.

AuNP allows increased sensitivity by several orders of magnitude; light scattered from one nanoparticle is equivalent to the light emitted from 5×10^5 (500,000) fluorophores. Tests that employ gold nanoparticles functionalized with antibodies, for example, are 2-3 orders of magnitude more sensitive than ELISA-based methods. Enables high-specificity for both nucleic acid and protein detection. For nucleic acid detection, single-base pair specificity is achieved due to assay reaction kinetics where gold nanoparticle probes, comprised of target-specific oligonucleotides, permit hybridization to target DNA over a very narrow temperature range. Reduces background noise (i.e. signal-to-signal) due to minimal non-specific binding of the gold nanoparticle probes, which in turn, create an enhanced assay signal. Requires little or no inventory control. The nanoparticle probes are extremely stable, have a long shelf-life and are non-toxic.

Carbon nanoparticles

Carbon nanoparticles are now used as coloured labels in lateral flow immuno assays (LFIA) and other diagnostic format such as micro-array. The use of carbon nanoparticles yields lines/spots that have a high signal-to-noise ratio, that is black/grey on a white membrane. Carbon nanotubes (CNTs) are well-ordered, all-carbon hollow graphitic nanomaterials with a high aspect ratio, lengths from several hundred nanometres to several micrometres and diameters of 0.4–2 nm for single-walled (SW-CNTs) and 2–100 nm for coaxial multiple-walled (MW-CNTs) carbon nanotubes. Conceptually, nanotubes are viewed as rolledup structures of single or multiple sheets of graphene to give SWCNTs and MWCNTs, respectively. These one dimensional carbon allotropes are of high surface area, high mechanical strength but ultra-light weight, rich electronic properties, and excellent chemical and thermal stability. Ever since the discovery of carbon nanotubes, researchers have been exploring their potential in disease diagnosis applications.

Quantum dots

The QDs, also known as nanocrystals, are inorganic fluorophores with a typical diameter of 2–10 nm. A QD consists of a core semiconductor covered with another shell semiconductor that has a larger spectral band gap (separation between electronic energy levels; specifically valence band and conduction band). The shell serves to increase QD's quantum yield as well as photo stability. They have broad-range excitation (a QD of any size can be excited by UV light), strong narrow emission bands, and high photo stability. The emission properties of QDs can be controlled by varying their size and composition. The QDs can be adapted to specific target recognition by conjugating them to a variety of biomolecules, such as antibodies, streptavidin, and oligonucleotides. When a QD absorbs a photon higher in energy than the spectral band gap of core semiconductor, an exciton (electron–hole pair) is produced. The absorption has an increased probability at shorter wavelengths and results in a broadband absorption spectrum; this is in contrast to conventional fluorophores. The excitation has a long life time of 10–40 ns as opposed to a few nanoseconds for a typical organic dye such as fluorescein. On returning back to lower energy state, a photon is emitted producing a narrow symmetric energy band, thus giving a strong fluorescence signal. When QD has a radius smaller than Bohr exciton radius, which is the natural separation distance between an electron and its hole (positive) in an exciton, the energy levels for a photon are quantized. A direct relationship between values of emitted energy quanta and size of QD exists. This phenomenon is known as quantum confinement effect, which gives QDs their name, and it is also the origin of size tuneable emission spectra of QDs. The fluorescent signal generated by QDs can be detected using different techniques including confocal microscopy, total internal reflection microscopy, fluorescence microscopy, wide field epi fluorescence microscopy as well as fluorometry.

Types of Nanoparticles Based Diagnostic

i. Nucleic Acid Lateral Flow Immuno-Assay

ii *Immuno based Nano-diagnostic*

Diagnostics using Nano particles

The areas of application of nanomolecules include therapeutics, cancer detection, diagnosis and treatment of infectious disease. Nanoparticles, such as gold nanoparticles and quantum dots, are most widely used for molecular detection purpose. It produces stronger and more intense signals than conventional fluorescent tags and also is more stable when exposed to light. Edgar et al. (2006) performed the diagnosis of bacterial pathogens by using host-specific bacteriophage and conjugation of phage to streptavidin-coated quantum dots. The phages were genetically modified to produce a specific protein (i.e. biotin) on their surface which selectively attracts the nanoparticles. In another study done by Huang et al. (2008), bovine serum albumin (dBSA)-coated water soluble cadmium telluride (CdTe) quantum dots (QDs), were successfully conjugated to an anti-*Escherichia coli* antibody via a cross-linking reaction and were then used to detect *E. Coli* and *Listeri*a monocytogenes using fluorescence microscopy.

Zhang et al. (2010) developed an ultrasensitive method for detection of porcine circovirus type 2 based on gold (III) enhanced chemiluminescence immunoassay. The gold (III) served as an analyte for the indirect measurement of the virus. The method has good sensitivity and reliability in analysis of 36 serum samples than the conventional polymerase chain reaction (PCR). Lin et al. (2005) used AuNPs for development of a microchannel immunoassay that detected *E. coli* and *H. pylori* antigens with a detection limit of 10 ng. The AuNPs were conjugated to secondary antibodies specific to biotinylated primary antibodies against pathogens. Although the assay sensitivity is comparable to that of conventional dot ELISA, the microchannel system is highly amenable to miniaturization. Duan et al. (2005) developed an assay for simultaneous detection of hepatitis B virus (HBV) and hepatitis C virus (HCV) using a protein chip, AuNP based detection, and silver enhancement. Antigens from HBV and HCV were immobilized on a glass chip to capture HBV and HCV antibodies in human sera. AuNP-labelled staphylococcal protein A was then added to detect captured antibodies and silver staining applied to amplify the detection signals. This produced black spots on the assay chip which were visible with the naked eye. Multimembrane composites (test strips) with immobilized polyclonal antibodies against viruses and colloidal gold-conjugated antibodies were used for bean mild mosaic virus (BMMV), carnation mottle virus (CarMV), rod-shaped tobacco mosaic virus (TMV), and filamentous potato viruses X and Y (PVX, PVY).

Detection of bacteria through nanoparticles

Fluorescent nanoparticles for detection of bacteria: Using bioconjugated dye-doped silica nanoparticles a bioassay was developed for accurate determination of single bacteria within 2 minutes. Fluorescent silica nanoparticles were used for detection of *Xanthomonas axonopodis* pv. *vesicatoria* that causes bacterial spot disease in Solanaceae plant.

Nanotechnology based detection for phytoplasma pathogen

A Lateral Flow Strip Test can be developed for nanotechnology-based quick detection of phytoplasma from infected samples. For this we need a 'detector antibody,' which is an antibody-specific for phytoplasma, a 'capture antibody' that is anti-rabbit antibody and 'gold nanoparticles'. First, the gold nanoparticle is linked to detector antibody. The infected plant sample has the antigen which will bind specifically to this antibody. The plant extract is taken and mixed with the antibody and if phytoplasma is present in the extract, phytoplasma specific proteins will bind to the antibody. This solution is then transferred to a filter stick which has the capture antibody imbedded in it. These antibodies will bind to a different part of protein and they do not have gold label. As the solution is wicked up the strip, the phytoplasma-specific proteins encounter the capture antibody and are adsorbed. All other proteins are washed away. The gold linked to detector antibody (which is still linked to the antigen) will now become concentrated and a pink colour appears on filter stick. This is because gold is pinkish colour in its natural state. If no phytoplasma is present in plant sample, no colour appears.

Detection of virus through nanoparticles

Aptamer-based biochips for label-free detection of plant virus: Aptamers are *in vitro* generated, short, single stranded nucleic acids that selectively bind to target compounds ranging from small molecules to macromolecules. Utilization of aptamers in label-free techniques such as surface plasmon resonance (SPR), SPR imaging, quartz crystal microbalance, microelectromechanical sensing, nano-field effect transistors (nanoFETs), and electrochemical impedance.

Conclusion

Early detection of pest ad diseases, nutrient deficiencies and heavy metal toxicities as utmost essential to protect the crops from catastrophe. The chapter summarized the role of nanotechnology in developing diagnostic kits for addressing both biotic and abiotic stress conditions. Since nanoparticles are extremely small but possessing extensive surface area, each and every molecule of analyte is taken into account while detecting the maladies. This is one of the most potential areas in agriculture that is given more impetus to advance early diagnosis of field problems.

26

Nanotechnology for Early Detection of Plant Pathogens

R. Selvarajan

Introduction

Plant diseases are one of the causes of concern in agriculture and horticulture industry. Estimated global crop loss due to plant diseases exceeds hundred billion US$ worldwide (Orke et al. 1994; Narayanasamy, 2010). Plant diseases are caused by microorganism such as fungi, bacteria, viral, viroids and phytoplasma. Most of the plant pathogens are transmitted through seed or planting material, which can spread secondarily through vectors and lead to severe loss to the crop. Among plant diseases, viral diseases cause serious crop losses and affect the quality of products, while the use of virus- free seeds and planting stocks results in a substantial increase in agricultural crop production (Waterworth & Hadidi, 1998; Strange& Scott, 2005). Due to the lack of effective treatment protocols, the main approach for the production of virus -free seeds and planting stocks is selection of healthy plants and reject the infected. The effective ELISA and PCR-based methods have been developed for laboratory detection of most of the commercially important phytopathogens including viruses. In recent years, the DNA microarray diagnostics method has become widely adopted, providing the capability of parallel detection of all pathogens in a single sample of the crop culture tested (Hadidi et al., 2004; Boonham et al 2007). All these methods of instrumental analysis possess both high sensitivity and productivity making them fit for wide use in planting material certification and quarantine control, as well as in monitoring viral infections. However, these methods are complicated and laborious; they require qualified personnel and expensive equipment, limiting their use to well-equipped facilities and laboratories. There is no on-site detection kits available for all the plant pathogens. In the field of clinical diagnosis, the advancement has already taken place by employing the nanotechnology and there is a scope to modify them to suit in plant virus diagnosis need to be explored.

Diagnostics of plant pathogens is vital for the production of quality planting material especially vegetatively propagated crops. Banana, Citrus, Potato, Cassava, Sugarcane, Taro, Amorphophalus, Black pepper and Cardamom are propagated vegetatively and are affected by many viral pathogens causing severe yield loss to the crop. Nucleic acid based diagnostics techniques such as PCR, real time PCR and nucleic acid spot hybridization (NASH) methods are available for most of these plant pathogens in India. ELISA is known popular detection technique used for the efficient detection of plant viruses and this robust, high-throughput technique can be handled for the detection of large number of samples at a time. Though this ELISA based techniques are robust, it cannot be used on-site or in the field and again this technique is less sensitive and require special skill and require at least a day in the laboratory. Of late, nanotechnology based detection tools are becoming popular in medical and agriculture sectors. In this article the nanotechnology based detection techniques available for the detection of plant pathogens are discussed in detail.

Nanotechnology

Nanotechnology is a collective term for a wide range of relatively novel technologies; the main unifying theme is that it is concerned with matter on the nanometre scale (Greek n˜anos means dwarf). Technology is generally regarded as the utilization or application of science to benefit society. Nanotechnology is an emerging technology, seeking to exploit distinct technological advances of controlling the structure of materials at a reduced dimensional scale, approaching individual molecules and their organised aggregates or supra molecular structures. Nanoparticles are typically in the size range of 11–100 nm (1 nm= 10^{-9} m) (Liu, 2006) and can have different shapes and compositions. Their very small size imparts physical and chemical properties that are very different from those of the same material in the bulk form. These properties include a large surface to volume ratio, enhanced or hindered particle aggregation depending on the type of surface modification, enhanced photoemission, high electrical and heat conductivity, and improved surface catalytic activity (Liu, 2006; Garg McNeil, 2005; Rosi and Mirkin, 2005, Shrestha et al., 2006). Nanoparticles are also structurally robust and their physical properties are tailorable by variation of particle size, shape and composition. Their nano-size is within the typical size range of biomolecules and cellular organelles. This would allow a nearly one-on-one interaction between the nanoparticle and the biomolecule of interest (Azzazy et al., 2006, 2007; Jain, 2005). The properties of nanoparticles make them of high potential for use in *in vitro* diagnostics where they promise increase increased sensitivity, speed and cost effectiveness.

Nanotechnology in diagnostics

There is a constant need to improve the performance of current diagnostic assays as well as develop innovative testing strategies to meet new testing challenges. The use of nanoparticles promises to help promote in vitro diagnostics to the next level of performance. There are three major areas where nanotechnology has been integrated into the next generation diagnostic techniques: (i) to improve assaysensitivity, specificity and limits of detection (ii) to increase sample throughput; and (iii) to reduce assay complexity and cost. Nanoparticles often require less analyte to register a response, which is largely due to the small area of the sensing surface. Smaller sensing areas generally allow higher-density arrays to be fabricated and this feature maximizes the number of analytes (e.g., pathogens or biomarkers) that can be interrogated in a single test without increasing sample requirements. Assay complexity and cost can also be significantly reduced by nanoparticles /nanosensor that eliminate sample processing steps, such as nucleic acid amplification (e.g., PCR), and/or provide "labelfluorescent dyes) detection platforms. However, it is important to note that several such devices self-described as "label free" often require a recognition element, such as an antibody or oligomers for selective detection.

The broad range of nanotechnologies applied to pathogen detection can be categorized according to the technologies, i.e., nanoarrays, nanofluidics, nanotransducers. Each of these types of nanotechnologies, while fundamentally different, has been shown to improve sensitivity, specificity, and throughput while decreasing analysis time and sample volume. Quantum dots (QDs), gold nanoparticles (AuNPs) and superparamagnetic nanoparticles are the most promising nanostructures for in vitro diagnostic applications. These nanoparticles can be conjugated to recognition moieties such as antibodies or oligonucleotides for detection of target biomolecules. Nanoparticles have been utilized in immunoassays, immune histochemistry and DNA diagnostics.

Nanotechnology based Lateral flow immune assay (LFIA) detection systems

Lateral flow immune assay (LFIA) are used for qualitative or semi-quantitative detection and monitoring of pathogens in non-laboratory environments. This LFIA works with chromatographic principle coupled with immunological recognition system. LFIA is based on the interaction between the target virus and immune- reagents (antibodies and their conjugates with colored colloidal particles or nanoparticles) applied on the membrane carriers (test strips). When the test strip is dipped into the sample being analyzed, the sample liquid flows through membranes and triggers immunochemical interactions resulting in visible coloration in test and reference lines (von Lode, 2005; Price & Kricka, 2007). A typical LFIA format consists of a surface layer to carry the sample from the sample application pad via the conjugate release pad along the strip encountering the detection zone up to the absorbent pad (Fig. 1).The membrane is often thin and fragile, so it is attached to a plastic or nylon basic layer to allow cutting and handling. In addition, robustness is achieved by housing the strips in a plastic holder, where only the sample application window and a reading window are exposed. Current membrane strips are produced from nitrocellulose, nylon, polyethersulfone, polyethylene or fused silica. At one end of the strip a sample application pad is provided. The sample application pad is usually made of cellulose or cross- linked silica. In close contact with the strip material and the sample application pad is the conjugate release pad, made of cross linked silica. Labelled analyte or recognition element(s) (depending on the application) are dried on this pad and after addition of the sample, this material will interact with the fluid flow; specific interactions will be initiated here and will continue during the chromatographic process. The strip based lateral flow immune-assay is widely being used in the detection of human pathogens (Edgar et al., 2006; Al-Yousif et al., 2002; Doering et al., 2007), toxic compounds like pesticides residues (Zhang et al., 2006) drugs, metabolites in biomedical field, phytosanitory, veterinary, feed/food (Delmulle et al., 2005), detecting many metabolic disorders (Cho and Paek, 2001; Choi et al., 2004) and ascertaining pregnancy (Tanaka et al., 2006). In agriculture, this technology has not yet been exploited and very limited information is available. LFIA strips have been employed in detect BT-cotton (Kumar et al., 2010; www.cicr.org.in) and pesticide quality control assessment in agriculture (Kranthi et al., 2009). The LFIA strip technique has been reported for the detection of few plant viruses (Tanaka et al., 1998; Danks and Barker, 2000; Saomone and Roggero, 2002; Salomone et al., 2002; Choi et al., 2001; Salomone et al., 2004 and Kusano et al., 2007). The use of nano particles as labels in conjunction with novel detection technologies has led to improvements in sensitivity and multiplexing capabilities (Jain, 2005; Rosi and Mirkin, 2005). Metallic nano particles composed of gold or silver have many optical and electronic properties, derived from their size and composition (Nath et al., 2008). When coupled to affinity ligands, these nanomaterials have found important applications as chemical sensor. For example gold nanoparticles conjugated with specific oligonucleotides can sense complementary DNA strands, detectable by color changes (Eghanian et al., 1997; Mirkin et al., 1996). Other nanoparticles including fluorescent quantum dots and carbon nanotubes have been used in various applications including DNA detection, and the development of immunoassays for the detection of pathogens (Bruchez et al., 1998; Edgar et al., 2006; Baptista et al., 2006; Alivisatos et al., 2005). Express immune-chromatographic test strip assays were developed for detection of five plant viruses varying in shape and size of virions: spherical carnation mottle virus, bean mild mosaic virus, rod shaped tobacco mosaic virus, and filamentous potato viruses X and Y. Multi-membrane composites (test strips) with immobilized polyclonal antibodies against viruses and colloidal gold-conjugated antibodies were used for the analysis. The immune-chromatographic test strips were shown to enable the detection of viruses both in purified preparations and in leaf extracts of infected plants with sensitivity from 0.08 to 0.5ìg/ml for 10 min. The test strips may be used for express diagnostics of plant virus diseases in field conditions (Byzovza et al., 2009).

A major challenge for field diagnostics that employ antibodies as targeting ligand is the need to maintain the antibody's structure and prevent thermal denaturation. Considering this challenge, gold nanoparticles and fluorescent conjugated polymer constructs have been used for the fluorescent- based identification of microorganisms without the need of antibodies (Phillips 2008). In this technique, the electrostatic interaction between cationic gold nanoparticles and anionic polymers led to fluorescence quenching (Phillips et al., 2008) (Fig. 2). However, in the presence of bacteria, the negatively charged bacterial cell wall caused displacement of the nanoparticles' polymer (Phillips et al., 2008). Hence, the interaction between the nanoparticles and bacteria and the concomitant dissociation of the polymer from the nanoparticles induced the release of the polymer's quenched fluorescence, leading to enhanced fluorescence emission. A library of three nanoparticle preparations was prepared and distinct fluorescence emission patterns were observed for each organism, including *E. coli, B. subtilis, L. lactis* and *S.coelicolor* (Phillips et al., 2008). Subsequent quantitative analysis for pattern recognition through linear discriminant analysis led to the construction of a signature plot, having each microorganism's characteristic fluorescence emission (Phillips et al., 2008) consequently, this approach can be utilized for the affordable and robust identification of microorganisms without the need for heat labile antibody-conjugated probes. It should be noted that as this method is at its infancy, the detection threshold was high (OD600=1; 1×10^9 colony forming units).

Nucleic acid lateral flow (immuno) assay

A combination of antigen-antibody interaction and specific tagged doubled-stranded amplicon (ds- amplicon) detection after PCR is also possible, and is called "nucleic acid lateral flow immunoassay" (NALFIA). Specific nucleic acid hybridization of amplicons with immobilized complementary probes is another option and is called "nucleic acid lateral flow assay" (NALF).

The NALFIA and the NALF set-ups are usually designed for testing the presence or absence of pathogens in food, feed or the environment. In the NALFIA set-up the analyte is an amplified double- stranded nucleic acid sequence (dsamplicon) specific of the organism using primers with two different tags; recognition of the analyte is done by binding to a tag- specific antibody. In a typical layout developed for the detection of pathogenic bacteria the nucleic acid was amplified using PCR with two tagged primers. A ds- amplicon was obtained with one strand labelled with biotin and the other strand labeled with, e.g., fluorescein isothiocyanate or digoxigenin. A solution of antibodies raised against the tag was sprayed at the test line. The biotin will bind to the avidin-labelled nano particles and the other tag will bind to the antitag antibody, resulting in the coloured signal. The response is directly proportional to the amount of analyte. For the NALF set- up several formats have been published.

1. Nanoparticle-labelled reporter probe and biotin- labelled immobilised capture probe via avidin, single-stranded amplicon (ss-amplicon) hybridises with complementary reporter and capture probes.
2. Nanoparticle-labelled reporter probe and bovine serum albumin labelled capture probe, immobilised through passive adsorption, ss-amplicon hybridises with complementary reporter and capture probes.
3. Nanoparticle-labelled reporter probe; capture probe is immobilised at the test line through passive adsorption, ss-amplicon hybridises with complementary reporter and capture probes. The response is directly proportional to the amount of analyte.

In DNA biosensors, molecular recognition is achieved via hybridization of the target sequence with complementary oligonucleotide probes. Detection is carried out by electrochemical, optical, or gravimetric transduction. Based on that principle, disposable dipstick-type biosensors can be fabricated for visual detection of DNA without the use of instruments. The dipstick-type dry-reagent format eliminates multiple incubation and washing steps, which are often necessary in DNA assays, and minimizes the requirements for highly qualified personnel. The presence of the analyte (target DNA sequence) triggers the formation of a biotinylated hybrid, which is captured at the test zone of the sensor by immobilized streptavidin and linked to oligonucleotide-functionalized, strongly colored nano- or microparticle reporters (e.g. gold nanoparticles and polystyrene microspheres).

Quantum dots

A quantum dot is a semiconductor whose exciton is confined in all three spatial dimensions. As a result, they have properties that are between those of bulk semi-conductors and those of discrete molecule. The conducting characteristics of quantum dot are closely related to the size and shape of the individual crystal. Generally, the smaller the size of the crystal, the larger the band gap, the greater the difference in energy between the highest valence band and the lowest conduction band. In fluorescent dye applications, this equates to higher frequencies of light emitted after excitation of the dot as the crystal size grows smaller, resulting in a color shift from red to blue in the light emitted. The main advantages in using quantum dots is that because of the high level of control possible over the size of the crystals produced, it is possible to have very precise control over the conductive properties of the material. Quantum dots (QDS) have been introduced as a promising new tool in life sciences, because of their unique optical properties. They are highly stable during excitation and have characteristic absorption and emission spectra. The emission peak of these nanoparticles is comparatively narrow and the dots fluorescence brighter than organic fluorescent dyes. Thus particle visibility is enhanced and weaker laser intensity is required for the imaging process. QDs can be excited using different wavelengths from UV up to the emission wavelength. Hence it is possible to excite simultaneously QDs emitting of different wavelengths potentially facilitating a simpler handing of multicolor- labelled samples. The core of the quantum dot particle is composed of a mixture of cadmium and selenide. This sphere, having a diameter of 20 to 55 Å, is coated with 1-2 monolayers of ZnS measuring 3,1 Proteins, antibodies, DNA or other molecules of interest can be attached to QDs allowing a wide range of applications in life sciences. The complete QD-streptavidin conjugate has a diameter of 10–15 nm. Hence quantum dots have been employed in live cell imaging, diagnostic and therapeutic purposes, immunohistochemistry and in fluorescence in situ hybridization (FISH) experiments. However, until now there have been only few reports of applications of QDs in plant research.

QDs are few nm in diameter, roughly spherical (some QDs have rod like structures), fluorescent, crystalline particles of semiconductors whose excitons are confined in all the three spatial dimensions. Their potential application in diverse fields can be attributed to the property of quantum confinement. In 1998, the first use of QDs for biological detection and about its photochemical properties was reported (Bruchez et al., 1998; Chan and Nie, 1998). Diagnostics using colloidal QDs has got tremendous hoist from this milestone finding. QDs are robust and very stable light emitters and can be broadly tuned through size variations. In the past two years, wide range of protocols for bio- conjugating

QDs have been developed in QDs have been developed in diverse areas of applications: cell labeling, cell tracking, *in vivo* imaging, DNA detection (Xu et al., 2003). Diagnosis of a disease in its very early stage can play important role in treatment. Due to phenomenal advancement in nanotechnology, QDs have emerged as pivotal tool for detection of a particular biological marker with extreme accuracy. QDs being very photo-stable and optically sensitive can be used as labeling and can be easily traced with ordinary equipment. Early detection of pathogens causing plant diseases using quantum dots would prove to be boon in agriculture.

Commercialized nanotechnology based detection kits

A number of foreign companies produce test strips for detection of plant viruses: Spot Check LF (Adgen Ltd., UK), Pocket Diagnostic (Forsite Diagnostics Ltd., UK), and Immunostrips (Agdia, United States). Results obtained with these within a few minutes. Among the substantial advantages of this approach are its high sensitivity, ease of both sample preparation, and the analysis itself.

Conclusions

Nanotechnology has the potential to make significant contributions to disease detection, diagnosis and prevention. Of late viral disease epidemics are very common in horticulture crops and the loss caused by these viruses are significant and affects the livelihood of growers. Potato, sugarcane, and banana plants propagated through the tissue culture process plants have been brought under seed act, 1966. As these vegetatively propagated plants can harbor viruses and transmits to their progeny in the process of tissue culture mass propagation, Dip stick or LFIA strips for banana and potato pathogens would boost the production and productivity in India. Crops like banana, cassava, amorphophalus and taro contain a lot of secondary metabolites, the purification of viruses from these plants are hampered by the presence of secondary metabolites. So, the viral coat protein gene can be cloned in a suitable expression vector and the over expressed viral protein can be purified to raise a high quality polyclonal antiserum for the development of LFIA strips. Since enormous losses are being incurred in different crops due to viral infection or different kinds of plant pathogens, quick and effective disease diagnosis should be available for field level diagnosis. In order to utilize the advancement taken place in the clinical diagnosis using nanotechnology, efforts should be taken to use the same for plant virus or plant pathogens diagnosis.

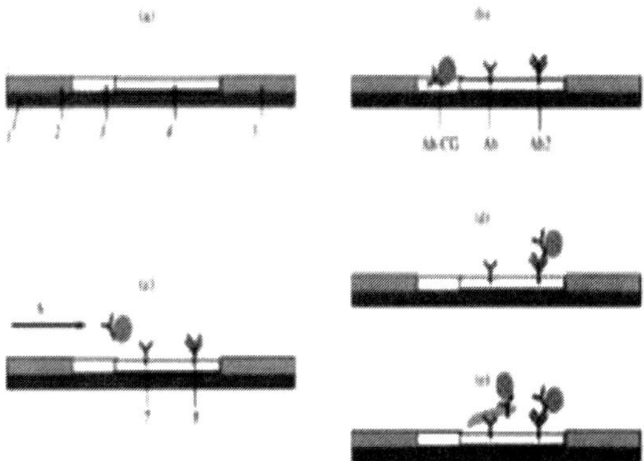

Fig. 1: Lateral flow device for virus detection a—test-system composition: 1—plastic mount; 2—absorption membrane; 3— conjugate moun; 4 —workingmembrane; 5—absorption pad. b—reagents introduced: Ab—virus-specific antibodies; Ab2— antibodies against rabbit IgG Ab-CG; gold colloidal conjugated to virus specific antibodies. c-A schematic representation of analysis: 6—spreading of liquid front; 7—test line; 8—control line d- test result in the absence of the virus: one colored band in the control line area, e—Test result in the presence of the virus: two colored bands in both test and control line areas (Byzova et al., 2009)

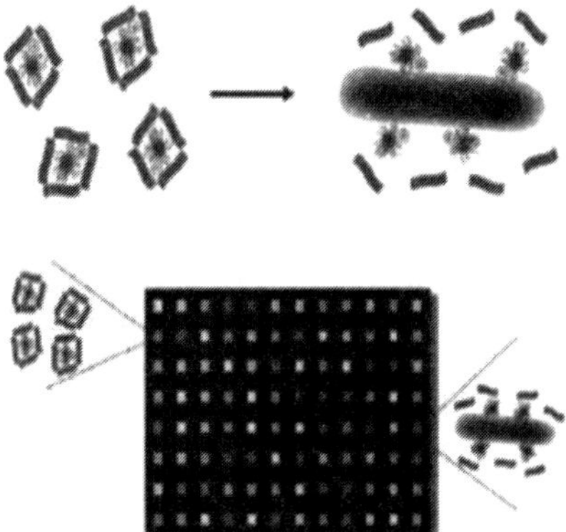

Fig. 2: Fluorescence-based detection of bacteria using cationic gold nanoparticles and anionic fluorescent conjugated polymers. In the presence of the bacterial anionic cell wall, there is displacement of the polymer leading to fluorescence emission. Discrete fluorescence emission patterns corresponding to different microorganisms can be obtained in a high throughput format. (Phillips et al., 2008)

References

Alivisatos, A.P., Gu, W. and C. Larabell. 2005. Quantum dots as cellular probes. *Biomed. Eng.* **7**:55-76.

Azzazy, H.M., M.M.Mansour and S.C. Kazmierczak. 2006. Nanodiagnostics: a new frontier forclinical laboratory medicine.

Azzazy, H.M., M.M.Mansour and S.C. Kazmierczak. 2007. From diagnostics to therapy: prospects of quantum dots.

Baptista, P.V., M. Koziol-Montewka, J. Paluch - Oles,G. Doria and R. Franco. 2006. Gold-nanoparticle- probe based assay for rapid and direct detection of Mycobacterium tuberculosis DNA in clinical samples, *Clin. Chem.* **52:**1433–1434.

Boonham, N., Tomlinson, J., and Mumford, R., 2007. Micro arrays for rapid detection of plant viruses. *Annu. Rev. Phytopathol.* **45:**307–328.

Bruchez Jr, M., M. Moronne, P. Gin, S. Weiss, A.P and Alivisatos. 1998. Semiconductor nanocrystals as fluorescent biological labels,

Byzova. N.A., I.V.Safenkova., S.N.Chirkov., A.V.Zherdev., A.N.Blintsov., B.B.Dzantiev and I.G.Atabekov. 2009. Development of Immunochoromatographic test systems for express detection of plant viruses. *Appl.Biochem. and Microbiol.* **45:**204-209.

Chan, W.C.W and S.Nie. 1998. Quantum Dot Bioconjugates for Ultrasensitive Nonisotopic Detection. *Science* **281**(5385), 2016-201.

Edgar, R., M. McKinstr y, J. Hwang, A.B. Oppenheim, R.A. Fekete, G. Giulian, C. Merril, K. Nagashima and S. Adhya. 2006. High sensitivity bacterial detection using biotin tagged phage and quantum-dot nanocomplexes. *Proc. Natl. Acad. Sci. U. S. A.* **103**:4841–4845.

Garg.J, B.Poudel, M.Chiesa J.B.Gordon., J.J.Ma., J.B.Wang., Z.F.Ren., Y.T.Kang., H.Ohtani., J.Nanda., G.H.McKinley and G.Chen. 2008. Enhanced thermal conductivity andviscosity of copper nanoparticles in ethylene glycol nanofluid. *J Appl Phys.***103**:074301.

Geertruda.A, Posthuma-Trumpie, Jakob Korf and Aart van Amerongen. 2009 Lateral flow (immuno) assay: its strengths, weaknesses, opportunities and threats. A literature survey. *Anal Bioanal Chem.*, **393**:569–582.

Hadidi, A., Czosnek, H. and Barba, M., J. 2004. DNA microarrays and their potential applications for the detection of plant viruses, viroids, and phytoplasmas. *J.Plant Pathol.*, **86**:97–104.

Jain, K.K. 2005. Nanotechnology in clinical laboratory diagnostics. *Clin Chim Acta,* **358**:37–54.

Liu, W.T. 2006. Nanoparticles and their biological and environmental applications. *J Biosci Bioeng.,* **102**:1– 7.

McNeil, S.E. 2005. Nanotechnology for the biologist. *J. Leukoc Biol:***78**:585–94.

Mirkin, C.A., R.L. Letsinger, R.C. Mucic and J.J. Storhoff. 1996. ADNA-based method forrationally assembling nanoparticles into macroscopic materials. *Nature,* **382**:607–609

Nath, S., C. Kaittanis, A. Tinkham and J.M. Perez. 2008. Dextran coated gold nanoparticles for the assessment of antimicrobial susceptibility. *Anal. Chem.* **80**: 1033–1038.

Narayanasamy, P. 2010. Pathogens-detection and Disease diagnostics. Bacterail and Phytoplasma pathogens. Vol 2; Spinger Publications, ISBN 978-90- 481-9768-2. DOI; 10.1007/978-90-481-9769-2.

27

Detection of Aflatoxins in Crop Produce using Nanotechnology

R. Velazhahan

Introduction

Spoilage in agricultural products due to mould growth can occur at various stages of production and storage and is significant in terms of trade economics, food safety and public health (Shephard, 2008). It has been reported that 5–10 % of agricultural products in the world are spoiled by mould contamination to the extent that they cannot be consumed by humans or animals. Mould growth decreases the quality of food and also creates a potential risk for human health because of the production of toxic secondary metabolites known as "mycotoxins". Foods contaminated by mycotoxins, when consumed by humans and animals causes a disease called "mycotoxicosis", resulting in death (Wagacha and Muthomi, 2008). Mould species belonging to genus Aspergillus, Penicillum, Fusarium and Alternaria produce most of the mycotoxins (Kabak et al., 2006). The most important groups of mycotoxins that contaminate foods are: aflatoxins (produced by *Aspergillus flavus* and *A. parasiticus*), deoxynivalenol (DON, a trichothecene mycotoxin) (produced primarily by Fusarium graminearum), fumonisins (produced prim arily by *Fusarium verticillioides* and *F. proliferatum*) and Zearalenone, produced primarily by *F. graminearum* (Vinod Kumar et al., 2008). Each of these toxins is associated with a specific set of health detriments that have been documented in domestic animals and humans. Mycotoxins are generally very stable, resistant against temperature, storage and processing conditions. International risk assessments performed by Joint FAO/WHO Expert

Committee on Food Additives (JECFA) for aflatoxin B1, aflatoxin M1, DON, fumonisins, ochratoxin A, T-2 toxin and HT-2 toxin indicate that health risks from mycotoxins are generally orders of magnitude lower in developed countries than that for populations from developing regions. In developing countries it is likely that consumers will be confronted with a diet that contains a low level of toxin, and in many cases, there may be other mycotoxins present. For example, aflatoxins, fumonisins, DON and zearalenone may occur together in the same grain since many fungi produce several mycotoxins simultaneously, especially *Fusarium* species. Co-occurrence of mycotoxins is of special concern, for instance, in the case of fumonisins and aflatoxin where a complimentary toxicity mechanism of action occurs (Bryden, 2007).

Aflatoxins

The sudden appearance of "Turkey-X" disease in 1960, which resulted in the death of over 1,00,000 turkeys in the United Kingdom, led to the discovery of a family of structurally related metabolites called "aflatoxins" (Blount, 1961; Vander Zijden, 1962; Hartley et al., 1963). Aflatoxins are a group of closely related heterocyclic compounds produced predominantly by two filamentous fungi, *Aspergillus flavus* and *A. parasiticus* when they grow on groundnut, maize, cotton, chilli and many other agricultural commodities. Foods and feeds, especially in warm climates, are susceptible to invasion by aflatoxigenic *Aspergillus species* and subsequent production of aflatoxins during pre-harvesting, transportation or storage (Vijayasamundeeswari et al., 2009). Several reports have indicated that some strains of *A. nominus* and *A. tamarii* are also producing aflatoxin, of which *A. nominus* is phenotypically similar to A. flavus (Kurtzman et al., 1987; Goto et al., 1996). Ito et al. (2001) isolated a strain of A. pseudotamarii that can produce aflatoxin. Aflatoxins are named after their fluorescence as blue or green under UV light and other analytical characteristics. Aflatoxin metabolites found in mammalian milk are named "AFM," where "M" denotes milk or mammalian metabolites. There are two broad categories of aflatoxins according to their structures. Aflatoxins $B_{1,2}$ (AFB_1, AFB_2) and aflatoxins parent compound, AFM1 has hepatotoxic and carcinogenic effects (Van Egmond, 1989). This toxin, initially classified as a Group 2B agent, has now been reclassified as Group 1 by the International Agency for the Research on Cancer (IARC). AFM1 is relatively stable during the pasteurisation, storage and preparation of various dairy products and therefore, AFM1 contamination poses a significant threat to human health, especially to children, who are the major consumers of milk. The legal regulations concerning AFM1 levels in milk and dairy products vary from country to country. EU regulations allow a maximum $M_{1,2}$ (AFM_1, AFM_2) are within the difuro- level of 0.05 μg L^{-1} (ppb) AFM1 in milk. coumarocyclopentenone series. Aflatoxins $G_{1,2}$ (AFG_1, AFG_2) are of the difuro-coumarolactone series. Presently, 18 different types of aflatoxins have been identified, with aflatoxin B_1, B_2, G_1, G_2, M_1, and M_2 being the most common. Of these, AFB_1 and AFG_1 occur most frequently in the agricultural commodities, with AFB_1 being the most potent (Mishra and Das, 2003). Among the 400 known mycotoxins, aflatoxins are the most dangerous to human health because of their highly toxic nature, hepatocarcinogenic (capable of causing liver cancer), mutagenic (capable of causing mutation) and teratogenic (capable of causing deformities in developing embryos) (Cullen and Newberne, 1994) and have been classified as class I human carcinogens by the International Agency for Research on Cancer (Williams et al., 2004).

Human exposure to AFB1 can arise from direct consumption of contaminated commodities and the milk of farm animals previously exposed to AFB1 in their feed (Concon, 1988). When aflatoxin B_1 (AFB1) is ingested by cows through contaminated feed, it is transformed into aflatoxin M_1 (AFM_1) through enzymatic hydroxylation of AFB_1 at the 9a-position and has an approximate overall conversion rate equal to 0.3 to 6.2% (Applebaum et al., 1982). AFM1 is secreted in milk by the mammary gland of dairy cows (Cathey et al., 1994). This transmission rate was shown to vary from animal to animal, from day to day, and from one milking to the next. Even though it is less toxic than its Aflatoxin contaminated diet has been linked with the high incidence of liver cancer (Bababunmi et al., 1978). The risk posed by aflatoxin depends on the level and type of aflatoxin in diet, the strain of animal and its nutritional status. Aflatoxins are detrimental to human health and their role in hepatocarcinogenesis often in conjunction with hepatitis B is well established (Wild and Hall, 1998). There is some evidence for associations with Reye's syndrome, Kwashiorkor and acute hepatitis (Wild and Hall, 1996). Aflatoxin exposure early in life has been associated with impaired growth, particularly stunting (Gong et al., 2002). Aflatoxins cause a variety of effects in poultry, including poor performance, liver damage, immunosuppression and changes in relative organ weights (Kubena et al., 1990). These compounds owe their toxicity to their ability to form irreversible adducts with nucleic acids with the concomitant inhibition of DNA replication and DNA-dependent transcription (Roebuck and Maxuitenko, 1994). AFB_1 has been shown to produce G-T transversions at codon 249 of the p53 tumor suppressor gene, whose altered sequence has been associated with a number of human cancers (Eaton and Gallagher, 1994). Due to concern for the potential effects of aflatoxins on human health even low level of contamination is important and hence most countries have extensive monitoring programmes and legislation that restrict marketing of aflatoxin contaminated grains and products (Van Egmond, 1989). The United States Food and Drug Administration (FDA) has set an aflatoxin tolerance limit of 20 ppb for foods and for most feeds and feed ingredients. The European Union has enacted a very stringent aflatoxin tolerance threshold of 2 μg/kg AFB_1 and 4 μg/kg total aflatoxins for nuts and cereals for human consumption (Bankole and Adebanjo, 2003).

The economic impacts attributed to aflatoxin are incurred directly by loss in crops, livestock, dairy and indirectly by a recurring expenditure in quality control programmes, research, education and lower foreign of other aflatoxins viz., AFB1 (61%), AFB2 (54%) and AFG2 (46%) by the dialyzed T. ammi extract was also reported. Mass spectral analysis of the degradation products of AFG1 suggests the modification of lactone ring structure. Variability in aflatoxin production potential of *A. flavus* isolates have been reported (Karthikeyan et al. 2009). The variability in aflatoxin production of *A. flavus* isolates might be due to their genetic makeup. Mohankumar et al. (2010) while evaluating isolates of Aspergillus flavus from maize collected from different agro-ecological zones of Tamil exchange earnings, increased storage and packaging costs of vulnerable commodities. The main approaches of Tamil Nadu, India for their ability to produce aflatoxin B_1 (AFB_1) for pre-harvest prevention of mycotoxin formation include good agricultural practices, such as crop rotation, time of irrigation, planting and harvesting, breeding for resistance to toxigenic fungi, genetically modified crops resistant to insect penetration, and competitive exclusion by using of non-toxigenic strains in the field. Prevention through pre-harvest and harvest management is the best method for controlling mycotoxin contamination in agricultural commodities, however if contamination still occurs, post-harvest decontamination/detoxification procedures can be used in order to remove or reduce the amount of toxin in agricultural products contaminated with unacceptable levels of mycotoxins (Chu, 1992). Sandosskumar et al. (2007) identified a medicinal plant, Zimmu (*Allium sativum* L. x *Allium cepa* L.) capable of detoxifying aflatoxin B1. Vijayasamundeeswari et al . (2010) demonstrated that intercropping of the medicinal plant zimmu (*Allium sativum* L. x *Allium cepa* L.) and seed and soil application of antagonistic bacterium Burkholderia sp. strain TNAU-1 significantly controlled *A. flavus* infection and aflatoxin B1 contamination in groundnut under field conditions. Velazhahan et al (2010) reported that the seed extract of Ajowan (Trachyspermum ammi) showed degradation of AFG1 up to 65%. The dialyzed T. ammi extract was more effective than the crude extract, capable of degrading >90% of the toxin. The aflatoxin detoxifying activity of the T. ammi extract was drastically reduced upon boiling at 100 °C for 10 min. Significant levels of degradation *in vitro* and their molecular variability by using restriction fragment length polymorphism (RFLP) analysis of the PCR-amplified internal transcribed spacer (ITS) regions of ribosomal DNA demonstrated that no association existed between clustering of A. flavus isolates based on ITS- RFLP and the production of aflatoxin.

Fumonisins

Fumonisins are a group of mycotoxins produced mainly by Fusarium vert icillioides and Fusarium proliferatum. They frequently contaminate corn and corn- based products and when ingested with food or feed cause several severe diseases in humans (Sydenham et al., 1990) and animals (Smith et al., 1996). To date, six different fumonisins (FA1, FA2, FBI, FB2, FB3, and FB4) have been described. Among them Fumonisin B1 (FB1) is the most prevalent fumonisin and holds the highest risk for human and animal nutrition. FB1 was shown to be carcinogenic (Lemmer et al., 1999), teratogenic (Marasas et al., 2004) and is suggested to be linked with the etiology of esophageal cancer (Voss et al., 2002) and neural tube defects in humans (Sydenham et al., 1990). FB1 is associated with a number of mycotoxicoses in animals, e.g. liver cancer in rats (Lemmer et al., 1999), leukoencephalomalacia in horse (Marasas et al., 1988b) and pulmonary edema in swine (Harrison et al., 1990). In swine, fumonisins in the feed can cause reduced average daily weight gain and immunosuppression. (Rotter et al., 1996). A carry- over of fumonisins from feed to meat has been reported in beef cattle (Smith and Thakur, 1996). FB1, the most abundant of the fumonisin analogues, has been classified by the International Agency for Research on Cancer (IARC) in Group 2B as a possible carcinogen to humans (IARC, 2002). The recommended maximum level of fumonisins in human foods is 2–4 mg/kg according to the particular corn-based product, while in animal feeds it is from 5–100 mg/kg depending on the animals that the feed is intended for (FDA, 2001). The European Union also regulated fumonisins (B1+B2): the maximum levels are 1 mg/kg for maize-based foods and 4 mg/kg for unprocessed maize. Heinl et al. (2010) isolated two genes that are responsible for degradation of fumonisin B1 from the bacterium Sphingopyxis sp. MTA144. The first gene encodes a protein which shows similarity to carboxylesterases, type B. The second gene encodes a polypept ide homologous to aminotransferases, class III. The two genes expressed heterologously in P. pastoris and purified. The effect of the recombinant enzymes on fumonisin B1 and hydrolyzed fumonisin B1 was determined. The recombinant carboxylesterase was shown to catalyze the deesterification of fumonisin B1 to hydrolyzed fumonisin B 1. The heterologously expressed aminotransferase was shown to deaminate hydrolyzed fumonisin B1 in the presence of pyruvate and pyridoxal phosphate. Azcona-Olivera et al. (1992) developed antibodies for detection of fumonisins using Cholera Toxin as the Carrier-Adjuvant. The antiserum cross- reacted with fumonisins B2 and B3 but not with the hydrolyzed backbone of fumonisin B, and tricarballylic acid and the detection limit for fumonisin B1in the ELISA was 100 ng/ml.

Zearalenone

Zearalenone (ZEN) is a non-steroidal estrogenic mycotoxin that is produced by numerous Fusarium species in pre- or post-harvest cereals. Fusarim graminearum (Gibberella zeae) and F.culmorum are the major zearalenone-producing species and are distributed worldwide. Zearalenone is insoluble in water and heat-stable, and it persists in both human foods and animal feeds prepared from contaminated grains. Tinyiro et al. (2011) recently reported that Bacillus subtilis 168 and Bacillus natto scavenged 95 and 78% ZEN, respectively, after incubation at 30° C for 48 h and suggested that these two Bacillus strains could be exploited for ZEN decontamination of food and feeds.

Deoxynivalenol

In addition to zearalenone, Fusarium species can produce deoxynivalenol, nivalenol, diacetoxyscirpenol, and other trichothecenes. Deoxynivalenol (DON) is a trichothecene secondary metabolite produced by Fusarium species infecting cereal crops. Deoxynivalenol is stable, survives processing (milling), and does occur in food products and feeds prepared from contaminated corn and wheat. The most common producer of DON is Fusarium graminearum (Marasas et al., 1984). In addition, F. culmorum has been demonstrated to produce significant levels of DON in laboratory cultures.

Biosensors for detection of mycotoxins

Analysis of mycotoxins can be accomplished by many techniques that range from determinative tests in which the presence of the toxin is confirmed, to presumptive tests in which the presence of the toxin is inferred from the presence of markers (Maragos, 2009). Aflatoxins, zearalenone, deoxynivalenol, fumonisins, and their respective metabolites require specific procedures for their determination because of their diverse chemistry and occurrence in complex matrices of food and feedstuffs. Major sources of error in the analysis of these mycotoxins arise from inadequate sampling and inefficient extraction and cleanup procedures. The determinative step in the assay for each of these toxins is sensitive to levels below those that are considered detrimental to humans and animals.

Because of the occurrence of mycotoxins in wide range of agricultural commodities and their wide range of toxicological effects, it is essential to monitor human and animal exposure via foods. Analytical methods such as thin-layer chromatography, Enzyme-Linked Immunosorbent Assay (ELISA), gas chromatography, high-performance liquid chromatography (HPLC), gas chromatography-mass spectroscopy (GC-MS) have been widely used for detection of mycotoxins. HPLC and ELISA methods require skilled personnel, expensive equipments, and complex sample preparations and cleanup procedures for detection of mycotoxins. Due to public health concerns, there is a current need in the food industry for a sensitive, specific and rapid method to monitor the level of mycotoxins in a variety of foods. van der Gaag et al. (2003) developed a biosensor for simultaneous detection of Aflatoxin B1, zearalenone, ochratoxin A, DON and fumunosin B1 based on the principle of surface plasmon resonance (SPR). Ngundi et al. (2005) developed a rapid and highly sensitive array biosensor for the detection and quantitation of ochratoxin A (OTA) in a variety of food and beverage samples. The array biosensor utilizes a competitive immunoassay format in which immobilized OTA derivatives compete with toxin in solution for binding to fluorescent anti-OTA antibody spiked into the sample. This competition was quantified by measuring the formation of the fluorescent immunocomplex on the waveguide surface. The fluorescent signal is inversely proportional to the concentration of OTA in the sample. The limit of detection for OTA in several cereals ranged from 3.8 to 100 ng/g, while in coffee and wine, detection limits were 7 and 38 ng/g, respectively. Sapsford et al. (2006) developed an Array Biosensor to measure both large pathogens, such as the bacteria Campylobacter jejuni (*C. jejuni*), and small toxins, including the mycotoxins ochratoxin A, fumonisin B, aflatoxin B1 and deoxynivalenol. Sandwich immunoassays were used to measure C. jejuni in buffer and a number of food matrices, while competitive immunoassays, taking only 15Dmin, were developed for the simultaneous detection of multiple mycotoxins. The combination of sandwich and competitive immunoassay formats on a single substrate was demonstrated, allowing the simultaneous detection of both large (*C. jejuni*) and small (aflatoxin B1) food contaminants. Yao et al (2006) developed an electrochemical biosensor f or sterigmatocystin based on the immobilisation of aflatoxin detoxifizyme on multi-walled carbon nanotubes. Wide working ranges and low limits of detection have been favoured by the high specific surface area available for enzymatic immobilisation. As an alternative, taking advantage of the effect of some mycotoxins on certain enzymes, enzymatic inhibition has also been used for biosensor development. Several studies revealed that the enzyme acetylcholinesterase (AChE) can be inhibited by aflatoxins. Based on this principle, an enzymatic biosensor for the assessment of aflatoxin B_1 (AFB1) in olive oil was developed (Ben Rejeb et al., 2009). In this case, the sample pretreatment step is very important to avoid matrix effects mainly caused by phenolic compounds and pesticides, also able to inhibit the enzyme. Kaushik et al. (2009) developed Cerium oxide-chitosan based nanobiocomposite for food borne mycotoxin detection. Cerium oxide nanoparticles (Nano CeO_2) and chitosan (CH) based nanobiocomposite film deposited onto indium-tin-oxide coated glass substrate has been used to coimmobilize rabbit immunoglobin (r-IgGs) and bovine serum albumin (BSA) for detection of ochratoxin- A. Välimaa et al. (2010)

developed a method for detecting zearalenone and its metabolites α-zearalanol, β-zearalanol, α-zearalenol and β-zearalenol in milk utilizing bioluminescent whole-cell biosensors. Ko et al. (2010) constructed a novel fusion protein by genetically fusing gold binding polypeptides (GBP) to protein A (ProA) and surface plasmon resonance (SPR) immunosensor was fabricated by using this fusion protein as a crosslinker for effective immobilization of antibodies for detection of AFB1. Consequently, a low detection limit (10 ng/ml) has been achieved for detection of mycotoxin. The results indicated that the GBP-ProA protein could be a valuable crosslinker for simple and oriented immobilization of antibodies onto SPR gold surfaces, and this SPR immunosensor could be a useful analytical tool for rapid and real-time detection of AFB1. Paniel et al. (2010) developed an electrochemical immunosensor for the detection of ultra-trace amounts of aflatoxin M1 (AFM1) in food products. The sensor was based on a competitive immunoassay using horseradish peroxidase (HRP) as a tag. The developed immunosensor had a low detection limit (0.01 ppb) and had good reproducibility.

Conclusions

Mycotoxins are a food safety risk worldwide. The most commonly encountered mycotoxins in foods and feeds are aflatoxins, zearalenone, deoxynivalenol and fumonisins. Mycotoxins can be assayed in a variety of matrices, but the methods of analysis are dependent on the matrix and the mycotoxin in question. It is important to ensure that a representative sample of the matrix is taken for analysis, especially if quantification is important. Qualitative analysis may be accomplished with a variety of immunologically based tests using limited laboratory facilities, but most quantitative tests employ sophisticated laboratory equipments and methods such as high performance liquid chromatography and gas chromatography. Current methods of analysis are capable of detecting mycotoxins in food and feedstuffs far below those concentrations considered to be harmful to animals and humans. Biosensors merit special mention due to their sensitivity, accuracy, cost-effectiveness and simplicity. Furthermore, biosensor arrays offer additional advantages, such as the possibility to measure multiple samples and provide multimycotoxin profiles in one assay. Biosensors and arrays for mycotoxins are promising biotechnological tools for mycotoxin detection in foods and feeds (Prieto-Simon and Campas, 2009).

References

Applebaum, R.S., Brachett, R.E., Wiseman, D.W. and Marth, E.H. 1982. Aflatoxin: Toxicity to dairy cattle and occurence in milk and milk products: A review. *J. Food Protection,* **45**: 752.

Azcona-Olivera, J.I., Abouzied, M.M., Plattner, R.D., Norred, W.P. and Pestka, J.J. 1992. Generation of Antibodies reactive with Fumonisins B1, B2, and B3 by Using Cholera Toxin as the Carrier-Adjuvant. *Appl. Environ. Microbiol.,* **58**: 169-173.

Bababunmi, E. A., Uwaifo, A. O. and Bassir, O. 1978. Hepatocarcinogens in Nigerian foodstuffs. *World Rev. Nut. Dietet.,* **28**: 188-209.

Bankole, S. A. and Adebanjo, A. 2003. Mycotoxins in food in West Africa: current situation and possibilities of controlling it. *Afr. J Biotechnol.*, **2**: 254-263.

Ben Rejeb, F., Arduini, A., Arvinte, A., Amine, M., Gargouri, L., Micheli, C., Bala, D., Moscone and Palleschi, G. 2009. Biosens. *Bioelectron*., **24**: 1962-1968.

Blount, W. P. 1961. Turkey "X" disease. *J. Brit. Turkey,* **9:**52-54.

Bryden, W.L. 2007. Mycotoxins in the food chain: human health Implications. Asia Pac. J. Clin. Nutr., **16** (Suppl 1):95-101.

Cathey, C.G., Huang, G., Sarr, A.B., Clement, B.A. and Phillips, T.D. 1994. Development and evaluation of a minicolumn assay for the detection of aflatoxin MI in milk. *J. Dairy Sci.,* **77**: 1223-1231.

Chu, F. S. 1992. Recent progress on analytical techniques for mycotoxins in feed stuffs. *J. Anim. Sci.,* **70**: 39-50.

Concon, J. M. 1988. Contaminants and additives. Food Toxicology, Part-B. New York: Marcel Dekker Inc. pp. 667–743.

Cullen, J. M., and Newberne, P. M. 1994. Acute hepatoxicity of aflatoxins. Pages: 3-26 in: The Toxicology of Aflatoxins: human health, veterinary, and agricultural significance. Eaton, D. L., Groopman, J. D. eds. Academic Press, San Diego.

Eaton, D. L., and Gallagher, E. P. 1994. Mechanism of aflatoxin carcinogenesis. *Annu. Rev. Pharmacol. Toxicol.*, **34**: 135-172.

FDA (Food and Drug Administration). 2001. Fumonisin levels in human foods and animal feeds. Available from http://www.cfsan.fda.gov/~dms/ fumongu2.htm.

Gong, Y. Y., Cardwell, K., Hounsa, A., Egal, S., Turner., Hall, A. J., and Wild, C. P. 2002. Dietary aflatoxin exposure and impaired growth in young children from Benin and Togo: cross sectional study. *British Medical J.*, **325**:20-21.

Goto, T., Wicklaw, D. T., and Ito, Y. 1996. Aflatoxin and cyclopiazonic acid production by a sclerotium producing starin of A. tamarii. *Appl. Env. Microbiol.*, **62**:4036-4038.

Harrison, L., Colvin, B., Greene, J., Newman, L., Cole, J.J., 1990. Pulmonary edema and hydrothorax in swine produced by fumonisin B1, atoxic metabolite of Fusarium moniliforme. *J. Vet. Diagn. Invest.*, **2:** 217– 221.

Hartley, R. D., Nesbitt, B. F., and O'Kelly, J. 1963. Toxic metabolites of Aspergillus flavus. *Nature*, **198**:1056-1058.

Heinl, S., Hartinger, D., Thamhesl, M., Vekiru, E., Krska, R., Schatzmayr, G., Moll, W.D., Grabherr, R. 2010. Degradation of fumonisin B1 by the consecutive action of two bacterial enzymes. *J. Biotechnol.* **145**: 120-129.

International Agency for Research on Cancer. 2002. Monograph on the Evaluation of Carcinogenic Risk to Humans. Some Traditional Herbal Medicines, Some Mycotoxins, Naphtalene and Styrene; IARC: Lyon, France.

Ito Y., Peterson S. W., Wicklaw D. T., and Goto T. 2001. Aspergillus pseudotamarii, a new aflatoxin producing sp. in Aspergillus section Flavi. *Mycol. Res.,* **105**: 233-239.

Kabak, D.B., Alan, D.W. and Dobson, V. 2006. Strategies to prevent mycotoxin contamination of food and animal feed: A Review. *Critical Rev. Food Sci. and Nut.*, **46**: 593-596.

Kart hikeyan, M., Sandosskumar, R., Mathiyazhahan, S., Mohankumar, M., Valluvaparidasan, V., Sangit Kumar and Velazhahan, R. 2009. Genetic variability and aflatoxigenic potential of Aspergillus flavus isolates from maize. *Arch. Phytopathol Plant Protect,* **42**: 83-91.

Kaushik, A., Solanki, P.R., Pandey, M.K., Ahmad, S. and Malhotra, B.D. 2009. Cerium oxide-chitosan based nanobiocomposite for food borne mycotoxin detection. *Appl. Phys. Lett.,* **95**: 173703.

Sapsford, K.E., Ngundi, M.M., Moore, M.H., Lassman, M.E., Shriver-Lake, L.C., Taitt, C.R. and Ligler,

F.S. 2006. Rapid detection of foodborne contaminants using an Array Biosensor. Sensors and Actuators B: *Chemical,* **113**: 599-607.

Ko, S., Kim, A.R., Kim, C.J. and Kwon, D.Y. 2010. Detection of aflatoxin B1 by SPR biosensor using fusion proteins as a linker. *Nanotech,* **3**: 129-132.

Kubena, L. F., Harvey, R. B., Huff, W. E., Corrier, D. E., Phillips, T. D. and Rottinghaus, G. E. 1990. Efficacy of a hydrated sodium calcium aluminosilicate to reduce the toxicity ofaflatoxin and T-2 toxin. *Poultry Sci.,* **69**:1078- 1086.

Kurtzman, C. P., Horn, B. W., and Hesseltine, C. 1987. A. nominus, a new aflatoxin producing species related to A. flavus and A. tamarii . *Antonnie Van Leewenhoek,* **53**:147–158.

Lemmer, E., de la Motte Hall, P., Omori, N., Omori, M., Shephard, E., Gelderblom,W., Cruse, J., Barnard, R., Marasas, W., Kirsch, R., Thorgeirsson, S., 1999. Histopathology and gene expression changes in rat liver during feeding of fumonisin B1, a carcinogenic mycotoxin produced by Fusarium moniliforme. *Carcinogenesis,* **20:** 817–824.

Maragos, C.M. 2009. Biosensors for mycotoxin analysis: recent developments and future prospects. *World Mycotoxin Journal,* **2:** 221-238.

Marasas, W., Kellerman, T., Gelderblom, W., Coetzer, J., Thiel, P., van der Lugt, J., 1988. Leukoencephalomalacia in a horse induced by fumonisin B1 isolated from Fusarium moniliforme. *Onderstepoort J. Vet. Res.,* **55**: 197-203.

Marasas, W., Riley, R., Hendricks, K., Stevens, V., Sadler, T., Gelineau-van Waes, J., Missmer, S., Cabrera, J., Torres, O., Gelderblom, W., Allegood, J., Martínez, C., Maddox, J., Miller, J., Starr, L., Sullards, M., Roman, A., Voss, K., Wang, E., Merrill, A.J., 2004. Fumonisins disrupt sphingolipid metabolism, folate transport and neural tube development in embryo culture and in vivo: a potential risk factor for human neural tube defects among populations consuming fumonisin contaminated maize. *J. Nutr.,* **134**: 711–716.

Mishra, H. N. and Das, C. 2003. A review on biological control and metabolism of aflatoxin. *Crit. Rev. Food Sci. Nut.,* **43:** 245-264.

Mohankumar, M., Vijayasamundeeswari, A., Karthikeyan, M., Mathiyazhagan, S., Paranidharan, V. and Velazhahan, R. 2010. Analysis of molecular variability among isolates of Aspergillus flavus by PCR- RFLP of the ITS regions of rDNA. *J. Plant Protect. Res.,* **50**: 446-451.

Ngundi, M.M., Shriver-Lake, L.C., Martin H. Moore, Lassman, M.E., Ligler, F.S. and Taitt, C.R. 2005. Array Biosensor for Detection of Ochratoxin A in Cereals and Beverages. *Anal. Chem.,* **77**: 148-154.

Paniel, N., Radoi, A. and Marty, J. L. 2010. Development of an electrochemical biosensor for the detection of Aflatoxin M1 in Milk. *Sensors,* **10**: 9439- 9448.

Prieto-Simon, B. and Campas, M. 2009. Immunochemical tools for mycotoxin detection in food. *Monatsh Chem,* **140:** 915-920.

Roebuck, B. D., and Maxuitenko, Y. Y. 1994. Biochemical mechanisms and biological implications of the toxicity of aflatoxins as related to aflatoxin carcinogenesis. Pages 27–43 in: The Toxicology of Aflatoxins. Eaton, D. L., Groopman, J. D., eds. Academic Press. San Diego.

Rotter, B., Thompson, B., Prelusky, D., Trenholm, H., Stewart, B., Miller, J., Savard, M., 1996. Response of growing swine to dietary exposure to pure fumonisin B1 during an eight-week period: growth and clinical parameters. *Nat. Toxins,* **4:** 42-50.

Sandosskumar, R., Karthikeyan, M., Mathiyazhagan, S., Mohankumar, M., Chandrasekar, G. and Velazhahan, R. 2007. Inhibition of Aspergillus flavus growth and detoxification of aflatoxin B1 by the medicinal plant zimmu (Allium sativum L. x Allium cepa L.). *World J. Microbiol. Biotechnol.,* **23**: 1007-1014.

Shephard, G. S. 2008. Impact of mycotoxins on human health in developing countries. *Food. Add. And Cont.,* **25:**146-151.

Smith, G., Constable, P. and Haschek, W., 1996. Cardiovascular responses to short-term fumonisin exposure in swine. *Fundam. Appl. Toxicol.,* **33**: 140–148.

Smith, J. and Thakur, R., 1996. Occurrence and fate of fumonisins in beef. *Adv. Exp. Med. Biol.,* **392:** 39-55.

Sydenham, E.W., Thiel, P.G., Marasas, W.F.O., Shephard, G.S., Van Schalkwyk, D.J. and Koch, K.R., 1990. Natural occurrence of some Fusarium mycotoxins in corn from low and high esophageal cancer prevalence areas of the Transkei, Southern *Africa. J. Agric. Food Chem.,* **38:** 1900-1903.

Tinyiro, S.E., Yao, W., Sun, X., Wokadala, C. and Wang, S. 2011. Scavenging of Zearalenone by Bacillus Strains-in vitro. *Research Journal of Microbiology,* **6:** 304-309.

Välimaa , A.L., Kivistö, A.T., Leskinen, P.I. and Karp, M.T. 2010. A novel biosensor for the detection of zearalenone family mycotoxins in milk. *J. Microbiol. Meth.,* **80**: 44-48.

Van der Gaag, B., Spath, S., Dietrich, H., Stigter, E., Boonzaaijer, G., van Osenbruggen, T. and Koopal, K. 2003. Biosensors and multiple mycotoxin analysis. *Food Control,* **14**:251-254.

Van Egmond, H.P. 1989. Current situation on the regulations for mycotoxins. Overview of tolerances and status of the standard methods of sampling and analysis. *Food Add. Contam,* **6:**139-188.

Van Egmond, H.P. 1989. Introduction to Mycotoxins in Dairy Products; Applied Science Publishers: London, UK, pp. 11-55.

Vander Zijden, A. S. M., Koelensmid, W.A.A.B., Bolding, J., Barett, C. B., Ord, O. W., and Philip, J. 1962. Isolation in crystalline form of a toxin responsible for Turkey X disease. *Nature,* **195:** 1060-1062.

Velazhahan, R., V ijayanandraj, S., Vijayasamundeeswari, A., Paranidharan, V., Samiyappan, R., Iw amoto, T., Friebe, B. and Muthukrishnan, S. 2010. Detoxification of aflatoxins by seed extracts of the medicinal plant, Trachyspermum ammi (L.) Sprague ex Turrill - structural analysis and biological toxicity of degradation product of aflatoxin G1. *Food Control* **21:** 719-725.

Vijayasamundeeswari, A., Mohankumar, M., Karthikeyan, M., Vijayanandraj, S., Paranidharan, V. and Velazhahan, R. 2009. Prevalence of aflatoxin B1 contamination in pre- and post-harvest maize kernels, food products, poultry and livestock feeds in Tamil Nadu, *India. J. Plant Protec. Res.,* **49**: 221-224.

Vijayasamundeeswari, A., Vijayanandraj, S., Paranidharan, V., Samiyappan, R. and Velazhahan, R. 2010. Integrated management of aflatoxin B1 contamination.

28

Nano Based Electrochemical Sensors for Detection of Pesticide Residues

S. Manisankar

Introduction

Fungicides are biocidal chemical compounds or biological organisms used to kill or inhibit fungi or fungal spores. An insecticide is a pesticide used against insects. Pyrethroids are widely used insecticides in agriculture as they control pests and diseases effectively in plants. Because of their low biodegradability and high persistence, they are extensively found in environmental samples such as air, water, soil, sediments, food and biological tissues [1-4]. Moreover, these chemicals induce cancer and act as endocrine disrupters in several organisms due to their high degree of toxicity. Although the use of most pyrethroids has been prohibited or limited in developed countries, they are used in many developing and under developed countries for controlling of pests and insecticides [5-8].

Fungicides and insecticides are determined in different kinds of samples such as soil [9], liver samples of birds [10], agrochemicals [11,12] and water samples [13,14]. High-performance liquid chromatography (HPLC) and gas chromatography [15] are commonly used. In HPLC, UV [9,14] or amperometric [15] detection is employed. Fluorimetric method [13] was also described. Differential pulse polarographic determination of pesticides was reported and applied for pesticide determination in commercial formulations [11]. Stripping analysis is an extremely sensitive electrochemical technique for measuring trace level organics [9]. In particularly stripping analysis provides a highly sensitive route to the measurement of many electroactive organic compounds [18-25]. Use of modified electrodes as means of selectively preconcentrating organic analytes is also attracting interest [26-28]. Nano size conducting polymeric layers coated electrodes exhibit preferential accumulation of analytes on bound surface functionalities, which may improve the selectivity and stability of the electrode. Present paper aims at studying the electrochemical behavior of some synthetic pyrethroids and a fungicide in aqueous-ethanolic medium using bare GCE and modified systems, and to develop an electroanalytical method for the quantitative trace determination of these in soil samples.

Three commonly employed insecticides and one fungicide have been selected for this investigation. Table 1 shows the list of insecticides and fungicide employed in the present study and table 2 shows their toxicity/carcinogenicity levels.

Experimental

Instrumentation

Cyclic and stripping voltammetric studies were performed using CHI 760C electrochemical workstation (CH Instruments, USA) coupled with a conventional three-electrode cell. This instrument uses the latest analog and microcomputer design to provide high performance, better precision and greater versatility in electrochemical measurements.

Table 1: Structure and physical properties of selected insecticides and fungicide

Compound	Structure	Class and physical nature
Cypermethrin (CYP)	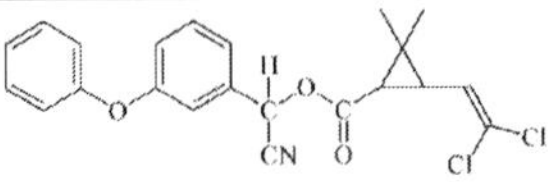	InsecticideYellow brown viscous semi solid M.P 60-80° C.
Deltamethrin (DEL)	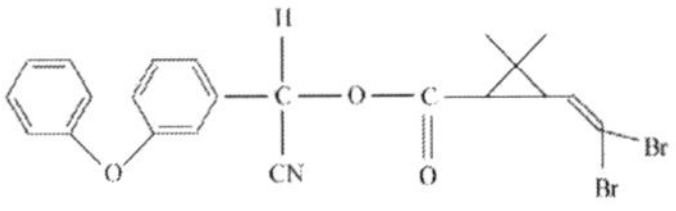	Insecticide, Crystalline, odourless, white powder M.pt. 98-101°C
Fenvalerate (FEN)	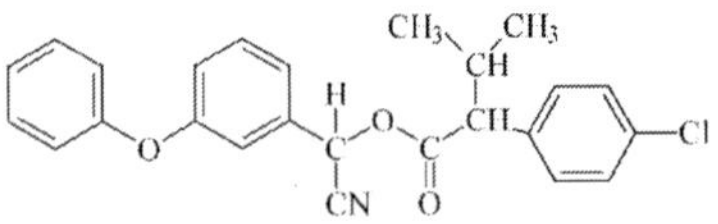	InsecticideClear viscous yellow liquid Vapor pressure 2.8x10^{-7} mm/Hg.
Carbendazim (CAR)	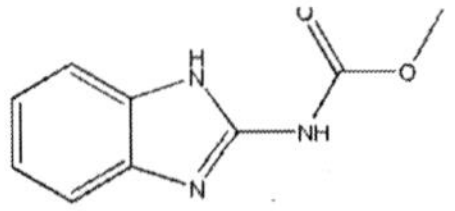	Benzamidazole fungicide Light gray powder M.pt. 302-307 °C (decomposes)

In order to identify a "most toxic" set of pesticides, Pesticide Action Network (PAN) and Californians for Pesticide Reform (CPR) created the term PAN

Table 2: Toxicity / carcinogenicity levels of Insecticides and Fungicide

Name of the selected Pesticide	PAN Bad Actor Chemical	Acute Toxicity	Carcinogen	Ground Water Contaminant	Developmental / Reproductive Toxin	Endocrine Disruptor
Cypermethrin	Not Yet Listed	Slight	Yes	Yes	Yes	Yes
Deltamethrin	Not Yet Listed	High	Yes	Yes	Yes	Yes
Fenvalerate	Not Yet Listed	High	Yes	Yes	Yes	Yes
Carbendazim	Not Yet Listed	Slight	Possible	Yes	Yes	Assesment in progress

Source: PAN (Pesticide Action Network) - Pesticides Database of US Environmental Protection Agency

Bad Actor pesticides. Pesticides with high acute toxicity, as designated by the World Health Organization (WHO), the U.S. EPA, or the U.S. National Toxicology Program.

Cell setup

The cell was made of glass, having a capacity of 10ml and the Teflon made cell top was comprised of three separate holes for the insertion of electrodes viz. working electrode, counter electrode and reference electrode. The cell top also has the purging and blanketing facilities of nitrogen gas with separate tubes to remove oxygen gas. This setup enables to maintain an inert atmosphere above the sample solution throughout the experiment.

Working electrode (GCE)

The pre-treated glassy carbon electrode (A=0.0707 cm^2) and various modified electrodes were employed as working electrodes for CV and stripping studies. Electrode modifications are carried out on glassy carbon electrode of area 0.0707 cm^2.

Modified electrodes

Fabrication of MWCNT/GCE

1 mg MWCNT was dispersed in 1 ml of 0.1 M sodium dodecyl sulphate using an ultrasonicator to give black suspensions. Cast films were prepared by placing 5 ?L of MWCNT/surfactant suspensions on GCE and then evaporating it in an oven at 50 °C.

Fabrication of PANI/MWCNT/GCE

Polyaniline was deposited on MWCNT/GCE by electrooxidation of 0.1 M aniline using 1 M sulphuric acid as a supporting electrolyte by applying a potential between -0.2 and 0.8 V.

Fabrication of PPY/MWCNT/GCE

Polypyrrole was deposited on MWCNT/GCE by the electrooxidation of 0.1 M pyrrole in acetonitrile medium containing 0.1 M lithium perchlorate as a supporting electrolyte by applying a potential between 0 and 1 V.

Fabrication of PEDOT/MWCNT/GCE

PEDOT was deposited by electrooxidation of 0.01 M EDOT in acetonitrile medium containing 0.1 M lithium perchlorate as a supporting electrolyte and cycling the potential between -0.2 V and 1.2 V.

Fabrication of PEDOT(CTAB) / MWCNT/ GCE AND PEDOT(SDS)/ MWCNT/ GCE

PEDOT was electrodeposited on MWCNT/GCE from 0.01 M EDOT in acetonitrile medium, in presence of 0.01 M CTAB or 0.01 M SDS and 0.1M lithium perchlorate supporting electrolyte and cycling the potential between -0.2 V and 1.2 V.

Fabrication of P3MT/MWCNT/GCE

P3MT was deposited by electrooxidation of 0.05 M 3-methyl thiophene in acetonitrile medium containing 0.05 M sodium perchlorate as a supporting electrolyte and cycling the potential between 0 V and 1.6 V.

Reference and Counter electrodes

Ag/AgCl reference electrode was used as reference electrode in this investigation. A platinum wire was used as counter electrode. Pt electrode was cleaned successively with detergent solution, isopropyl alcohol and sodium hydroxide solution. Finally it was rinsed with distilled water. All electrochemical experiments were performed at 25 ± 0.1°C.

Results and Discussions

Cyclic voltammetric studies of CYP

Figs. 1: 1(A) - 1(G) exhibit the cyclic voltammograms of CYP at plain GCE and all modified systems chosen. Effect of pH was studied and analyzing the peak characteristics, pH 13 was chosen for the electrochemical studies of CYP, since CV with higher peak current and lower reduction potential was obtained only at pH 13. In CV, it showed one well defined reduction peak at -1.448 V (25.00 μA), -1.445 V (56.30 μA), -1.399 V (61.82 μA), -1.403 V (74.59 μA), -1.39V 132.7 μA), -1.37 V (161.1μA), -1.38V (145.9μA) and -1.381 V (312.9 μA) on plain GCE, MWCNT/GCE, PANI/MWCNT/GCE, PPY/ MWCNT/GCE, PEDOT/MWCNT/GCE, PEDOT(SDS)/MWCNT/GCE, PEDOT(CTAB)/MWCNT/GCE and P3MT/MWCNT/GCE respectively at pH 13 and the scan rate of 0.1 Vs^{-1}.

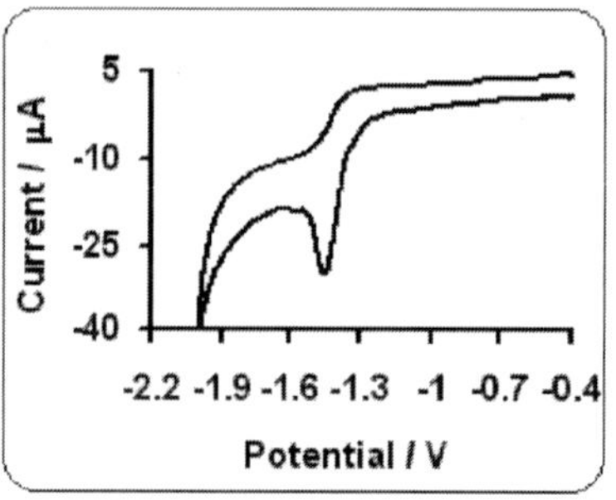

Fig. 1: CV of 1 mg L^{-1} CYP in pH 13 on bare GCE at scan rate of 0.1 V s^{-1}

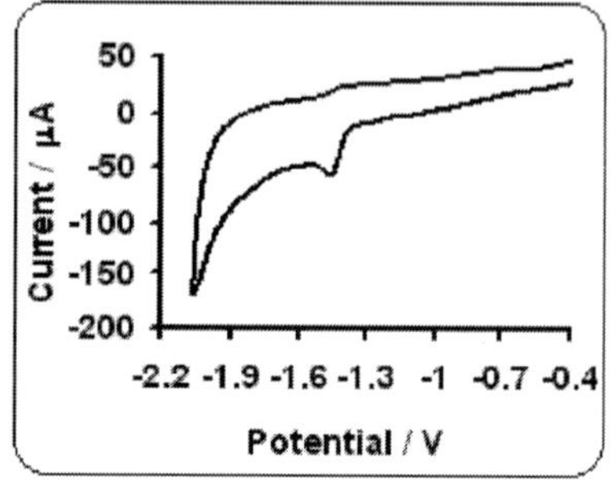

Fig. 1(A): CV of 1 mg L^{-1} CYP in pH 13 on MWCNT/GCE at scan rate of 0.1 V s^{-1}

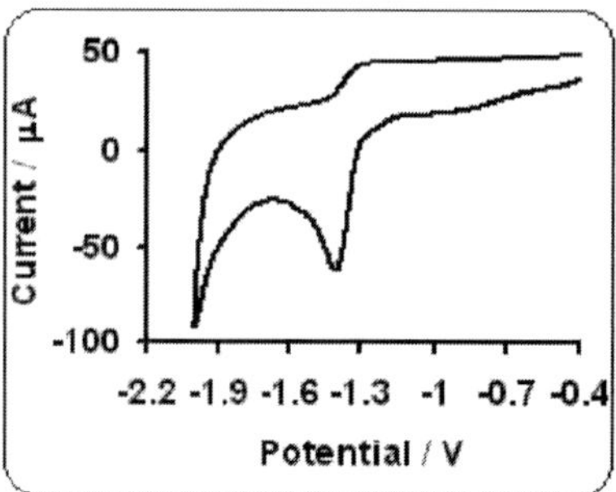

Fig. 1(B): CV of 1 mg L^{-1} CYP in pH 13 on PANI/MWCNT/GCE at scan rate of 0.1 Vs^{-1}

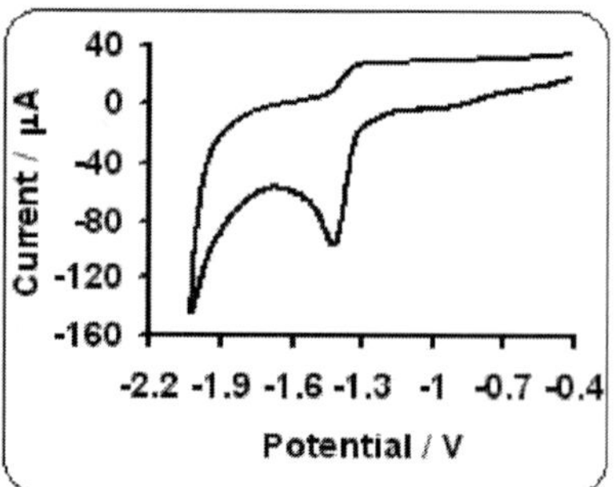

Fig. 1(C): CV of 1 mg L^{-1} CYP in pH 13 on PPY/MWCNT/GCE at scan rate of 0.1 V s^{-1}

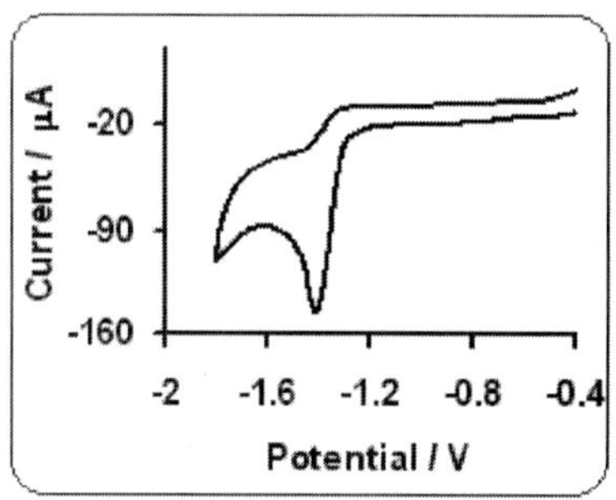

Fig. 1(D): CV of 0.99 mg L^{-1} CYP in pH 13 on PEDOT/MWCNT/GCE at scan rate of 0.1Vs^{-1}

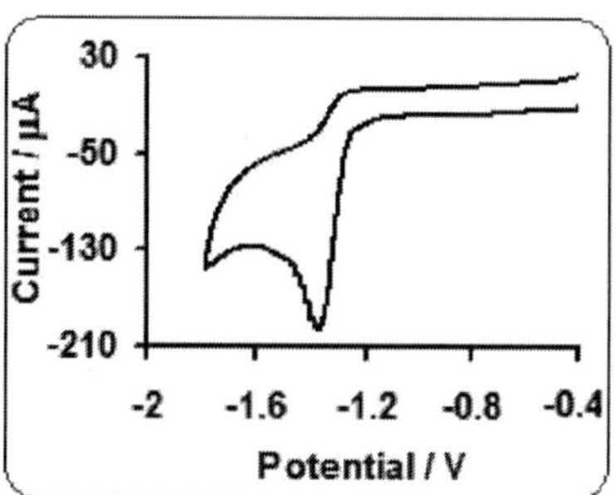

Fig. 1(E): CV of 0.99 mg L^{-1} CYP in pH 13 on PEDOT(SDS)/MWCNT/GCE at scan rate of 0.1 Vs^{-1}

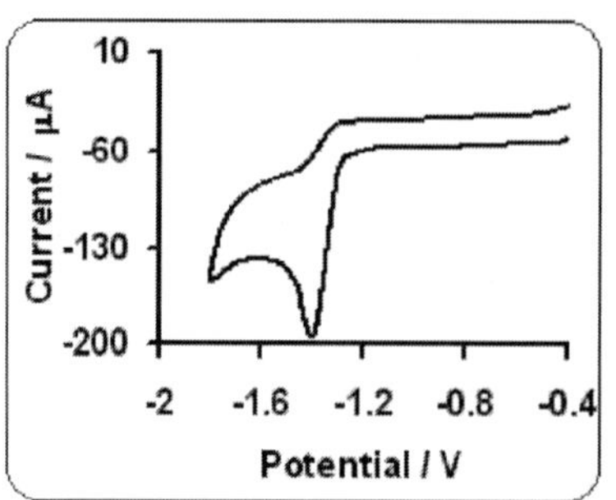

Fig. 1(F): CV of 1 mg L^{-1} CYP in pH 13 on PEDOT(CTAB)/MWCNT/GCE at scan rate of 0.1 V s^{-1}

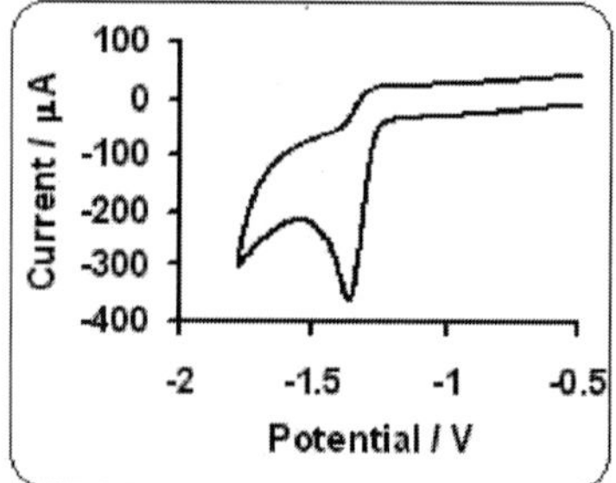

Fig. 1(G): CV of 1 mg L^{-1} CYP in pH 13 on P3MT/MWCNT/GCE at scan rate of 0.1 Vs^{-1}

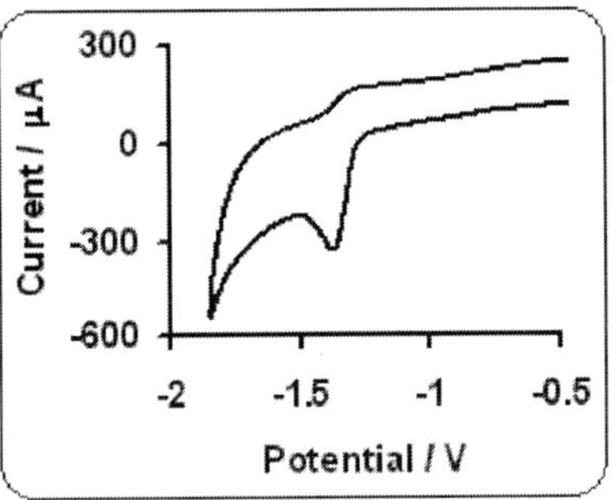

Fig. 2.1 (A): CV of 1 mg L^{-1} DEL in pH 13 on P3MT/MWCNT/GCE at scan rate of 0.1 V s^{-1}

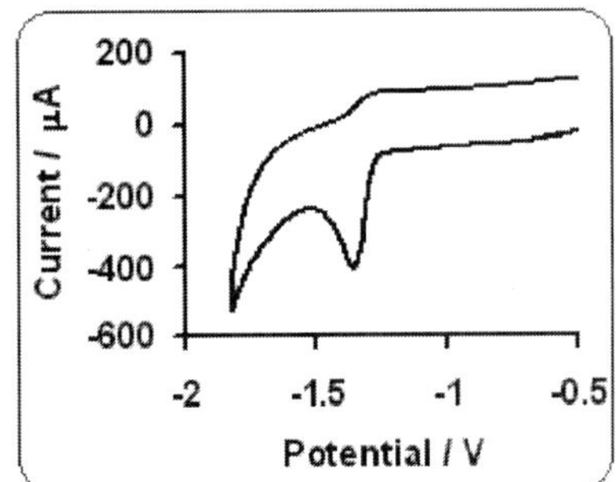

Fig. 2.2 (A) CV of 1mg L^{-1} FEN in pH 13 on P3MT/MWCNT/GCE at scan rate of 0.1 V s^{-1}

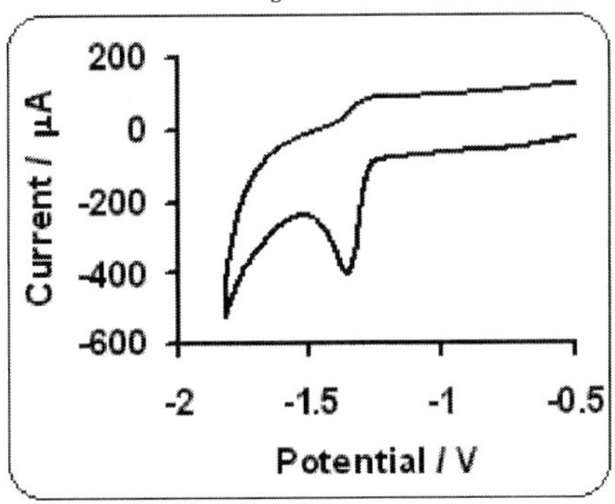

Fig. 2.3 (A) CV of 1mg L^{-1} FEN in pH 13 on P3MT/MWCNT/GCE at scan rate of 0.1 V s^{-1}

Adsorption studies

Adsorption of CYP/DEL/FEN/CAR was performed on best performing system. Herein adsorption studies were done with an accumulation potential of 0.1 V and an accumulation time of 10 s. To study the surface morphological changes during adsorption, SEM and XRD of adsorbed surfaces were made. Uniform adsorption of molecules in between the network of the modified system can be well understood from the Figs. 3.1 – 3.4. XRD pattern shows additional peaks due to the adsorption of insecticide/fungicide particles (Fig.3.5 – 3.8). These studies clearly illustrate the adsorption of CYP/DEL/FEN/CAR on the modified system.

Fig. 3.1: SEM micrograph of CYP adsorbed on P3MT/MWCNT/GCE modified surface

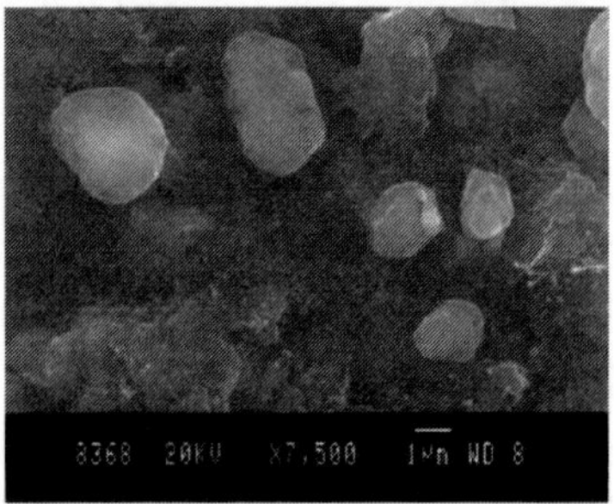

Fig. 3.2: SEM micrograph of DEL adsorbed on P3MT/MWCNT/GCE modified surface

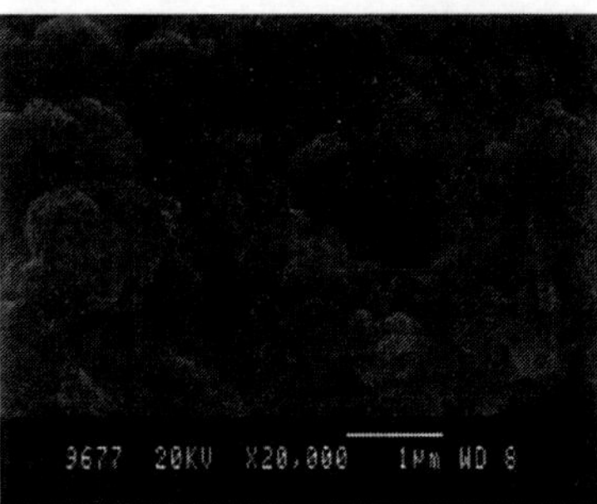

Fig. 3.3: SEM micrograph of FEN adsorbed on P3MT/MWCNT/GCE modified surfaces

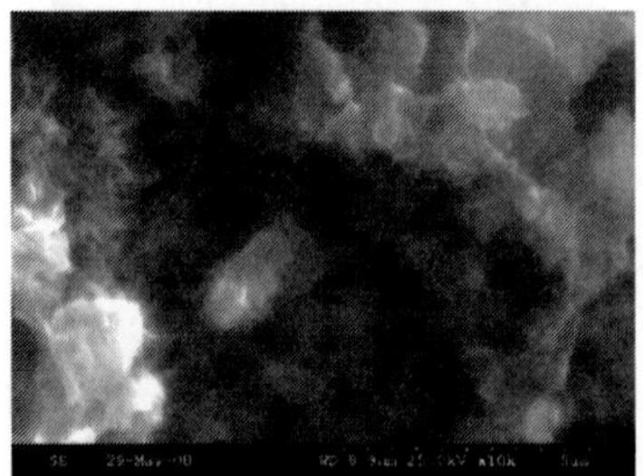

Fig. 3.4. SEM micrograph of CAR adsorbed on P3MT/MWCNT/GCE modified surfaces

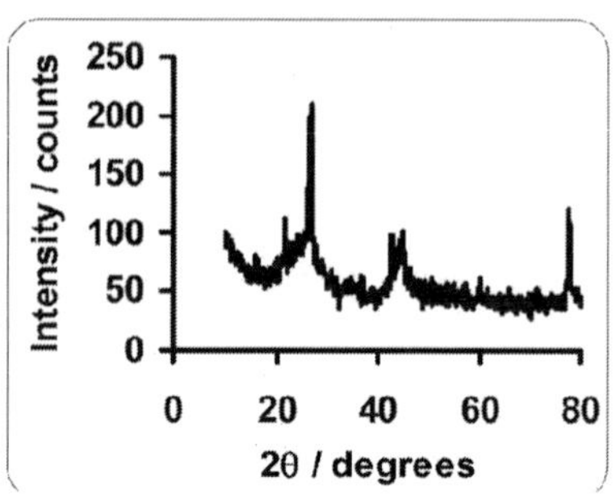

Fig. 3.5: XRD pattern of CYP adsorbed on P3MT/MWCNT/GCE modified surface

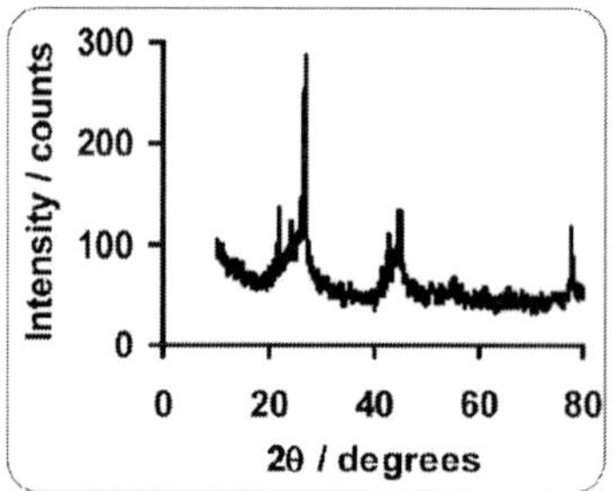

Fig. 3.6: XRD pattern of DEL adsorbed on P3MT/MWCNT/GCE modified surface

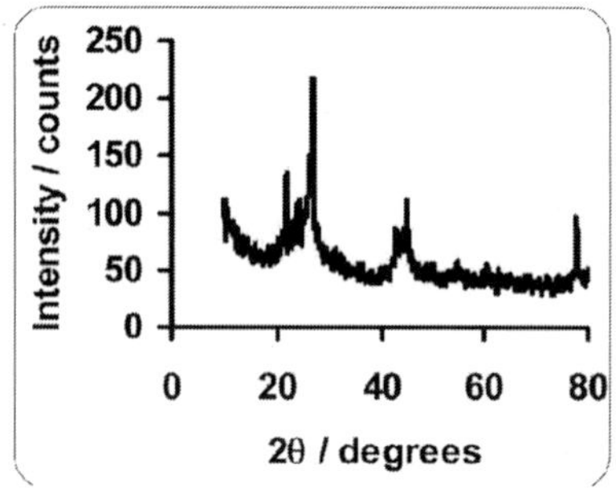

Fig. 3.7: XRD pattern of FEN adsorbed on P3MT/MWCNT/GCE modified surfaces

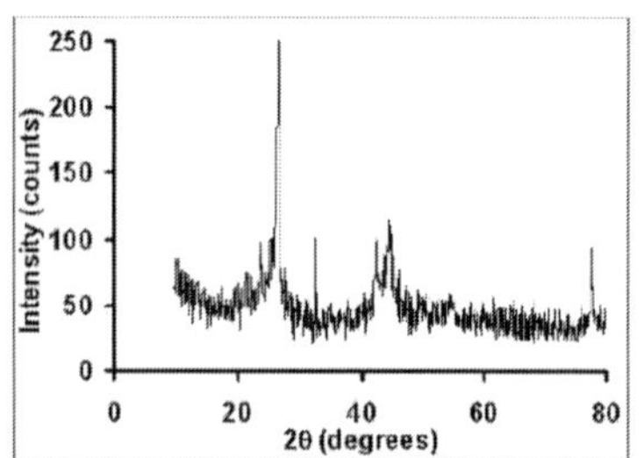

Fig.3.8: XRD pattern of CAR adsorbed on P3MT/MWCNT/GCE modified surfaces

Stripping studies

For the striping studies of CYP, differential pulse mode was preferred. Stripping parameters such as Initial scan potential (E_{is}), amplitude (PA) and pulse width (PW) are also influencing the current response. These parameters are interrelated and have a combined effect on the signal. E_{is} varied from -0.200 V to -1.0 V and the optimum values were found to be -0.800 V. An increase in peak current was identified, when pulse amplitude was changed from 0.025 to 0.100 V, hence amplitude of 0.100 was chosen. Pulse width was changed from 0.025 to 0.125s. Maximum peak current was observed at 0.05 s. Under these optimum experimental conditions, the effect of pesticide concentration on the stripping signal was studied. The representative differential pulse stripping voltammograms are given in Figs. 4(A) – 4(G).

Calibration plots were made and it shows linear dependence of peak current with concentration and it was found to be good from 0.01 to 100 g L^{-1} on P3MT/MWCNT/GCE modified system.

Similar studies were carried out for DEL/FEN/CAR and the range of determination and LOD values are presented in Table 3.

Reproducibility

Cyclic voltammetric studies were carried out for six separately prepared modified electrodes in pH 13. The reproducibility is fairly good with a standard error of about 2.5 %. Modified electrodes were prepared and stored in an atmosphere at room temperature for four weeks, after that CV's were recorded. Peak potential was unchanged and current signals showed only less than 2.2 % decrease of initial response. Also, the reproducibility of stripping signal was understood from relative standard deviation (1.8 to 2.9 %) calculated for six identical measurements at a concentration level of 25 □g L^{-1} in DPSV.

Residue analysis in soil sample

The suitability of this modified electrode was tested for the determination CYP in soil samples. Soil sample to be analyzed, was collected from a paddy field near by Karaikudi, Tamil Nadu, India and it was washed repeatedly with water and exposed to atmosphere. Approximately 50 g of sieved soil was spiked with 25 ml of 1 mg L^{-1} CYP solution and shaken well in a closed bottle for about 30 min and it is extracted using dichloromethane. This extract was filtered and evaporated to dryness by gentle heating on a water bath. This residue was transferred into 250 ml calibrated flask, dissolved in ethanol and made up to the mark. A 10 ml portion of this solution was transferred into a 50 ml calibrated flask and 0.1 M L^{-1} NaOH containing 50 % aqueous ethanol was used to dilute

Table 3: Determination ranges and detection limits of insecticides at various modified systems

S. No	Modified systems	CYP		DEL		FEN		CAR	
		A	B	A	B	A	B	A	B
1	MWCNT/GCE	2.5 to 100	2	10 to 100	5	1.5 to 100	1	-	-
2	PANI/MWCNT/GCE	1.5 to 100	0.8	1.5 to 100	1	1 to 100	0.5	-	-
3	PPY/MWCNT/GCE	0.8 to 100	0.35	1.5 to 100	0.9	0.5 to 100	0.1	-	-
4	PEDOT/MWCNT/GCE	0.15 to 100	0.10	0.3 to 100	0.21	0.25 to 100	0.18	-	-
5	PEDOT(SDS)/MWCNT/GCE	0.15 to 100	0.015	0.1 to 100	0.063	0.1 to 100	0.061	-	-
6	PEDOT(CTAB)/MWCNT/GCE	0.15 to 100	0.07	0.3 to 100	0.14	0.25 to 100	0.12	-	-
7	P3MT/MWCNT/GCE	0.01 to 100	0.0015	0.01 to 100	0.0019	0.01 to 100	0.0061	0.009 to $5x10^4$	0.0032

A - Determination range, □g L^{-1}
B - Detection limits, □g L^{-1}

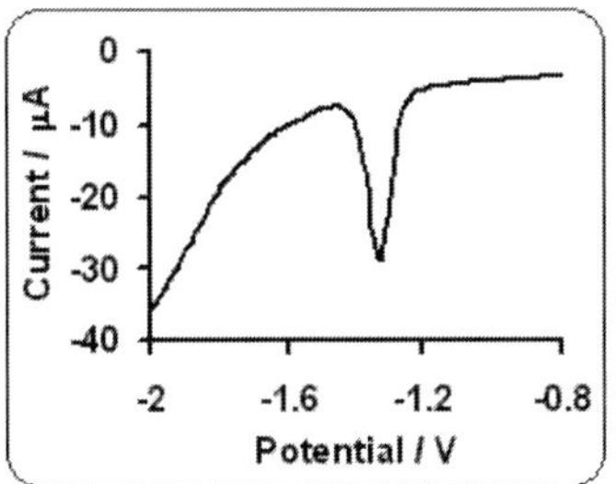

Fig. 4(A): DPSV of 50 □g L^{-1} CYP under optimum experimental conditions pH 13 on MWCNT/GCE

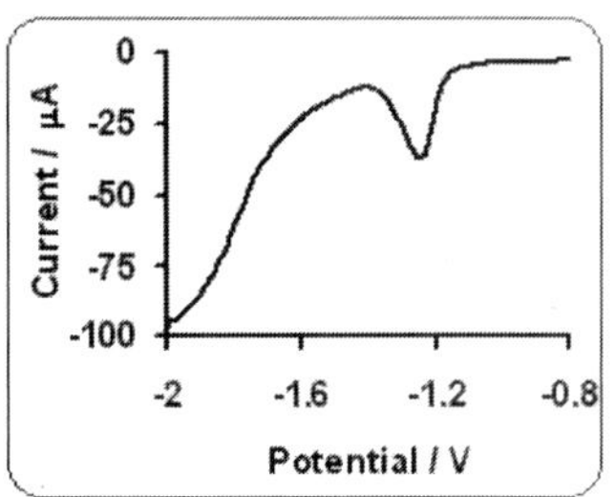

Fig. 4(B): DPSV of 50 □g L^{-1} CYP under optimum experimental conditions in pH 13 on PANI/MWCNT/GCE

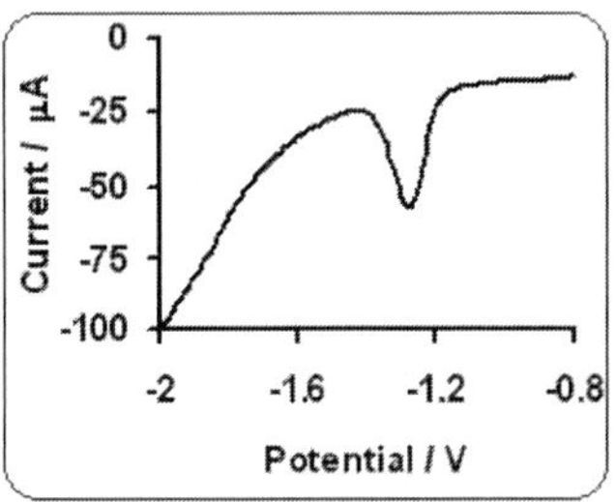

Fig. 4(C): DPSV of 50 □g L^{-1} CYP under optimum experimental conditions in pH 13 on PPY/MWCNT/GCE

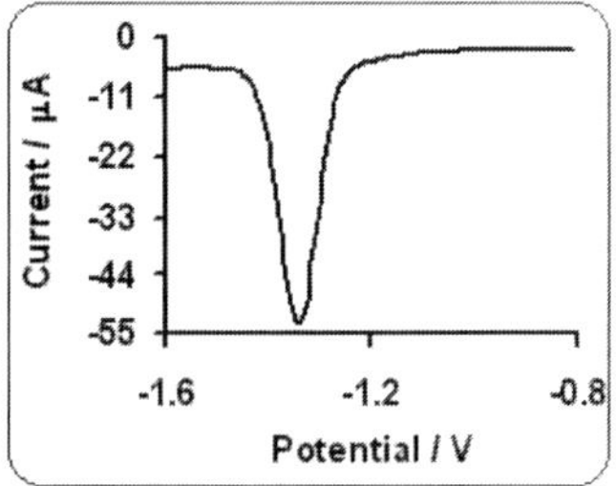

Fig. 4 (D): DPSV of 50 □g L^{-1} CYP under optimum experimental conditions in pH 13 on PEDOT/MWCNT/GCE

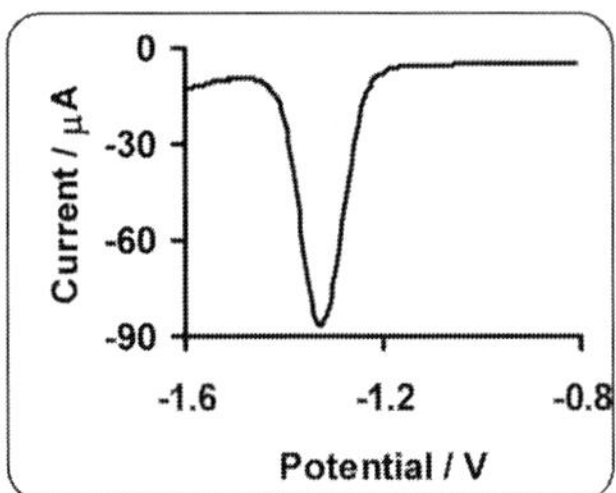

Fig. 4 (E): DPSV of 50 □g L^{-1} CYP under optimum experimental conditions in pH 13 on PEDOT(SDS)/MWCNT/GCE

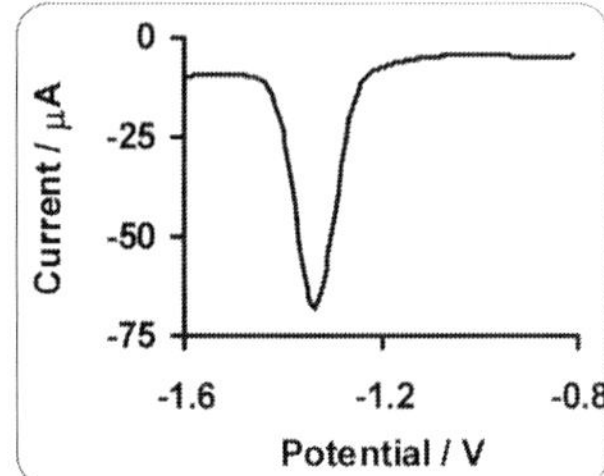

Fig. 4(F): DPSV of 50 □g L^{-1} CYP under optimum experimental conditions in pH 13 on PEDOT(CTAB)/MWCNT/GCE

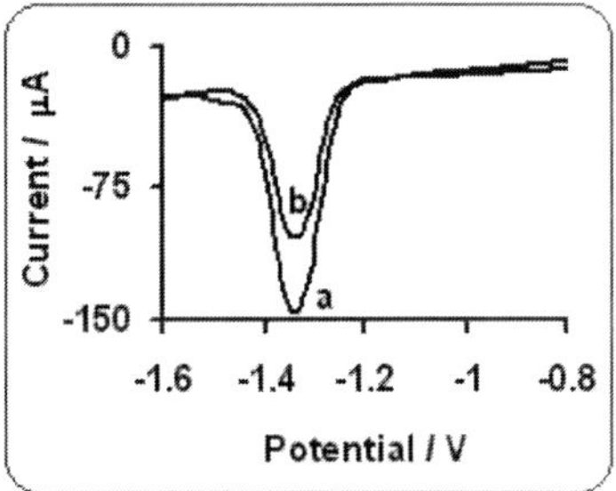

Fig. 4(G): DPSV of 50 □g L^{-1} CYP under optimum experimental conditions pH 13 on P3MT/MWCNT/GCE (a) conventional sample (b) real sample

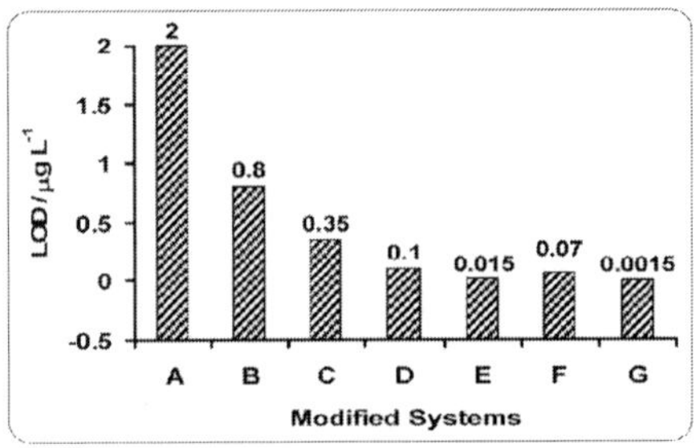

Fig. 4a: Comparison of LOD values of CYP on different sensors

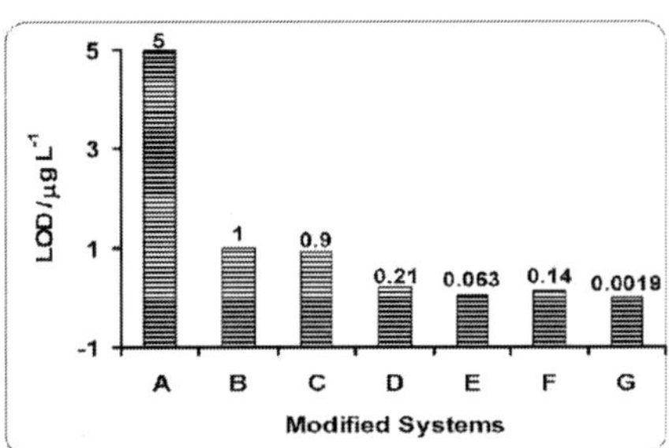

Fig. 4b: Comparison of LOD values of DEL on different sensors

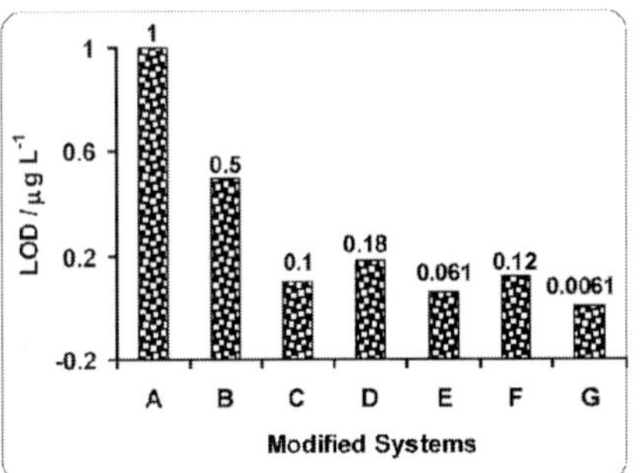

Fig. 4c: Comparison of LOD values of FEN on different sensors

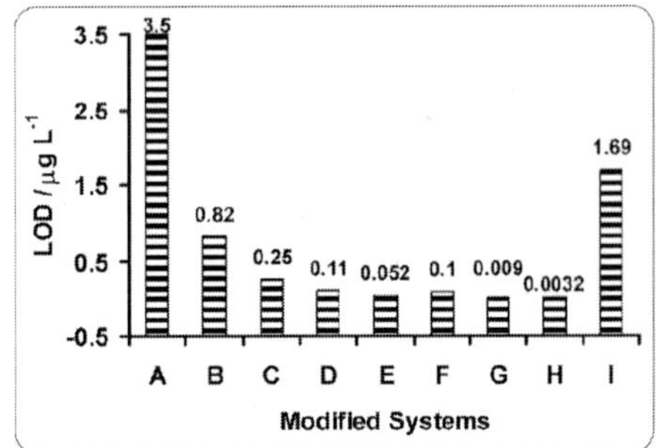

Fig. 4d: Comparison of LOD values of CAR on different modified systems

Fig. 4a-4d: Show LOD values of CYP, DEL, FEN and CAR obtained at various modified systems.

the contents of flask to required volume. Standard addition method was used. 0.05 ml aliquot of the 1 mg L^{-1} CYP stock standard solution was added to the solution prepared as above. This solution was taken in the cell and adsorptive differential pulse stripping voltammetric studies were carried out. The amount of CYP in the residue was determined and the percentage recoveries with standard error were arrived at. Precision of the method was found out from six repeated analyses at different intervals. The stripping voltammogram thus obtained is given in Fig. 4 (G) (b) on P3MT/MWCNT/GCE under similar experimental conditions. Similar studies were carried out for DEL, FEN and CAR and the percentage recoveries were determined. The percentage of recovery in all cases is presented in Table 4.

Table 4: Percentage recovery of CYP from soil sample

Pesticide	Spiked in □g L^{-1}	Found	Recovery in %	RSD*
CYP	25	22.49	89.96	1.5
	100	89.04	89.04	2.5
	200	185.9	181.9	1.9
DEL	25	21.94	87.76	1.7
	100	86.95	86.95	2.0
	200	172.0	86.02	2.1
FEN	25	22.13	88.52	1.6
	100	87.15	87.15	3.0
	200	179	89.50	2.2
CAR	10	8.45	84.50	2.8
	50	45.26	90.52	3.4
	100	94.76	94.76	1.9
	200	186.5	93.25	2.1
	300	281.2	93.73	3.2

Conclusions

All the reported insecticides and fungicide are electroactive and responded similarly in cyclic voltametry and their electrochemical behavior was also similar. All the insecticides undergo hydrolysis during electrolysis and the resulted 3-phenoxy benzaldehyde undergoes reduction. Of all the modified systems studied, P3MT/MWCNT sensor showed better response over the other systems, at which lowest LOD achieved for the three insecticides and one fungicide.

References

Berny, P.J., F. Buronfosse, B. Videmann, T. Buronfosse, J. Liq. Chromatogr. 1999. *Relat. Technol.* **22**:1547.

Dong, C., Z. Zeng, X. Li, 2005. *Talanta* **66**:721.

Dong, S., Baskaran, R.S. Kookana, R. Naidu, *J. Chromatogr*. 1997. 787:271.

F. Ye, Z. Xie, X. Wu, X. Lin, 2006. *Talanta,* **69**:97.

Fernandez-Alba, A.R., A. Valverde, A. Aguera, M. Contreras, S. Chiron, A 1996. *J. Chromatogr.* **721**:97.

Garrido, E.M., J.L. Costa Lima, C. Delerue-matos, M.F.M. Borges, A.M. Oliveira Brett, 2001. *Electroanalysis* **13**:199.

Goncalves, C., M.F. Alpendurada, A. 2002. *J. Chromatogr*, **968**:177.

Guiberteau, T., N. Galeano, P. Mora, F. Parrilla, Salinas, 2001. *Talanta* **53**:943.

Hernandez, S., I. Palchetti, M. Mascini, 2000. *Int. J. Environ. Anal. Chem.* **78**:263.

Ibrahim, M.S., K.M. Almagboul, M.M. Kamal, 2001. *Anal. Chim. Acta* **432**:21.

Leone, A.D., E.M. Ulrich, C.E.Bodnar, R.L. Falconer, R.A. Hites, 2000. Atmos. *Environ.* **4**:4131.

MacDonald, L.M., T.R. Meyer, 1998. *J. Agric. Food Chem*. **46**:3133.

Manisankar, P., C. Vedhi, S. Viswanathan, H.G. Prabu, 2004. *J. Environ. Sci. Health,* Part B **39**:89.

Manisankar, P., H.G. Prabu, 1995. *Electroanalysis* **7**:594.
Manisankar, P., S. Viswanathan, H.G. Prabu, 2002. *Int. J. Anal. Chem.* **82**:331.
Manisankar, P., S. Viswanathan, H.G. Prabu, 2004. *Int. J. Anal. Chem.* **84**:389.
Mart´ynez-Galera, M., A. Garrido Frenish, J.L. Mart´ynez-Vidal, P. Parrilla-Vazquez, A 1998. *J. Chromatogr*, 799 149.
Oudou, H.C., R.M. Alonso, R.M. Jimenez, 2001. *Electroanalysis* **13**:72.
Porazzi, E., M.P. Martinez, R. Fanelli, E. Benfenati, 2005. *Talanta* **68**:146.
Sreedhar, N.Y., K. Samatha, D. Sujatha, 2000. *Analyst* **125**:1645.
Tanabe, S., H. Iwata, R. Tatsukawa, 1994. *Sci. Total Environ.* **154**:163.
Vilchez, J.L., R. El-Khattabi, R. Blanc, A. Navalon, 1998. *Anal. Chim. Acta* **371**:247.
Wania, F., D. Mackay, 1996. *Environ. Sci. Technol.* **30**:390A.
Zen, J.M., A.S. Kumar, M.R. Chang, 2000. *Electrochim. Acta* **45**:1691.
Zhou, Q., J. Xiao, W. Wang, G. Liu, Q. Shi, J. Wang,. 2006. *Talanta* **68**:1309.
Zuman, P., *Anal. Lett.* 2000. **33**:163.

29

Encapsulation of Functional Foods

K. Thangavel

Introduction

Fruits are important sources of vitamins, carbohydrates, fiber and sugar. A large variety of fruits are grown in India, of which mango, banana, citrus, guava, grape, pineapple and apple are the major ones. In India, grapes occupies fifth position amongst fruit crops with a production of 16.77 million tonnes from an area of 0.64 million ha (FAO, 2009). The total grape export from India during the year 2007-08 was 96,723 tonnes worth Rs 317.84 crores. (APEDA, 2009). Grape is an excellent source of many nutrients and phytochemicals, able to contribute to a healthy diet. Grapefruit is a good source of vitamin C and pectin fiber. The major varieties of grapes grown in India are Thomson seedless, Sonaka, Anab-e-Shahi, Perlette, Banglore blue, Pusa seedless, Beauty seedless *etc.* (NRCG, 2009). Red grapes are grown in large quantities, especially for wine and juice production. Grapes are used for making jam, juice, jelly, vinegar, wine, grape seed extracts, raisins, grape seed oil and some kinds of candy.

Approximately 71 per cent of world grape production is used for winemaking and the rest 27 per cent is used as fresh fruit for table purpose and 2 percent as dried fruit (FAO, 2005). Grape pomace and other solid winery waste have been used for the recovery of food ingredients, nutraceuticals and functional foods (anthocyanin, grape seed oil, β-glucans, antioxidants, *etc.*), which provide demonstrated physiological benefits or reduce the risk of chronic disease (Hang,

1998). Grape skin pigment is added as a colorant in wine making. The recovery of pigment from the grapes varies from 9 to 12 kg/t (red grape pomace), the final product in the form of a liquid concentrate with 30 g of pigments per kilogram of solution (Leber, 2004). The continuing reduction in the number of coal tar dyes allowed as food additives has recently renewed the interest in natural pigments as food colourants. Grape anthocyanins are not quantitatively transferred to the wine or juice and considerable quantities of them are left in pomace which may eventually become an important source for the extraction of pigments.

Grape skin anthocyanin

Anthocyanins are natural, nontoxic and water-soluble pigments. Anthocyanins are part of a larger class of molecules known as flavonoids, responsible for the attractive red colour in fruits, vegetables and grape skin.

Grapes owe their attractive red to purple coloration to the water soluble flavonoids. Generally, the colouring matter of grapes is found only in the cells of the skin and is a good source of anthocyanins. Grape skin contains a great number of ployphenolics compounds. The most abundant of these compounds in grapes are anthocyanins, mainly 3- glycosides, 3-acetylglycosides and 3-p-coumaroylglycosides of malvidin (Mv), peonidin (Pn), delphinidin (Dp), petunidin (Pt) and cyanidin (Cy).

Anthocyanins and other pigment chemicals of the larger family of polyphenols in red grapes are responsible for the varying shades of purple and red colour (Walker et al., 2007).

Basic structure of anthocyanin

Anthocyanins belong to the flavonoid group of polyphenols. They have a C6 (A-ring) C3(C-ring) C6 (B- ring) -skeleton typical of flavonoids. Anthocyanins are glycosylated polyhydroxy and polymethoxy derivatives of 2-phenylbenzopyrylium cation, *ie.* the flavylium cation. The main part of anthocyanins is its aglycone, the flavylium cation, which contains conjugated double bonds responsible for the absorption of light around 500 nm causing the pigments to appear red to human eye. The aglycones are called anthocyanidins, which are usually penta- (3,5,7,3′,4′) or hexa-substituted (3,5,7,3′,4′,5′). The most important anthocyanidins are pelargonidin, cyanidin, peonidin, delphinidin, malvidin, and petunidin. The structure of anthocyanin is shown in Fig.1. These aglycones differ in the number of hydroxyl and methoxyl groups in the B-ring of the flavylium cation.

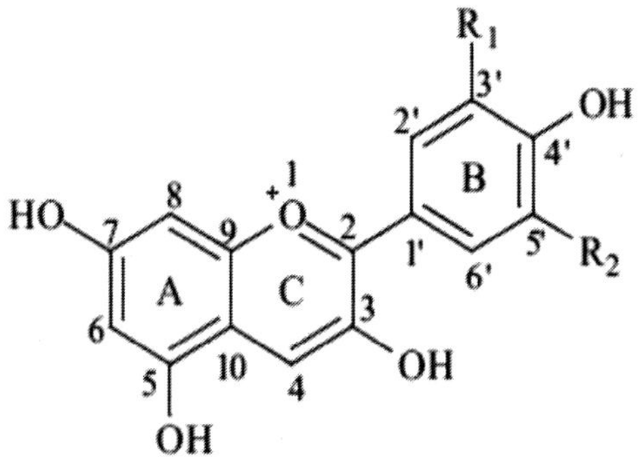

Fig.1: Structure of anthocyanin **(Francis, 1989).**

Name	R_1	R_2	Colour
Delphinidin	OH	OH	Bluish-red
Petunidin	OCH_3	H	Bluish-red
Cyanidin	OH	H	Orange-red
Pelargonidin	H	H	Orange
Peonidin	OCH_3	H	Orange-red
Malvidin	OCH_3	OCH_3	Bluish-red

Distribution of anthocyanins

Amrani and Glories (1995) reported that the majority of the phenols in red grape berries are located in the inner cells of grape skin.

The accumulation of anthocyanins in grapes begins after the phenological stage known as veraison and is affected by weather conditions, especially light intensity and temperature, which limit the cultivation of red grape cultivars in regions too cool or too warm during grape maturation. When grapes are harvested, the content of anthocyanins in grapes is as high as 1000 mg/kg (Boulton et al., 1995).

Bridle and Timberlake (1997) stated that grapes owe their attractive red to purple colouration to the water – soluble flavonoids. Generally, the colouring matter of grapes is found only in the cells of the skin and is a good source of anthocyanins. It was reported that grape pomace was a relatively inexpensive source of anthocyanins, since it is a by-product of the wine industry.

Harborne (1998) stated that anthocyanins were responsible for many of the attractive colors of flower, fruits, leaves and storage organs, from scarlet to blue.

Anthocyanin is an important source of the naturally occurring colorants of red fruits (Lepidot et al., 1999). An account of the anthocyanin content in some common fruits and vegetables is given in Table.1.

Table 1: Anth ocyanin content in selected common fruits and vegetables

Commodity	Total anthocyanin concentration (mg/kg)	References
Apple (peel)	100 – 21,600	Eder (2002)
Bilberry	4600	Eder (2002)
Blackberry	820 -1,800	Clifford (2000); Eder (2002)
Blueberry	825-5,030	Clifford (2000); Eder(2002)
Boysenberry	1,609	Clifford (2000)
Cherry (sweet)	3,500 – 4,500	Eder(2002)
Cherry (tart)	288	Eder (2002)
Chokeberry	5,060 -10,000	Clifford (2000)
Cranberry	460-2,000	Timberlake (1988)
Elderberry	2,000-15,600	Eder(2002)
Grape (red)	300 – 7,500	Timberlake (1988) ; Eder (2002)
Grape (blue)	80- 3,880	Eder (2002)
Loganberry	774	Clifford (2000)
Marion berry	237	Prior (2004)
Orange (juice)	2,000	Clifford (2000)
Plum	19-250	Timberlake (1988)
Raspberry (red)	100-600	Clifford (2000)
Raspberry (black)	763 -4,277	Timberlake (1988); Clifford (2000)
Strawberry	127-360	Timberlake (1988); Eder(2002)
Cabbage (red)	250	Timberlake(1988)
Cabbage (black)	1,300 – 4,000	Timberlake (1988); Eder(2000)
Currant (red)	119- 186	Eder (2002)
Eggplant	7,500	Clifford (2000)
Radish (red)	110-600	Giusti and Wrolstad (2003)
Potato (red)	150-450	Rodriguez- saona et al.(1998)
Onion	Up to 250	Timberlake (1988)

Among different plants or even cultivars in the same plant, the total anthocyanin content varies considerably, affected by genetic make–up, light, temperature and agronomic factors (Shahidi and Naczk, 2004).

Nutraceutical use of anthocyanin

A great number of literatures on anthocyanins indicate that anthocyanins are strong antioxidants and free radicals scavengers. Epidemiological data associate anthocyanins with reduction of risk of various diseases such as visual and vascular diseases, obesity, tumours, neurological dysfunctions and some types of cancer.

Mazza (1995) stated that nutraceutical health benefits *viz.* enhancement of visual activity, neurological dysfunctions, antimutagenic and antioxidative properties were associated with extracted grape skin anthocyanins.

Hou (2003) reported that anthocyanins were the chemical components that gave the intense color to many fruits and vegetables, such as blueberries, red cabbages and purple sweet potatoes. Epidemiological investigations indicated that the consumption of anthocyanin products such as grape skin extract and billberries extract was associated with a lower risk of cardiovascular disease and improvement of visual functions.

Waterhouse and Teissedre (1995) reported that Grape skin contains a high amount of secondary metabolites including phenolic acids, flavanols and anthocyanins which are reported to possess antibacterial, antiviral, antioxidant, anti-inflammatory, anti-cancerogenic properties and can prevent cardiovascular diseases.

Anthocyanins reduce the risk of cardiovascular diseases for people who consume wine, berry, and grape. The mechanism postulated was that anthocyanins act as antioxidants by donating hydrogen atoms to highly reactive free radicals, breaking the free radical chain reaction (Rice-Evans et al., 1996).

Hollman and Katan (1999) stated that the biological effects of flavonoids in reducing the risk of cardiovascular diseases are possibly associated with their antioxidant properties. These effects also include protection of tissues against free radical attack and lipid peroxidation. The effects of flavonoids on atherosclerosis, cancer, inflammation, and plasma cholesterol level have been investigated by many researchers. The results have been reported in epidemiological studies which seek to establish an inverse relationship between incidence of cancer and dietary flavonoid intake.

Hagiwara et al. (2001) stated that anthocyanins are not only nontoxic and nonmutagenic, but have positive therapeutic properties, such as, antioxidant , anti-inflammatory ,anticarcinogenic , antiviral and antibacterial properties.

Shi et al. (2001) studied the Cancer preventing activities of flavonoids. Components of tea, grape seeds and skins were reported to have anticarcinogenic properties. Antitumor, antiplatelet, antiallergic, antiischemic, and antiinflammatory activities of flavonoids are mostly associated with the antioxidative properties of plant flavonoids.

Bazzano et al. (2002) reported that regular consumption of anthocyanin containing fruits and vegetables is associated with reduced risks of chronic diseases such as cancer, cardiovascular disease, stroke, Alzheimer's disease, cataract and age-related functional decline.

Folts (2002) found that consumption of red wine (5 ml/kg) and purple grape juice (5–10 ml/kg) resulted in high blood antiplatelet activity. Consumption of purple grape juice offered further protection to patients against the oxidation of low-density lipoprotein (LDL) cholesterol. The flavonoids present in purple grape juice and red wine may inhibit the initiation of atherosclerosis.

An in vitro study indicated that catechin and quercetin have a property to inhibit LDL oxidation. Moreover, tannins (tannic acid), flavonols (quercetin and rutin), cinnamic acids (caffeic and ferulic acid), stilbenes (resveratrol), benzoic acids (gallic acid), and anthocyanidins (malvidin) also inhibit the in vitro oxidation of LDL in a dose-dependent manner, and this inhibition was better than that of common antioxidants like vitamins E and C (Manthey et al., 2002).

Passamonti et al. (2002) reported that anthocyanins display an array of beneficial actions on human health and well-being. They were shown to be absorbed as intact molecules in the stomach or possibly supported by a bilitranslocase mediated mechanism, or in the intestinum resisting enzymatic activity of the gut flora and subsequently detected unchanged in plasma and urine. Though their absorption rate is far below 1%, anthocyanins after being transported to sites of high metabolic activity may reach levels to show systemic activity such as antineoplastic, anticarcinogenic, antiatherogenic, antiviral, and anti-inflammatory effects, decrease of capillary permeability and fragility, inhibition of platelet aggregation and immune stimulation all of which are mainly based on the antioxidant activities of anthocyanins .

Grapes constitute one of the major sources of phenolic compounds among fruits. Antioxidative phenolic compounds like phenolic acids, polyphenols and flavonoids have been shown to scavenge free radicals such as superoxide, peroxyl, and hydroxyl radicals, and thus impact on the redox mechanisms that may lead to degenerative diseased conditions *i.e.* atherosclerosis, diabetes and cancers (Alonso et al., 2003).

Yoo et al. (2003) found that grape contains flavonoids with antioxidant properties which are believed to be protective against various types of cancer. This antioxidative protection is possibly provided by the effective scavenging of reactive oxygen species (ROS), thus defending cellular DNA from oxidative damage and potential mutations. These results indicate that the consumption of grape anthocyanin may increase plasma antioxidant capacity resulting in reduced DNA damage in peripheral lymphocytes achieved at least partially by a reduced release of ROS.

Greenspan et al. (2005) reported the anti-inflammatory properties in both *in vitro* and *in vivo* studies powder prepared from muscadine grape skins.

Dryden et al. (2006) reported that anthocyanins were strong antioxidants and free radical scavengers with anticarcinogenic, anti-inflammatory and antibacterial activities; thus, they are related with the reduction of risk of coronary heart disease, circulatory disorders, some types of tumours and chronic diseases.

Principles of microencapsulation

Microencapsulation is defined as a technology of packaging solids, liquids or gaseous materials in miniature, sealed capsules that can release their contents at controlled rates under specific conditions.

Shahidi and Han (1993) stated that microencapsulation enabled transforming liquid ingredients into flowable dry powders. Compound powders with "soft" cores provided means for protecting sensitive components with pharmaceutical benefits.

Onwulata et al. (1995) stated that microencapsulation was a process where droplets core were coated by thin films. During microencapsulation the entrapment of sensitive ingredients within a continuous film or coating could protect them from environmental factors such as moisture, air or light.

Jain et al. (1997) defined microencapsulation as a technology of packaging solids, liquids or gaseous material in miniature sealed capsules that could release their contents at controlled rates at specific conditions. The miniature packs in packages, called 'microcapsules', may range from sub micron to several millimeters in size and was ideally spherical.

Microencapsulation is the technique by which one material or a mixture of materials is coated with or entrapped within another material or system. The coated material is called active or core material, and the coating material is called shell, wall material, carrier or encapsulant (Augustin et al., 2001).

Microencapsulation is a technique in which a membrane encloses small particles of solid, liquid or volatile compounds with the objective of offering protection to the core material from adverse environmental conditions such as undesirable effects of light, moisture and oxygen, thus contributing to an increase in the shelf life of the product and thereby promoting a controlled liberation of the core material (Bertolini and Siani, 2001).

Tari and Singhal (2002) reported that microencapsulation can protect the flavour from undesirable interactions with food, minimize loss against light-induced reactions and oxidation, increase the flavour shelf-life, allow a controlled release and retain aroma in a food product during storage. A schematic illustration of microencapsulation of the core material as reported by Tari and Singhal (2002) is shown in Fig.2.

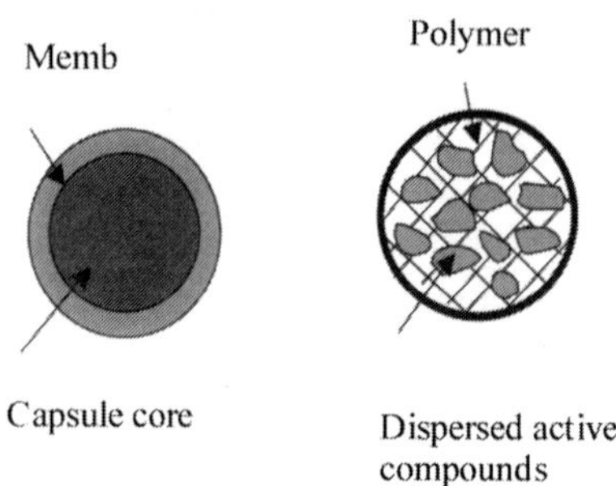

Fig. 2: Schematic illustration of microcapsule

Microencapsulation is a technique successfully used to improve the survival of microorganisms in dairy products, protect sensitive food components, ensure against nutritional loss and incorporate unusual or time-release mechanisms into the formulation (Ko et al., 2008).

Materials for microencapsulation

In microencapsulation, usually a material or a mixture of materials is coated with or entrapped within another material or system. Commonly, food constituents like fats, oils, aroma compounds, oleoresins, vitamins, minerals, colorants and enzymes are used for microencapsulation purposes.

Core material

Core material is the substance to be microencapsulated. The different types of core material that have been microencapsulated are shown in Table .2.

Table 2: Microencapsulated food ingredients

S.No	Type of ingredient
1	Flavouring agents such as oils, spices, seasonings
2	Sweeteners
3	Acids, alkalies, buffers
4	Lipids
5	Redox agents (bleaching, maturing)
6	Enzymes or microorganisms
7	Artificial sweeteners
8	Leavening agents
9	Antioxidants
10	Preservatives
11	Colorants
12	Essential oils, amino acids, vitamins and minerals

Source: (Kirby, 1991)

Madene et al. (2005) reported that the core material composed of one or several droplets. Core material had low water solubility with low rates of dissolution, which led to decrease in particle size of suspended core material. Very small core particle will give rise to aggregation problems during production because of their increased surface attractive forces. Large particles can cause problems because of their rapid sedimentation. Spherical cores of very narrow size are easier to deposit with uniform coating.

Soottitantawat et al. (2005) stated that coating the core enabled it to act as a semi permeable membrane protecting the core from severe conditions.

Ersus and Yurdagel (2007) microencapsulated black carrot anthocyanins (core material) by spray drying using maltodextrin as wall material. Microencapsulation protected the loss of black carrot anthocyanin from light, temperature and oxygen during storage.

Amuthaselvi (2009) microencapsulated grape skin anthocyanins (core material) by spray drying using Gum Arabic and maltodextrin as wall material. Microencapsulation protected the loss of anthocyanin to environmental factors during storage.

Wall materials and its properties

The wall material is designed to protect the core material from factors that may cause its deterioration, to prevent a premature interaction between the core material and other ingredients, to limit volatile losses and to allow controlled or sustained release under desired conditions.

For microencapsulation of the food compounds, the wall material must not have reactivity with the core material; be present in a form that is easy to handle, *i.e.* with low viscosity at high concentrations, give the maximum protection of the active ingredient against the external factors, ensure good emulsion-stabilization properties and effective redispersion behaviour in order to release the flavour at the desired time and place (Reineccius, 1988).

A suitable material that can be used for microencapsulation should possess high emulsifying activity, high stability and tendency to form fine and dense network during drying. It should not permit lipid separation from the emulsion during dehydration (Tanimoto and Matsuno, 1992).

Reineccius et al. (1995) stated that wall material used in the food industry includes gum arabic, alginates, carbohydrates, gums, lipids, proteins (milk, whey and soy protein), gelatin, waxes, gelatin and some chemically modified natural polymers .

McNamee et al. (2002) reported that the role of microencapsulation wall material is two fold. First, the material must have surface- active and film-forming properties for the encapsulation of lipid-based ingredients. Second, the encapsulating agent had to form a dry matrix around the lipid droplets, which binds them together within the encapsulate so as to prevent volatile loss and contact with the atmosphere.

Lucchesi et al. (2003) stated that wall material should have good properties of emulsification, film- formation, water-solubility, edibility and biodegradability.

The most important step in encapsulation of any "core" material by spray drying was the selection of suitable "wall" material which should form a continuous thin film and prevent deterioration and evaporation of the core material. The material should be low in cost, bland in taste and stable during storage and must possess required criteria of mechanical strength, be compatible with the food product with appropriate particle size, thermal properties and dissolution release (Perez- Alonso et al., 2003).

Astaxanthin pigment was microencapsulated by spray drying using chitosan as wall material. Results showed that the microencapsulated pigment was stable under ambient storage conditions at room temperature (Ciapara et al., 2004).

Madene et al. (2005) reported that the wall materials were selected according to the nature of the core material. If the core material was lipophilic, a hydrophilic substance was used as the wall material. When an aqueous solution was used as the core material, water insoluble substance was used as the wall material. The characteristics of different wall materials were also reported and it is given in Table 3.

Table 3: Characteristics of different wall materials

S.No	Wall material	Characteristics
1.	Maltodextrin (DE < 20)	Film forming
2.	Corn syrup solid (DE > 20)	Film forming, reductability
3.	Modified starch	Very good emulsifier
4.	Gum arabic	Emulsifier, film forming
5.	Modified cellulose	Film forming
6.	Gelatin	Emulsifier, film forming
7.	Cyclodextrin	Encapsulant, emulsifier
8.	Lecithin	Emulsifier
9.	Whey protein	Good emulsifier
10.	Hydrogenated fat	Barrier to oxygen and water

Gharsallaoui et al. (2007) reported that the solubility, film forming ability and retention of core compounds during microcapsule preparation by spray-drying were mainly dependent upon the wall material. Microencapsulation efficiency and microcapsule stability during storage were largely dependent on wall material composition.

Maltodextrin

Maltodextrins are cost effective, are bland in flavor, have low viscosity at high solids ratio. This allows for blending to create different wall densities, which provides protection against oxidation of the encapsulated ingredient. Higher DE maltodextrins formed a denser and more oxygen impermeable matrix providing longer shelf life for orange oil (Anandaraman and Reineccius, 1986).

Characterization of maltodextrin was performed by Raja et al. (1989) and it was reported that maltodextrin with dextrose equivalence between 10 and 20 were fit to be used as wall materials for microencapsulation of flavoring agents. The maltodextrin encapsulated samples showed the highest retention of flavor since they could be dispersed in water up to 35.5 per cent of the solution without clods formation.

Mckernan (1992) observed that maltodextrins are being increasingly used in microencapsulation studies as they had acceptable food grade qualities. They are found to be non- reactive with core materials with low viscosity at high solid content and are relatively inexpensive.

Maltodextrins as wall materials provided good oxidative stability to encapsulated oil with good emulsion stability (Kenyon, 1995). These materials are considered as good encapsulating wall materials, as they exhibited low viscosities at high solids contents with good solubility.

Whorton and Reineccius (1995) evaluated the mechanism associated with the controlled release of the flavouring material (a mixture of several odd-numbered carbon aldehydes, C-3 to C-11) encapsulated in maltodextrin of different dextrose equivalent (DE). The flavour retention during storage improved with an increase in the DE value of maltodextrin.

Usage of maltodextrin as wall material for production of *Amaranthus* betacyanin pigment significantly reduced the hygroscopicity of the betacyanin extracts enhancing their storage stability (Cai and Corke, 2000)

Maltodextrin have low flavour and are non-reactive with the core component (anthocyanin) used for microencapsulation. It improved the shelf life of black carrot anthocyanins and maintained the colour and stability during storage (Ersus and Yurdagel, 2007).

Gums

Gum arabic is a hydrocolloid produced by natural exudation of acacia trees and is an effective encapsulation material. It has high water solubility and low cost compared to other hydrocolloid gums. It is relatively less viscous and has the ability to act as an oil-in-water emulsifier (Glicksman, 1983).

Gums and thickeners are generally bland or tasteless, but they can have a pronounced effect on the taste and flavour of foods. In general, hydrocolloids decrease sweetness, with much of the effect being attributed to viscosity and hindered diffusion (Godshall, 1997).

Gum arabic is a polymer consisting primarily of D-glucuronic acid, L-rhamnose, D-galactose and L-arabinose with about 5 per cent protein. The protein fraction is responsible for the emulsification of the gum since it acts as an interface between oil and water (Re-Mi, 1998).

Bertolini and Siani (2001) stated that gum arabic has excellent emulsification properties. It acts as a semi permeable membrane during spray drying. The external surfaces had no apparent cracks, providing good protection to the core material.

Gum arabic is most often used as a flavour-encapsulating material. Its solubility, low viscosity, emulsification characteristics and its good retention of volatile compounds make it versatile for most encapsulation methods (Buffo et al., 2002).

Rodriguez – Huezo et al. (2004) conducted a study on microencapsulation of red chilly carotenoid pigments with gum arabic. It was reported that gum arabic gave better morphology and higher encapsulation efficiency.

Bixin pigment was encapsulated by spray-drying with gum arabic and maltodextrin and evaluated for its stability. The microcapsules containing gum arabic and maltodextrin as emulsifier showed the highest encapsulation efficiency of 86 per cent and 75 per cent, respectively. Bixin encapsulated with gum arabic was 3 to 4 times more stable than that encapsulated with maltodextrin (Barbosa et al., 2005).

Sheikh et al. (2006) reported that gum arabic was a suitable wall material for encapsulation of pepper oleoresin. It protected the oleoresin from oxidation and volatilization.

Kanakdande et al. (2007) reported that gum arabic was a better wall material for encapsulation of cumin oleoresin by spray drying technique when compared to the other wall materials such as maltodextrin and modified starch.

Grape skin anthocyanin was microencapsulated with gum Arabic and maltodextrin and analysed for its stability. Gum Arabic coated micoencapsules showed higher stability compared to maltodextrin coated materials (Amuthaselvi, 2009).

Spray drying Technique for Microencapsulation

Spray-drying is a unit operation by which a liquid product is atomized in hot air to instantaneously obtain its powder. The initial liquid feed is in the form of solution, emulsion or a suspension. Spray-drying depends upon the feed material and operating conditions to produce a very fine powder (10–50 µm) or large sized particles (2–3 mm).

Reineccius (1988) reported that production of microencapsulated powders by spray-drying generally involved the formation of a stable emulsion in which the wall material acted as a stabilizer for the core material. The emulsion was then spray-dried to yield the encapsulated powder product.

Spray drying has paved the way for the production of powder colorants with high storage stability, ease of handling for some applications, and minimize the weight for transportation, in comparison to liquid concentrates. Some advantages of spray drying include the ability to quickly produce a dry powder (e.g. as compared to lyophilization) and the ability to control the particle size distribution (Masters, 1991).

Shahidi and Han (1993) reported that microencapsulation by spray-drying involved four stages *viz*, preparation of the dispersion or emulsion; homogenization of the dispersion; atomization of the feed emulsion; and dehydration of the atomized particles.

Spray-drying is the most common and cheapest technique to produce microencapsulated food materials. Equipment is readily available and production costs are lower than most other methods. Compared to freeze-drying, the cost of spray-drying method is 30–50 times cheaper (Desobry et al., 1997).

Cai and Corke (2000) reported that spray drying was an economical method for preservation of natural colorants by entrapping the ingredients in coating materials.

Spray drying is the most common and cheapest technique to produce microencapsulated food materials. Equipment is readily available and production costs are lower than most other methods.

Tari and Singhal (2002) reported that the microencapsulation of sensitive compounds using spray drying consisted of two steps. The first was emulsification of a core material, such as the lipid system, with a dense solution of a wall material such as a polysaccharide or protein. The second was spray drying of the emulsions.

The spray drying process involves the dispersion of the substance to be encapsulated in a carrier material, followed by atomization and spraying of the mixture into a hot chamber (Watanabe et al., 2002). The resulting microcapsules are then transported to a cyclone separator for recovery.

Mathew (2003) has reported spray drying to be an excellent economical way to form particles containing suspended flavours and scents in a polymer matrix. The short contact time of this high volume process minimized loss of volatile cores.

Ersus and Yurdagel (2007) found that microencapsulation of black carrot anthocyanin using spray drying for preservation of natural colorants increased the stability and half life period of microencapsulated pigment stored at different storage temperatures 4 and 25° C and light intensity of 3000 lx.

Parize et al. (2008) microencapsulated the urucum pigment extracted from *Bixa orellana* seed by spray drying. It was observed that urucum pigment could be successfully incorporated into chitosan by spray-drying process, resulting in dry and colourful powders, which was water-soluble. Urucum pigment is widely used in the food industry and interest in its application in the pharmaceutical and cosmetics industry is growing.

Obon et al. (2009) reported that betacyanin pigment obtained from *Opuntia stricta* microencapsulated by co-current spray drying of *O.stricta* fruit juices with a bench-scale two fluid nozzle spray dryer showed high color strength when stored at room temperature for one month.

Ge et al. (2009) studied the microencapsulation of red pigments from a hybrid rose using spray drier. The stability of red pigment was maintained by microencapsulation against the variations in pH, light and heat.

Amuthaselvi, 2009 conducted a microencapsulation study for the encapsulation of anthocyanin using the spray drier with a capacity of 1500 g/h water evaporation. Anthocyanin concentration of 15, 20 and 25°Brix were emulsified with maltodextrin and gum arabic as wall material at 180 to 220 °C.

Type of atomizers for microencapsulation

Hang (1988) stated that in the food industry single fluid, high-pressure nozzle and centrifugal wheel / disc atomizers are used. Nozzle type, speed and pressure of the nozzles had significant effect on the particle size distribution of the resultant powder.

Atomization is typically accomplished by either a twin fluid high-pressure spray nozzles or centrifugal wheel. Centrifugal wheel atomizers have an advantage of handling very viscous and abrasive feed materials while the pressure spray systems offer greater flexibility in producing wide range of particle size powder (Reineccius, 2004).

Teunou and Poncelet (2005) stated that disintegration of a liquid by a rotating disc is a technique widely used in spray drying, cooling or freezing for atomisation of the liquid into fine droplets before drying or freezing. This technique is applied today for microencapsulation, which is a promising method to protect the core material, improve the product characteristics and functionalities. For this purpose, the structure and the quality of the microencapsules are determining factors in guaranteeing the properties of the microencapsulated particles.

Core to wall material ratio

Lee and Rosenberg (2000) investigated the effects of core-to-wall ratio ranging from 1:1.5 to 5:1.5 on formation, properties and core release from whey protein-based microcapsules containing theophylline (tea leaf extract). Core retention and microcapsules size increased with increase in core-to-wall ratio. Outer topography and inner structure of microcapsules were also influenced by core-to-wall ratio.

Ciapara et al. (2004) reported that *Astaxanthin* pigment was microencapsulated using chitosan at 1: 9 ratio of core and wall material. There was no pigment loss due to oxidation during storage.

Rodriguez – Huezo et al. (2004) conducted experiment on micro-encapsulation of red chilly carotenoid pigment with gum arabic at different core to wall ratios (1:2.6, 1:3.9 and 1:1.4). Results showed that 1:3.9 core to wall ratio gave highest encapsulation efficiency.

Ersus and Yurdagel (2007) conducted experiment on microencapsulation of anthocyanin pigment of black carrot by spray drying with core to maltodextrin ratio of 1:3. Storage at 4°C increased the half life of spray dried anthocyanin pigments three times when compared to 25°C storage temperature.

Temperature of hot air

Relatively high inlet temperature is desirable to allow rapid formation of semi permeable membrane on the droplet surface, provided no damage like 'ballooning' of the droplets was caused to the dry product (Reineccius, 1988). High exit air temperature might result in heat damage to some flavouring materials. Hence, a dryer outlet temperature of 80 to 100°C is suggested for the spray drying of flavouring materials.

Cai and Corke (2000) microencapsulated *Amaranthus* betacyanin extracts with maltodextrin and modified starches as coating materials at five inlet/outlet air temperatures and four feed solid contents. Higher inlet/outlet air temperature caused greater betacyanin loss during spray drying and slightly affected the pigment stability during storage. Temperature of inlet / outlet ranging from 170 to 200°C / 92 to 96°C and 20 to 40 per cent feed solid content were suitable for production of *Amaranthus* betacyanin extracts with less pigment degradation.

The acidified ethanol extracts of black carrot with a high anthocyanin content (125 ± 17.22 mg/100 g) were spray dried using maltodextrin as carrier and coating agents, at three different inlet/outlet air temperatures with constant feed solid content (20 per cent). Higher inlet/outlet air temperatures caused greater anthocyanin loss during spray drying (Ersus and Yurdagel, 2007).

Microencapsulated Powder characteristics

Microencapsulation Efficiency

Encapsulation efficiency is the amount of anthocyanin entrapped inside the wall material to the total anthocyanin present in surface anthocyanin powder. Rodriguez – Huezo et al. (2004) conducted experiments on microencapsulation of red chilly pigment emulsion with 25 per cent and 35 per cent solid content with spray drying. Emulsions at 35 per cent solid content showed the best morphology, highest microencapsulation efficiency during drying and highest storage stability.

Shu et al. (2005) prepared lycopene microcapsules by spray drying method using a wall system consisting of gelatin and sucrose. Encapsulation efficiency (EE) was significantly affected by the ratio of core and wall materials, ratio of gelatin and sucrose, homogenization pressure, inlet temperature, feed temperature, and lycopene purity. The ratio of gelatin: sucrose of 3:7 and the ratio of core and wall material of 1:4, feed temperature of 55°C, inlet temperature of 190°C, homogenisation pressure of 40 MPa were the optimal conditions reported.

Bixin was encapsulated by spray-drying using gum arabic and maltodextrin by Barbosa et al. (2005). The encapsulation efficiency and stability were evaluated. Encapsulation efficiency was found to be 86 and 75 percent for gum arabic and maltodextrin, respectively.

Xie et al. (2007) conducted studies on microencapsulation of vitamin A with gelatin-sucrose and gelatin peach gum-sucrose as wall material. The microencapsulation efficiency of gelatin-sucrose and gelatin peach sucrose varied from 94.77 to 96.70 per cent. It was reported that by adding peach gum, the microcapsules exhibited a spherical shape with a smooth surface.

Moisture content

Rodríguez – Huezo et al. (2004) conducted studies on microencapsulation of red chilly carotenoid pigment by spray drying. It was reported that maximum carotenoid degradation occurred at a water activity (a_w) of 0.628.

Erus and Yurdagel (2007) reported that increasing spray drying temperature reduced the moisture content of powders. Moisture content ranged from 1.09 to 3.76 per cent for powders where the flow rates were kept constant at 5 mL/min for the temperature ranges of 160 to 200° C.

Stability of the microencapsulated powders

Reineccius et al. (2002) evaluated the incorporation and retention on storage of a variety of flavor compounds spray-dried in α-, β- and γ-cyclodextrins (CyDs). CyD/flavour complexes were stored at 20 or 40°C at 65 or 80 per cent relative humidity, and losses during both the inclusion process and the subsequent storage period were monitored analytically. γ-CyD generally functioned best in terms of initial flavor retention. On storage, losses of volatiles were greatest for γ-CyD and least in the case of α-CyD. The results suggested that CyD encapsulation via spray-drying involved matrix entrapment as well as molecular inclusion.

Barbosa et al. (2005) studied the stability of microencapsulated bixin with maltodextrin and mixture of maltodextrin with sucrose. The bixin stability was observed to be 10 times greater for both wall materials when compared to the non-encapsulated bixin.

Erus and Yurdagel (2007) reported microencapsulation of anthocyanin pigment of carrot by spray drier using maltodextrin as wall material. The microencapsulated anthocyanin powder had more stability with anthocyanin content retaining up to 64 days at 25°C.

Amuthaselvi (2009) conducted studies on microencapsulation of anthocyanin from grape skin by spray drying, using maltodextrin and gum Arabic as wall materials. The microencapsules were evaluated for total content stability for 3 months. Gum Arabic coated microencapsulated offered greater protection to anthocyanin from the environmental factors than maltodextrin coated.

Packaging and Storage of Microencapsulated Powder

Beta carotene was microencapsulated in maltodextrin by spray drying and stored in petridishes at 25°C. There were slight changes in colour values during storage due to oxidation. Beta carotene content reduced to 11 per cent compared to initial content (80 per cent) when stored at 25°C (Desorby et al., 1997).

Hassan and Ahmed (1998) prepared pineapple powder using foam mat drying process and the dehydrated product was packaged in 300 gauge high-density polyethylene (HDPE) packages and aluminium foil and the samples were kept at room temperature. It was reported that the dried product was shelf stable for six months and the foam mat dried pineapple powder was better in colour, flavour, flowability, solubility and overall acceptability than the spray dried powder.

Shu et al. (2005) studied the microencapsulation of lycopene stored in plastic bags at 0°C. It was observed that lycopene was effectively stable from oxygen and light during storage.

Kanakdande et al. (2007) conducted experiments on microencapsulation of cumin oleoresin powder and packed in polyethelyene pouches for storage studies. Flavor components were found to be safe against oxidation as well as volatilization.

Ersus and Yurdagel (2007) performed storage studies on microencapsulated anthocyanin pigment powders stored in brown bottles with screw caps at 4 and 25° C to determine the effects of storage temperatures. Anthocyanin contents of encapsulated powders decreased by 33 per cent at the end of 64 days at 25°C and 11 per cent at 4° C during storage.

Amuthaselvi (2009) microencapsulated grape skin anthocyanin powder by spray drier and evaluated the powder characteristics during storage .There was no significant change in encapsulation efficiency during the storage of microencapsulated powder. The solubility of microencapsulated anthocyanin powders decreased from 93.7 to 90.9 per cent during storage. It was also seen that redness of anthocyanin ('a' value) decreased after three months of storage. This might be due to the oxidative degradation of active components present in the surface anthocyanin in the microencapsulated powder.

Conclusion

Encapsulation is one of the widely used technique to assist smart delivery of agricultural inputs. In this chapter the use of encapsulation technique in developing value added food product is elucidated. The country like India produces huge quantities of fruits, vegetables, species and medicinal plant but their nutritional qualities are less exploited. Encapsulation technique would relatively help to produce value added products by extracting functional qualities.

References

Alonso Borbalan, A. M., L. Zorro, D.A. Guillen and C. G. Barroso. 2003. Study of the polyphenol content of red and white grape varieties by liquid chromatographymass spectrometry and its relationship to antioxidant power. *Journal of Chromatography*,**1012(1)**:31–38.

Amrani, J. K. and Y. Glories. 1995. Tannins and anthocyanins of grape berries: Localization and extraction technique. Analytical, *Nutritional and Clinical Methods*. **153**:28–31.

Amuthaselvi, G. 2009. Studies on microencapsulation of anthocyanins from grape skin. Unpublished M.Tech thesis, Department of Food and Agricultural Process Engineering, Tamil Nadu Agricultural University, Coimbatore.

Anandaraman, S. and G.A.Reineccius. 1986. Stability of encapsulated orange peel oil. *Food Technology*. **40**: 88 – 93.

APEDA. 2009. Export and Import values of commodities in India. www.apeda.in. dt 24.3.2009.

Augustin, M.A., L.Sanguansri, C. Margetts and B. Young. 2001. Microencapsulation of food ingredients. *Food Australia*. **53**: 220–223.

Barbosa, M.I., C.D.Borsarelli and A.Z.Mercadante. 2005. Light stability of spray-dried bixin encapsulated with different edible polysaccharide preparations. *Food Research International*. **38(9):** 989-994.

Bazzano, L. A., J.,He, L. G. Ogden, C. M. Loria, S.Vupputuri, L. Myers and P.Whelton 2002. Fruit and vegetable intake and risk of cardiovascular disease in US adults: the first National health and nutrition examination survey epidemiologic follow up study. *American Journal of Clinical Nutrition*. **76**: 93–99.

Bertolini, A.C. and C.R.F.Siani. 2001. Stability of monoterpenes encapsulated in gum Arabic by spray drying. *Journal of Agricultural and Food Chemistry*. **49**:780 -785.

Boulton, R. B., V. L Singlcton, L. F.Bisson and R.E.Kunkee.1995. Principles and Practices of Winemaking. Chapman and Hall: New York.

Bridle, P. and C.F.Timberlake. 1997. Anthocyanins as natural food colours-selected aspects. *Food Chemistry*. **58**:103–109.

Buffo, R.A., K.Probst, G.Zehentbauer, Z.Luo and G.A.Reineccius. 2002. Effects of agglomeration on the properties of spray – dried encapsulated flavours. *Flavour and Fragrance Journal*. **17**: 292:299.

Cai,Y.Z. and H.Corke.2000. Production and properties of spray- dried *Amarnthus* Betacyanin Pigments. *Journal of Food Science: Sensory and Nutritive Qualities of Food*. **65(6):** 1248 – 1252.

Ciapara,I.H., L.F.Valenzuela, F.M.Goycoolea, W.A.Monal. 2004. Microencapsulation of astaxanthin in a chitosan matrix. *Carbohydrate Polymers*. **56**: 41–45.

Clifford, M.N. 2000. Anthocyanins- nature, occurance and dietary burden. *Journal of Science and Food Chemistry*. **80** : 1063 -1072.

Desobry, S., F.M.Netto and T.P.Labuza. 1997. Comparison of spray-drying, drum-drying and freeze-drying for â-carotene encapsulation and preservation. *Journal of Food Science*, **62**: 1158–1162.

Dryden, G.W., M.Song and C. McClain. 2006. Polyphenols and gastrointestinal diseases. *Current Opinion Gastroen*, **22(2)**: 165–170.

Eder. 2002. Pigments, In : Food analysis by HPLC . (Eds.) Nollet, M.L.L., Marcel Dekker, New York. 845 – 880.

Ersus, S. and U.Yurdagel. 2007. Microencapsulation of anthocyanin pigments of black carrot by spray drier. *Journal of Food Engineering*. **80**: 805 – 812.

FAO. 2005. Distribution and production of grapes in world. www.fao.com.dt 24.3.2009

FAO. 2009. Production of grapes in India. www.fao.com. 24.3.2009

Folts, J. D. 2002. Potential health benefits from the flavonoids in grape products on vascular disease. *Advances in Experimental Medicine and Biology*. **505**: 95–111.

Francis, F.J. 1989. Food colorants, anthocyanins. Critical Review of Food Nutrition. **28** : 273–314.

Ge, X., Z.Wan, N.Song, A.Fan and R.Wua. 2009. Efficient methods for the extraction and microencapsulation of red pigments from a hybrid rose. *Journal of Food Engineering*. 23.

Gharsallaoui, A., G.Roudaut, O.Chambin, A.Voilley and R.Saurel. 2007. Applications of spray - drying in microencapsulation of food ingredients: An overview. *Food Research International.* **40**:1107–1121.

Giusti, M.M. and R.E. Wrolstad. 2003. Acylated anthocyanins from edible sources and their applications in food systems. *Journal of Biochemical Engineering.* **14**: 217-225.

Glicksman, R.T. 1983. A method of analysis of non-volatile encapsulated substances. *Journal of American Chemical Society.* **5:**448 -450.

Godshall, M.A. 1997. How carbohydrates influence food flavour. Journal of Food Technology. **51:** 63-67.

Greenspan, P., J. D.Bauer, S. H. Pollock, J. D. Gangemi, E. P. Mayer and A.Ghaffar. 2005. Antiinflammatory properties of the muscadine grape (Vitis rotundifolia). *Journal of Agricultural and Food Chemistry.* **53**: 8481–8484.

Hagiwara, A., K,Miyashita, T. Nakanishi, M. Sano, S.TamanoT. Kadota. 2001. Pronounced inhibition by a natural anthocyanin, purple corn colour, of 2-amino-1-methyl-6-phenylimidazo[4,5-b]pyridine (PhIP)-associated colourectal carcinogenesis in male F344 rats pretreated with 1,2-dimethylhydrazine. *Cancer Letters*, **171(1)**: 17–25.

Hang, Y.D. 1998. Recovery of food ingredients from grape pomace. *Process Biochemistry.* **23**: 2-4.

Harbone, J.B.1998. Phenolic compounds. In : Phytochemical methods – a guide to modern techniques of plant analysis 3rd edition. Chapman and Hall, New York. pp.66 – 74.

Hassan, M. and J.Ahmed . 1998. Sensory quality of foam- mat dried pineapple juice powder. *Indian Food Pack.* **52** (7): 31- 33.

Hollman, P.C.H. and M.B. Katan. 1999. Dietary flavonoids, intake, health effects and bioavailability. *Food and Chemical Toxicology.* **379**: 937–942.

Hou, D.X. 2003. Potential mechanisms of cancer chemoprevention by anthocyanins. *Current Molecular Medicine.* **3:**149–59.

Jain, P., S.Arora and T.Rai. 1997. Flavour encapsulation and its application. *Beverage and Food World.* **24**(4): 21-24.

Kanakdande, D., R.Bhosale and R.S.Singhal. 2007. Stability of cumin oleoresin microencapsulated in different combination of gum arabic, maltodextrin and modified starch. *Carbohydrate Polymers.* **67**: 536–541.

Kenyon, M. M. 1995. Modified starch, maltodextrin and corn syrup solids as wall materials for food encapsulation. In S. J. Risch & G. A. Reineccius (Eds.), Encapsulation and controlled release of food ingredients ACS symposium series Washington, DC: *American Chemical Society.* **590**:42–50.

Kirby, C.J.1991. Microencapsulation and controlled delivery of food ingredients. Food Science and Technology. **5:**74–78.

Ko, J.A., S.Y.Koo and H.J. Park. 2008. Effects of alginate microencapsulation on the fibrinolytic activity of fermented soybean paste (Cheonggukjang) extract. *Food Chemistry.* **111**: 921–924.

Leber, E. 2004. Recovering value from winemaking by-products. ASEV 55th Annual Meeting San Diego, California.

Lee, S.L. and M.Rosenberg. 2000. Whey protein-based microcapsules prepared by double emulsification and heat gelation. *Food Science Technology* (LWT). **33**: 80–88.

Lepidot, T., S.Harel , B.Akiri , R.Granit and J.Kanner. 1999. pH-dependent forms of red wine anthocyanins as antioxidants. *Journal of Agriculture and Food Chemistry.* **47**: 67.

Lucchesi, M.B., M.Bruschi, M.L.Cardoso and M.P.D.Gremiao. 2003. Gelatin microparticles containing propolis obtained by spray drying technique preparation and characterization. *International Journal of Pharmaceutics.* **264**: 45–55.

Madene, A., M. Jacquot, J.Scher and S. Desobry. 2005. Flavour encapsulation and controlled release – a review. *International Journal of Food Science and Technology*. **41**: 1–21.

Manthey, J. A., B. S. Buslig and M. E. Baker. 2002. Flavonoids in cell function. *Advances in Experimental Medicine and Biology*. **505**:1–7.

Masters, K. 1991. Spray Drying Handbook. Longman Scientific and Technical, UK. pp.100-150.

Mathew, A.G. 2003. Some features of Indian spice extraction technology. Spice India. 9-12.

Mazza, G. 1995. Anthocyanins in grape and grape products. *Critical review of Food Science and Nutrition*. **35**: 341-371.

McKernan, W.M. 1992. Microencapsulation in the flavor industry. Part II. The flavour industry. **4:** 72.

McNamee, B.F., E.D. Riordan and M.Sullivan. 2002. Emulsification and microencapsulation properties of gum arabic. *Journal of Agricultural and Food Chemistry.* **46**:4551–4555.

NRCG. 2009. Production of Grape varieties in India. www.nrcgrapes.nic.in. dt 19.5.2009

Obon, J.M., M.R. Castellar, M. Alacid and J.A. Fernandez-Lopez. 2009. Production of a red–purple food colorant from Opuntia stricta fruits by spray drying and its application in food model systems. *Journal of Food Engineering*. **90**: 471–479.

Onwulata, C.I., P.W. Smith and V.H. Holsinger. 1995. Flow and compaction of spray–dried powders of anhydrous butter oil and high melting milk fat encapsulated in disaccharides. *Journal of Food Science*. **60**: 836-840.

Parize, A.L., T.C. Souza, B.Valfredo, T. Favere, C.M. Laranjeira, A. Spinelli and E.Longo. 2008. Microencapsulation of the natural urucum pigment with chitosan by spray drying in different solvents. *African Journal of Biotechnology*. **7 (17)**: 3107-3114.

Passamonti, S., U. Vrhovsek and F. Mattivi. 2002. The interaction of anthocyanins with bilitranslocase. *Biochemical and Biophysical Research Communications*. **296**: 631–636.

Perez-Alonso, C., J.G. Baez-Gonzalez, C.I. Beristain, E.J. Vernon-Carter and M.G. Vizcarra-Mendoza. 2003. Estimation of the activation energy of carbohydrate polymers blends as selection criteria for their use as wall material for spray-dried microcapsules. *Carbohydrate polymers*. **53**:197–203.

Prior, R.L. 2004. absorption and metabolism of anthocyanins: potential health effects in phytochemicals – mechanism of action. In: phytochemicals, mechanism of action. (Eds.) Meskin, M.S., W.R. bidlack, A.J. davies, D.S. Lewis and R.K. Randolph., CRC Press, Boca raton, pp 1-19.

Raja, K. C.M., B.Sankarikutty, M.Sreekumar, A.Jayalekshmy and C.S.Narayanan. 1989. Material characterization studies of maltodextrin samples for the use of wall material. Starch/ starke. **41**:298–303.

Reineccius, G.A.1988. Spray-drying of food flavors. In G.A.Reineccius and S.J.Risch (Eds.), Flavor encapsulation. Washington, DC: Amercian Chemical Society. 55–66

Reineccius, G.A., F.M. Ward, C. Whorten and S.A. Andon. 1995. Developments in gum acacias for the encapsulation of flavors. In S.J.Risch and G.A. Reineccius (Eds.), Encapsulation and controlled release of food ingredients. ACS symposium series. Washington, DC: *American Chemical Society.* **590**: 161–168.

Reineccius, T.A., G.A. Reineccius and T.L. Peppard. 2002. Encapsulation of flavors using cyclodextrins: comparison of flavor retention in alpha, beta and gamma types. *Journal of Food Science*. **67**: 3271–3279.

Reineccius, G.A. 2004. The spray drying of food flavour. Drying technology. 22 (6): 1289 –1324.

Re-Mi. 1998. Microencapsulation by spray drying. *Drying Technology*. **16**:1195 -1236.

Rice-Evans, C.A., N.J. Miller, G. Paganga.1996. Structure antioxidant activity relationships of flavonoids and phenolic acids, Free Radical Biological Medicine. **20:**933–956.

Rodriguez–Huezo, M.E., R.Pedroza-islas, L.A.Prado-barragan, C.I.Beristain and E.J.Vernon-carter. 2004. Microencapsulation by spray drying of multiple emulsions containing caroteinoids. Journal of food science. *Food Engineering and Physical properties*. **69**:351-359.

Rodriguez-saona, L. E., R.E. Wrolstad and C.Pereria. 1998. Glycoalkaloid content and anthocyanin stability to alkaline treatment of red fleshed potato extract. *Journal of Food Science.* **64**: 445 -450.

Rosenberg, M., I.J. Kopelman and Y. Talmon. 1990. Factors affecting retention in spray drying microencapsulation of volatile materials. *Journal of Agricultural Food Chemistry.* **38**:1288 -1294.

Shahidi, F. and X.Q. Han.1993. Encapsulation of food ingredients. *Critical Reviews in Food Science and Nutrition.* **33**: 501–547.

Shahidi, F. and M. Naczk. 2004. Food phenolics : sources, chemistry, effects and applications, technomic Publ. Co. Lancasyter, PA. pp 281–313.

Sheikh, J., R. Bhosale and R.Singhal. 2006. Microencapsulation of black pepper oleoresin. Food chemistry. 105-110.

Shi, H., N. Noguchi and E. Niki. 2001. Galvinoxyl method for standardizing electron and proton donation activity. *Methods in Enzymology.* **335**: 157–166.

Shu, B., Y. Wenli , Z.Yaping and X. Liu. 2005. Study on microencapsulation of lycopene by spray-drying. *Journal of Food Engineering.* **86**: 85-94.

Soottitantawat, A., F. Bigeard, H. Yoshii, T. Furuta, M. Ohkawara and P.Linko. 2005. Influence of emulsion and powder size on the stability of encapsulated D-limonene by spray drying. Innovative Food Science and Emerging Technologies. **6:** 107–114.

Tanimoto, M. and R.Matsuno. 1992. Properties of agents that effectively entrap liquid lipids. Bioscience Biotechnology and Biochemistry. **56**: 477–480

Tari, T.A. and R.S. Singhal. 2002. Starch based spherical aggregates: reconfirmation of the role of amylose on the stability of a model flavouring compound, vanillin. *Carbohydrates Polymers.* **50**: 279–282.

Teunou, E. and D.Poncelet.2005. Rotary disc atomisation for microencapsulation applications – Prediction of the particle trajectories. *Journal of Food Engineering.* **71**: 345–353.

Timberlake, C.F., 1988. The biological properties of anthocyanin compounds. NATCOL Quarterly report Bulletin. 1, pp. 4–15.

Walker, A.R., E. Lee, J. Bogs, D.A. McDavid, M.R.Thomas and S.P. Robinson. 2007. White grapes arose through the mutation of two similar and adjacent regulatory genes. *Journal of Plant.* **49 (5):** 772–85.

Watanabe, Y., X.Fang, Y.Minemoto, S.Adachi and R.Matsuno. 2002. Suppressive effect of saturated acyl L-ascorbate on the oxidation of linoleic acid encapsulated with maltodextrin or gum arabic by spray-drying. *Journal of Agricultural and Food Chemistry.* **50:** 3984–3987.

Waterhouse, A. L and V. Teissedre. 1995. Wine and heart disease. Chemical India.338-341.

Whorton, C. and G.A. Reineccius. 1995. Evaluation of the mechanisms associated with the release of encapsulated flavor from maltodextrin matrices. In: Encapsulation and Controlled Release of Food Ingredients (ed) Rish & G.A. Reineccius. pp. 143–160. Washington, DC: American Chemical Society.

Xie,Y., H. Zhou and Z.R. Zhang. 2007. Effect of relative humidity on retention and stability of vitamin A microencapsulated by spray drying. *Journal of Biochemistry.* **31**: 68 -80.

Yoo Kyoung Park, Eunju Park , Jung-Shin Kima and Myung-Hee Kanga.2003. *Mutation Research.* **529**:77–86.

30

Nano Food Packaging to Enhance Shelf Life of Crop Produce

M.R. Manikantan and N. Varadharaju

Introduction

Food packaging is a critical technology addressing the ever-increasing demands for convenience, freshness, ease, shelf-life, safety and security of food products.

The main purpose of food packaging is to protect the food from microbial and chemical contamination, oxygen, water vapor and light. The type of packaging used therefore has an important role in determining the shelf-life of the food. Nanotechnology can be intervened to achieve the purpose of food packaging and it will become one of the most powerful forces for innovation in the food packaging industry.

Nanotechnology already formed the basis for a number of commercially available products. The technology has been widely used by the electronics industry for many years. Even agricultural production and food industries have witnessed the use of nanotechnology that it can be integrated into number of food systems and food packaging products (Manikantan and Varadharaju, 2007). The use of nanotechnology in Food Science and Technology is shown in Fig. 1.

In the food packaging arena, nano-materials are being developed with enhanced mechanical and thermal properties to ensure better protection of food from exterior mechanical, thermal, chemical or microbiological effects. This would endow packaged food with an additional level of safety and functionality. Also it would offer advantages along the supply chain and potentially increase the shelf-life of food. Some of the potential uses of nanotechnology in food packaging include modifying the permeation behaviour of films, increasing barrier properties, improving mechanical and heat resistance properties, developing active anti-microbial and antifungal surfaces, sensing and signaling microbiological and biochemical changes. Materials exhibiting antimicrobial properties by nano particles of silver have already entered the market.

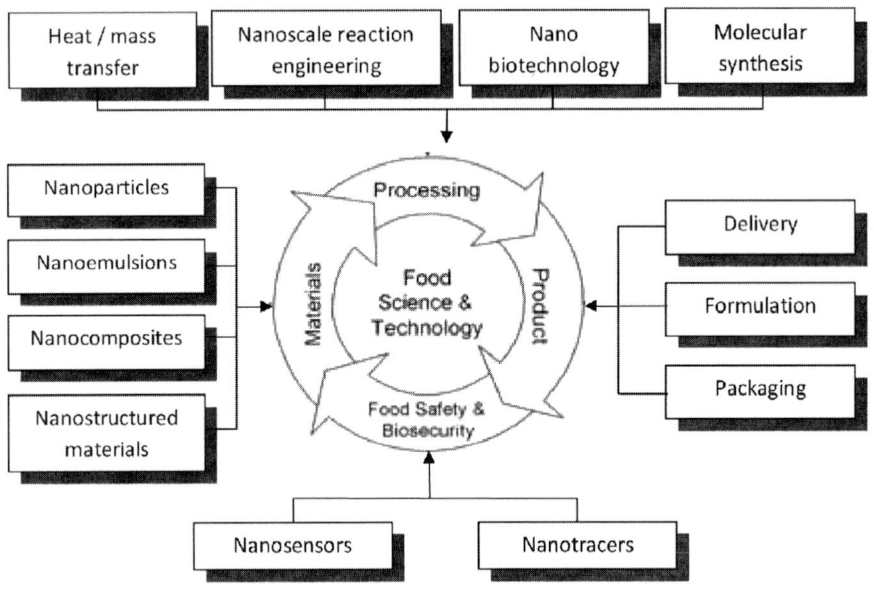

Fig. 1: Use of nanotechnology in Food Science and Technology

Nanotechnology enables the designers to alter the structure of the packaging materials on the molecular scale, to give the materials desired properties. With different nanostructure, the plastics can obtain various gas/water vapor permeability to fit the requirements of preserving fruit, vegetable, beverage, wine and other food. By adding nanoparticles, people can also produce bottles and packages with more light and fire-resistance, stronger mechanical and thermal performance and less gas absorption. These properties can significantly increase the shelf life, efficiently preserve flavour and colour and facilitate transportation and usage. Further, nanostructured film can effectively prevent the food from the invasion of bacteria and microorganism and ensure the food safety.

About 400-500 nano-packaging products are estimated to be in commercial use at the moment, while nanotechnology is predicted to be used in the manufacture of 25 percent of all food packaging within the next decade. The contributing factors to these developments include significant benefits in terms of lightweight but strong packaging materials and prolonged shelf life of packaged foodstuffs and the likely low risk to the consumer attributable to the fixed or embedded nature of engineered nano materials (ENMs) in plastic polymers. A number of nanotechnology-derived food contact materials (FCMs) are currently available worldwide, the main areas of application of which fall into the following broad categories:

- FCMs incorporating nanomaterials for improved packaging properties (flexibility, gas barrier properties, temperature/ moisture stability)
- Biodegradable polymer–nanomaterial composites, with enhanced mechanical and functional properties
- "Active" FCMs incorporating nanoparticles with antimicrobial or oxygen scavenging properties
- "Intelligent" and "Smart" food packaging, which incorporates nanosensors to monitor and report the condition of the food.

Nanoparticles reinforced materials for improved packaging properties

The most prominent products in the food industry's research and development pipeline include new polymer nanocomposites for packaging and wrapping. Nanocomposite technology has been described as the next great frontier of material science in packaging. This technology is developed to improve barrier performance pertaining to gases such as oxygen and carbon dioxide. It also enhances the barrier performance to ultraviolet rays, as well as adding strength, stiffness, dimensional stability, and heat resistance. (Manikantan et al., 2009).

Polymer composites with nanoclay

These are among the first nanocomposites to emerge on the market as improved materials for packaging (including food packaging). Nanoclay has a natural nanoscaled layer structure, which when incorporated into polymer composite restricts the permeation of gases. Nanoclay–polymer composites have been made from a thermoset or thermoplastic polymer reinforced with nanoparticles of clay. These include polyamides (PA), nylons, polyolefins, polystyrene (PS), ethylene-vinylacetate (EVA) copolymer, epoxy resins, polyurethane, polyimides and polyethyleneterephthalate (PET).

Polymer nanocomposites consist of resins (either thermoset or thermoplastics) that have fillers added with at least one dimension measured in nanometres. Because the nanoparticles are so small and their aspect ratios (largest dimension/smallest dimension) are very high, even at such low loadings, certain polymer properties can be greatly improved without the detrimental impact on density, transparency, and processability associated with conventional reinforcements like talc or glass. New plastics created with this technology demonstrate an increased shelf life and are less likely to shatter. These plastics may offer the improved characteristics at competitive price and more attractive for use in food and beverage packaging and pharmaceutical applications.

Preparations methods of nanocomposite are solution included intercalation, in-situ polymerization and melt processing. Melt processing is done simultaneously when the polymer is being processed through an extruder, injection molder, or other processing machine. The polymer pellets and clay are pressed together using shear forces to help with exfoliation and dispersion. With in-situ polymerization, clay is added directly to the liquid monomer during the polymerization stage. In the solution included intercalation method, clay is added to a polymer solution using solvents to integrate the polymer and clay molecules.

There are a number of nanoclay–polymer composites available commercially. Table 1 gives the details of commercial application of polymer clay nanocomposites. Known applications of nanoclay in multilayer film packaging include bottles for beer, carbonated drinks and thermoformed containers. Some large breweries are reported to be using the technology already in their beer bottles.

Chemical giant Bayer produces a transparent plastic film (called Durethan) containing nanoparticles of clay. The nanoparticles are dispersed throughout the plastic and are able to block oxygen, carbon dioxide and moisture from reaching fresh meats or other foods. The nanoclay also makes the plastic lighter, stronger and more heat-resistant.

Table 1: Commercial application of Polymer-Clay Nanocomposites

Supplier (Trade Name)	Base Resin	Reinforcement	Commercial Applications
Bayer AG, Germany (Durethan LDPU)	Nylon 6	Organo-clay	Barrier films
Clariant Technologies Packaging	Polypropylene (PP)		Organo-clay
General Motors R&D, Basell product	Thermoplastic Polyolefin (TPO)	Organo-clay Southern clay	Polyolefins and Auto body-side claddings, step assist for 2002 General Motors Safari and Astro vans
Ube Industries, Japan	Nylon 6 and Nylon 6/66 copolymer	Organo-clay	Barrier films, Timing belt cover for Toyota Motors
Honeywell (Aegis)	Nylon 6	Organo-clay	Bottles and film
Mitsubishi Motors GDI models	Nylon	Organo-clay	Engine cover on its
Kabelwerk Eupen of Belgium	Ethylene vinyl acetate copol ymer	Organo-clay	Wire and cable
Nanocor (Imperm)	Nylon 6	Organo-clay	Molding, PET beer bottles
RTP	Nylon 6	Organo-clay	Multi-purpose

Until recently, industry's quest to package beer in plastic bottles (for cheaper transport) was unsuccessful because of spoilage and flavour problems. Nanocor, a subsidiary of Amcol International Corp., is a world leader in the production of nanocomposites, which are polymers bonded with nanocrystals to provide the materials with enhanced properties. The company is currently producing nanocomposites for use in plastic beer bottles that give the contents a six-month shelf- life. The material works by introducing nanocrystals into the plastic that essentially create a maze from which oxygen molecules find it difficult to escape (Fig. 2). Nanocor and another company, Southern Clay, are working on improvements expected to increase shelf life up to 18 months.

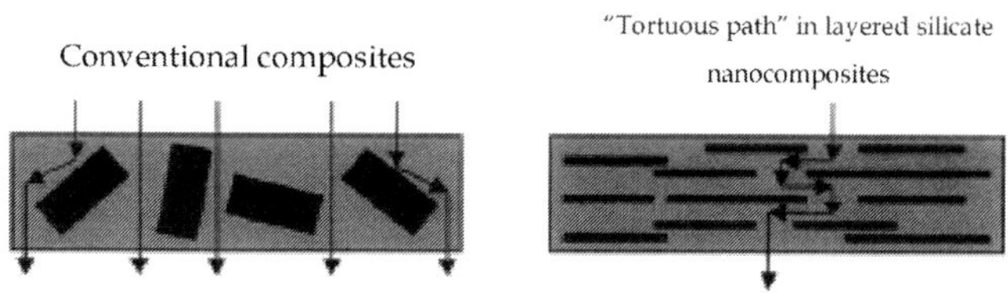

Fig. 2: Formation of tortuous path in polymer clay nanocomposite.

In addition to addressing spoilage and flavour issues, nanocomposites also offer other benefits like lighter weight and better recyclability, and suggest that producers may be able to significantly reduce transportation and production costs by reducing the amount of material used to package items. Nanocomposites, at the present time, are more expensive than most plastic resins but the price continues to drop and it is expected that nanocomposites materials will be very cost competitive in the near future.

Nano bio-degradable packaging

The use of nanomaterials to strengthen bio plastics (plant-based plastics) may enable bio based nanocomposite plastics to be used instead of fossil-fuel based plastics for food packaging and carry bags. Bio based nanocomposites are a new class of materials in food packaging industry with improved barrier and mechanical properties as compared to those of neat biopolymers. They are biodegradable and they are also produced from renewable resources. So, these make them environment friendly. It should be noted that barrier properties and especially mechanical properties of bio based nanocomposite films are stronger than edible films and synthetic polymeric films. Unlike edible films, they could not have been consumed as a part of food.

Bio based nanocomposites are produced from incorporation of nanoclay into biopolymers (or Edible films). Advantages of bio based nanocomposites are numerous and possibilities for application in the packaging industry are endless. Bio based nanocomposites can be used to extend the shelf-life of the fresh products such as fruits and vegetables by controlling respiratory exchange. Also it can improve the quality of fresh, frozen, and processed meat, poultry, and seafood products by retarding moisture loss, reducing lipid oxidation and discoloration, enhancing product appearance, and reducing oil uptake by battered and breaded products during frying.

Bio based nanocomposites are composed of biopolymer, nanoclay and usually compatibilizing agents. Major component of bio based nanocomposites is biopolymers. Biopolymers have great commercial potential for bio plastic and edible films, but some of the properties such as brittleness, low heat distortion temperature; high gas permeability and low melt viscosity for further processing restrict their use in a wide range of applications (Ray and Bousmina, 2005). As mentioned before, modification of biopolymers with nanotechnology is an effective way to improve their properties. Biopolymers derived from renewable resources are broadly classified according to method of production. This gives the following three main categories (Petersen et al., 1999):

i. Biopolymers directly extracted/removed from natural materials (mainly plants) such as hydrocolloids (polysaccharides and proteins). The most frequently utilized polysaccharides were cellulose and starch (and their derivatives), chitosan, seaweed extracts (carrageenans and alginates), exudates (arabic gum), seed (guar gum), xanthan and gellan gum and pectin. Proteins include collagen, gelatin, casein, whey proteins, corn zein, wheat gluten and soy proteins.

ii. Biopolymers produced by classical chemical synthesis from renewable bio derived monomers like polylactate.

iii. Biopolymers produced by microorganisms or genetically transformed bacteria like Polyhydroxyalkanoates.

Hence, biopolymers which can be used in bio based nanocomposites formulation are numerous.

The utilization of special compatibilizing agents (modifier) between the two basic materials (biopolymer and nanoclay) for the preparation of biobased nanocomposite is necessary. Layered silicates are characterized by a periodic stacking of mineral sheets with a weak interaction between the layers and a strong interaction within the layer. The space in-between the layers is occupied by cations. By cation exchange reactions between the clay and organic cations (such as alkyl ammonium salts), the layered silicate can be transformed into organically modified clay. The inter-layer distance will increase by using voluminous modifiers. If this modifier is compatible with biopolymer as well, a homogeneously and nanoscaled distribution (exfoliation) of the clay sheets can be effected in the polymer matrix. The pure clay shows an interlayer distance of 1.26 nm. It has been proven by XRD analysis that most of the layers are indeed *swollen* after the modification reaction. The inter-layer distance changes to 2.34 nm, an increase of nearly 100% compared to the pure clay.

Biopolymers like starch present some drawbacks, such as the strong hydrophilic behavior (poor moisture barrier) and poorer mechanical properties than the conventional non-biodegradable plastic films used in the food packaging industries (McGlashan and Halley, 2003; Park et al., 2003; Park et al., 2002). So, incorporation of nanoclay in biopolymers like starch can improve its properties such as barrier and mechanical properties (Vaia et al., 2000). The most commonly used nanoclays include montmorillonite, a 2:1 phyllosilicate (Chiou et al., 2005).

Antimicrobial nanopackaging materials

The incorporation of antimicrobial compounds into food packaging materials has received considerable attention. Films with antimicrobial activity could help control the growth of pathogenic and spoilage microorganisms. An antimicrobial nanocomposite film is particularly desirable due to its acceptable structural integrity and barrier properties imparted by the nanocomposite matrix, and the antimicrobial properties contributed by the natural antimicrobial agents impregnated within. Materials in the nanoscale range have a higher surface to volume ratio when compared with their microscale counterparts. This allows nanomaterials to be able to attach more copies of biological molecules, which confers greater efficiency. Nanoscale materials have been investigated for antimicrobial activity so that they can be used as growth inhibitors, killing agents or antibiotic carriers.

Polymer composites with nano-metals or metal oxides

Polymer nanocomposites incorporating metal or metal oxide nanoparticles are utilized mainly for their antimicrobial action, abrasion resistance, UV absorption, and strength. Some nano materials have been used to develop active packaging that can absorb oxygen and therefore keep food fresh. Other nano materials have been incorporated as UV absorbers to prevent UV degradation in plastics such as PS, PE and PVC. The commercially important nanomaterials in this respect include nanosilver and nanozinc oxide for antimicrobial action, nanotitanium dioxide for UV protection in transparent plastics, nanotitanium nitride for mechanical strength and as a processing aid, and nano silica for surface coating.

The most common nanocomposites used as antimicrobial films for food packaging are based on silver, which is well known for its strong toxicity to a wide range of microorganisms, with high temperature stability and low volatility. Some mechanisms have been proposed for the antimicrobial property of silver nanoparticles (Ag-NPs): adhesion to the cell surface, degrading lipopolysaccharides and forming "pits" in the membranes, largely increasing permeability; penetration inside bacterial cell, damaging DNA; and releasing antimicrobial $Ag+$ ions by Ag-NPs dissolution. The latter mechanism is consistent with findings by Kumar and Munstedt (2005), who affirmed that the antimicrobial activity of silver-based polymers depends on releasing of $Ag+$, which binds to electron donor groups in biological molecules containing sulphur,oxygen or nitrogen.

It is important to note that the surface biocides, such as nano silver, in packaging materials are not intended to have a preservative effect on the food.

Instead, the biocidal agent is intended to help maintain the hygienic condition of the surface by preventing or reducing microbial growth. Where the use of a nano material gives a preservative effect in the packaged product, there would be a requirement for additional regulatory authorization as a direct food additive in most countries. Based on the antimicrobial action of nano silver, a number of "active" FCMs have been developed that are claimed to preserve the food materials by inhibiting the growth of micro-organisms. Examples include food storage containers and plastic storage bags. Nano silver has also been incorporated into the inner surface of some domestic refrigerators to prevent microbial growth and maintain a clean and hygienic environment in the fridge. The discovery of antimicrobial properties of nano zinc oxide and nano magnesium oxide at the University of Leeds may provide more affordable materials for such applications in food packaging (Zhang et al., 2007). A plastic wrap containing nano zinc oxide is also available, which is claimed to sterilize under indoor lighting.

Nanoscale chitosan also has antibacterial activity. One possible antimicrobial mechanism by nanoscale chitosan involves interactions between positively charged chitosan and negatively charged cell membranes, increasing membrane permeability and eventually causing rupture and leakage of intracellular material. Another two antimicrobial mechanisms suggest: chelation of trace metals by chitosan, inhibiting enzyme activities; and, in fungal cells, penetration through the cell wall and membranes to bind DNA and inhibit RNA synthesis. Carbon Nano Tubes (CNTs) have been also reported to have antibacterial properties. Direct contact with aggregates of CNTs was demonstrated to be fatal for E. coli, possibly because the long and thin CNTs puncture microbial cells, causing irreversible damages.

Coatings containing nanoparticles

Coatings that contain nanoparticles are used to create antimicrobial, scratch resistant, anti-reflective, or corrosion-resistant surfaces. This involves the coating of nanoparticulate form of a metal, metal oxide or a film resin substance with nanoparticles. Examples of FCMs with nanocoating include antibacterial kitchenware, cutting boards and teapots. High-barrier nanocoatings have also been developed that contain numerous nanodispersed platelets per micron of coating thickness to increase the barrier properties of PET; this enhances the oxygen barrier when used in food and drink applications, ensuring longer shelf life. The coatings have been reported to be very efficient at keeping out oxygen and retaining carbon dioxide and can rival traditional active packaging technologies such as oxygen scavengers (Garland, 2004). Examples include a nanocoating which is an aqueous-based nanocomposite barrier coating that provides an oxygen barrier with a 1–2 micron coating for food packaging use, and plasma arc deposition of amorphous carbon inside PET bottles as a gas barrier.

Waxy coating is used widely for some foods such as apples and cheeses. Recently, nanotechnology has enabled the development of nanoscale edible coatings as thin as 5 nm thick, which are invisible to the human eye. These edible nano-coatings could be used on meats, cheese, fruits, vegetables, confectionery, bakery goods and fast foods. They could provide a barrier to moisture and gas exchange, act as a vehicle to deliver colours, flavours, antioxidants, enzymes and anti-browning agents and could also increase the shelf life of manufactured foods, even after the packaging is opened. The U.S. Company Sono-Tec Corporation announced in early 2007 that it has developed an edible antibacterial nano-coating, which can be applied directly to bakery goods. It is currently testing the process with its clients.

Another trend in this respect is the chemical release nano-packaging. This technique enables food packaging to interact with the food. The exchange can be processed in both directions. Packaging can release nanoscale antimicrobials, antioxidants, flavours, fragrances or nutraceuticals into the food or beverages to extend its shelf life or to improve its taste or smell.

In many instances chemical release packaging also incorporates surveillance elements, where the release of nano-chemicals will occur in response to a particular trigger event. Conversely, nano-packaging using carbon nanotubes are being developed with the ability to pump out carbon dioxide that would otherwise result in food or beverages deterioration. Nano-packaging that absorbs undesirable flavours is also in development stage.

The chemical release packaging is also designed to release biocides in response to the growth of a microbial population, humidity or other changing conditions. Other packaging and food contact materials incorporate antimicrobial nanomaterials, that are designed not to be released, so that the packaging itself acts as an antimicrobial agent. These products commonly use nanoparticles of silver, although some use nanosize zinc oxide or chlorine dioxide and other materials. Nano magnesium oxide, nano copper oxide, nano titanium dioxie and carbon nanotubes are also predicted for future use in antimicrobial food packaging.

Antimicrobial nano-emulsions

Nano-emulsions have been developed for use in the decontamination of food packaging equipment and in the packaging of food. A typical example is a nanomicelle-based product which is claimed to contain natural glycerine and removes pesticide residues from fruits and vegetables, as well as the oil/dirt from cutlery.

Intelligent packaging concepts based on nanosensors

Nanotechnology has also enabled the development of nanosensors that can be applied as labels or coatings to add an intelligent function to food packaging in terms of ensuring the integrity of the package through detection of leaks (for foodstuffs packed under vacuum or inert atmosphere), indications of time–temperature variations (e.g. freeze–thaw–refreezing), or microbial safety (deterioration of foodstuffs). Examples include an indicator that turns from transparent to blue, informing the consumer that air has entered the modified atmosphere of the packaged materials. For this type of application, nanotechnology-derived printable inks have been developed. One example is an oxygen detecting ink containing light-sensitive titanium oxide (TiO_2) nanoparticles, which only detect oxygen when they are "switched on" with UV light. Other conductive inks for ink jet printing based on copper nanoparticles have also been developed (Park et al., 2007).

Food safety also requires confirmation of the authenticity of products. This is where application of nano-barcodes incorporated into printing inks or coatings has shown the potential for use in tracing the authenticity of the packaged product (Han et al., 2001). Food quality indicators have also been developed that provide visual indication to the consumer when a packaged foodstuff starts to deteriorate. An example of such food quality indicators is a label based on detection of hydrogen sulphide, which is designed for use on fresh poultry products. The indicator is based on a reaction between hydrogen sulphide and a nanolayer of silver (Smolander et al., 2004). The nanosilver layer is opaque light brown, but when processed meat starts to deteriorate silver sulphide is formed and the layer becomes transparent, indicating that the food may be unsafe to consume.

Other materials developed for potential food packaging applications are based on nanostructured silicon with nanopores. The potential applications include detection of pathogens in food and variations of temperature during food storage. Another relevant development is aimed at providing a basis for intelligent preservative packaging technology that will release a preservative only when a packaged food begins to spoil (ETC Group, 2004).

Based on applied studies of the surface properties of materials, several types of gas sensors have been developed, which translates chemical interactions between particles on the surfaces into a response signal. Metal oxide gas sensors are one of the most popular type of sensors because of their high sensitivity and stability. Conducting polymers (CPs) or electro active conjugated polymers, which can be synthesized either by chemical or electrochemical oxidation, are very important because of their electrical, electronic, magnetic and optical properties, which are related to their conjugated Π electron backbones. Polyene and polyaromatic CPs such as polyaniline (PANI), polyacetylene, polypyrrole (PPy) has been widely studied. Electrochemically polymerized CPs have a remarkable ability to switch between conducting oxidized (doped) and insulating reduced (undoped) state, which is the basis of many applications. Food spoilage is caused by microorganisms, whose metabolism produces gases which can be detected by conducting polymer nanocomposites (CPC) or metal oxides, which can be used for quantification and/or identification of microorganisms based on their gas emissions. Sensors based on CPC consist on conducting particles embedded into an insulating polymer matrix. The resistance changes of the sensors produce a pattern that corresponds to the gas under investigation.

Nanotechnology is also enabling sensor packaging to incorporate cheap Radio Frequency Identification (RFID) tags. The nano-enabled RFID tags are much smaller, flexible and can be printed on thin labels. This increases the tags versatility and thus enables much cheaper production. Other nano-based track and trace packaging technologies are also being improved. For instance, a United States company Oxonica Inc, has developed nano-barcodes to be used for individual items or pellets, which must be read with a modified microscope. These have been developed primarily for anti- counterfeiting purposes.

Summary

Packaging has developed into an essential technology in the handling and commercialization of foodstuffs to provide, by maintaining or even increasing, the required levels of quality and safety. There are high expectations in food and packaging: longer shelf life, safer packaging, better traceability of products and healthier food is only a few of the expected improvements. Nanotechnology may be effectively applied in the various areas of food packaging such as designing the packaging material according to the barrier requirements of the food materials to be stored, anti microbial coatings to ensure the safety of the packaged food, sensing the reactions inside the packaging environment and utilizing the bio materials as packaging materials to reduce the pollution and intelligent packaging by using nanosensors.

References

Chiou, B.S., Yee, E., Glenn, G.M. and Orts, W.J. 2005. Rheology of starch–clay nanocomposites. *Carbohydrate Polymers*, **59**: 467–475.

ETC Group. 2004. Food Packaging Using Nanotechnology Methods: an Overview of 'Smart Packaging' and 'Active packaging', http://www.azonano.com/ Article ID=1317

Garland, A. 2004. Nanotechnology in plastics packaging. Commercial applications in nanotechnology (ed). Pira International Limited UK, 14 -63.

Han, M., Gao, X., Su, J.Z. and Nie, S. 2001. Quantum-dot-tagged microbeads for multiplexed optical coding of biomolecules. *Nat. Biotechnol.*, **19**:631–635.

Huang, M., Yu, J. and Ma, X. 2006. High mechanical performance MMT-urea and formamide-plasticized thermoplastic corn starch biodegradable nanocomposites. *Carbohydrate Polymers*, **63**:393–399.

Kumar, R. and Munstedt, H. 2005. Silver ion release from antimicrobial polyamide/silver composites. *Biomaterials*, **26**:2081–2088.

Manikantan, M.R., Arumuganathan, T., Balakrishnan, M. and Varadharaju, N. 2009. Role of Nanotechnology in Food Industry, *Beverage and Food World*, 41-44.

Manikantan, M.R. and Varadharaju. 2007. Nanotechnology in Food Preservation. In: Nanotechnology Applications in Agriculture. (eds.) Chinnamuthu, C.R., Chandrasekaran, B. and Ramasamy, C., TNAU, Coimbatore, 63-76.

McGlashan, S. A. and Halley, P. J. 2003. Preparation and characterization of biodegradable starch-based nanocomposite materials. *Polymer International*, **52**:1767–1773.

Park, B.K., Kim, D., Jeong, S., Moon, J. and Kim, J.S. 2007. *Thin solid films*, **515 (19):** 7706–7711.

Park, H. M., Lee, W.K., Park, C.Y., Cho, W.J. and Ha, C.S. 2003. Environmentally friendly polymer hybrids Part 1. Mechanical, thermal, and barrier properties of thermoplastic starch/clay nanocomposites. *Journal of Materials Science*, **38**: 909–915.

Park, H.M., Li, X., Jin, C.Z., Park, C.Y., Cho, W.J. and Ha, C.S. 2002. Preparation and properties of biodegradable thermoplastic starch/clay hybrids. *Macromolecular Materials and Engineering*, **287**:553–558.

Petersen, K., Nielsen, V., Bertelsen, G., Lawther, M., Olsen, M.B., Nilsson, N. H. and Mortensen, G. 1999. Potential of biobased materials for food packaging. *Trends in Food Science & Technology*, **10**:52-68.

Ray, S.S. and Bousmina, M. 2005. Biodegradable polymers and their layered silicate nanocomposites: In greening the 21st century materials world. *Progress in Materials Science*, **50**:962–1079.

Smolander, M., Hurme, E., Koivisto, M. and Kivinen, S. 2004. PCT International Patent Application, WO 2004/102185 A1.

Vaia, R. A., In Pinnavaia, T. J. and Beall, G. W. 2000. Polymer–clay nanocomposites (Eds.), New York, Wiley, 229–266.

Zhang, L., Jiang, Y., Ding, Y., Povey, M. and York, D. 2007. Investigation into the antibacterial behaviour of suspensions of ZnO nanoparticles (ZnO nanofluids). *J. Nanoparticle Res.*, **9(3)**:479–489.

31

Nano Remediation of Soil and Aquatic Pollutants

C. Udayasoorian, K. Vinoth Kumar and R.M. Jayabalakrishnan

Introduction

Maintaining and restoring the quality of air, water and soil is one of the great challenges of our time. Most countries face serious environmental problems, such as the availability of drinking water, the treatment of waste and wastewater, air pollution and the contamination of soil and groundwater. There are two main approaches to these problems: the first is to control pollution at source to minimize or eliminate waste production or harmful emissions; the second is to remedy the pollutants that accumulate in the environment. Considering the enormous scale of soil and groundwater contamination, the complexity of the task seems intractable. In many cases, conventional remediation and treatment technologies have shown only limited effectiveness in reducing the levels of pollutants, especially in soil and water (Rickerby and Morrison, 2007). Nanotechnology offers a number of promising solutions to these environmental issues and represents a new approach to the cleanup of contaminated sites and waste streams, particularly for those substances that are highly toxic, persistent, or difficult to treat using conventional means. It also includes green manufacturing and green energy production to reduce pollution at source. Nanotechnology promises the creation of improved, more specific, faster and more cost effective solutions for pollution treatment. The early impact of nanotechnology research has been seen mostly in remediation and treatment technologies (Masciangioli and Zhang, 2003).

Since the 1990s the use of engineered nano-materials (ENMs) in a variety of environmental applications, such as water purification, waste water treatment, indoor and outdoor air cleaning, and soil and groundwater remediation, has been investigated. Various applications have been successfully demonstrated at the laboratory scale, but most of them still require verification of their efficacy and safety in the field. It is thus not surprising that to date few nanotechnological applications for environmental use have been commercialized. However, some environmental applications of nanomaterials have entered the market or are under close investigation. These include,

- Nanoscale TiO_2 for the photocatalytic degradation of contaminants in air (e.g. NOx, VOCs) and water (e.g. microorganisms, organic materials)
- Nanofiltration for wastewater treatment and drinking water purification (e.g. removal of hardness and desalination)
- Nanoscale zero-valent iron for soil and groundwater remediation

Nano-remediation

Nanoremediation is the application of reactive nanomaterials for trans-formation and detoxification of pollutants. These nanomaterials have properties that enable both chemical reduction and catalysis to mitigate the pollutants of concern (Otto et al., 2008).

Soil and groundwater remediation

Contamination of soil and groundwater with carcinogenic organic substances and/or metals occurs all around the world. Groundwater sources are also frequently contaminated with pesticides or halogenated compounds. Landfill leakage, agriculture and chemical accidents are the main sources of these pollutants. Conventional remediation technologies include *ex-situ* soil washing and pump and treat operations and *in-situ* thermal treatment, chemical oxidation and use of reactive barriers with iron (Hodson, 2010). Soil and groundwater remediation is generally very expensive, and conventional methods are not always successful or they take a long time for the remediation to become effective. Environmental remediation methods can be classified as adsorptive or reactive and as *in-situ* or *ex-situ* (Hodson, 2010). The use of nanomaterials in all these scenarios has been investigated (Table. 1).

Table 1: Classification of remediation methods involving nanoparticles

	In-situ	*Ex-situ*
Adsorptive	*In-situ* sequestration of contaminants by adding binding agents, e.g. iron oxides	Extraction of contaminated solution, which is then treated with adsorbents, as in nanofiltration
Reactive	*In-situ* reaction of nanomaterial, e.g.n ZVI, with target contaminant	Extraction of contaminated solution, which is then treated with reactants, as in TiO_2 photo-oxidation

Ex-situ application

Ex situ or pump and treat systems operate on the basis of removing contaminated groundwater from the ground, downstream of the contamination site and then treating it before returning it to the ground. SAMMS (Self Assembled Monolayers on Mesoporous Supports) are created by self assembly of a monolayer of functionalized surfactants onto mesoporous ceramic supports, resulting in very high surface areas (~1000 m^2/g) with adsorptive properties that can be tuned to target contaminants such as mercury, chromate, arsenate, pertechnetate, and selenite. Dendritic polymers are another type of nanostructured material that has the potential for use in remediation. Both of these types of nanostructured adsorbants are most likely to be applied *ex situ*, where they can be recovered with the concentrated hazardous material that they adsorb. Nanotechnologies that affect remediation by contaminant degradation rather than adsorption are particularly attractive for organic contaminants. A well established approach for remediation of organic contaminants is photooxidation and the potential benefits of quantum sized photocatalysts have long been recognized for contaminant degradation applications. However, as with the absorptive technologies described above, photooxidation with nanostructured semiconductors is primarily an *ex-situ* strategy, in this case because effective illumination usually requires that treatment be done in a reactor that is designed for this purpose. With this technology "it takes a long time to achieve cleanup goals, it has been demonstrated to spread contamination in some cases, and it is expensive to operate and maintain," requiring continual energy input (USEPA, 2004).

In-situ application

In soil and groundwater remediation, *in-situ* applications seem to be most promising as they are in general less costly. For nanoremediation *in situ*, no groundwater is pumped out for above ground treatment, and no soil is transported to other places for treat-ment and disposal. Nanomaterials have highly desired properties for *in-situ* applications. Because of their minute size and innovative surface coatings, nanoparticles may be able to pervade very small spaces in the subsurface and remain sus-pended in groundwater, allowing the particles to travel farther than larger, macrosized particles and achieve wider distribution. However, in practice, current nanomaterials used for remediation do not move very far from their injection point (Tratnyek and Johnson, 2006).

For *in-situ* treatment, it is necessary to create either an *in-situ* reactive zone with relatively immobile nanoparticles or a reactive nanoparticle plume that migrates to contaminated zones. For applications in topsoil, nanoparticles can be worked into the surface of the contaminated soil using conventional agricultural practices. These different approaches are shown in Fig. 1.

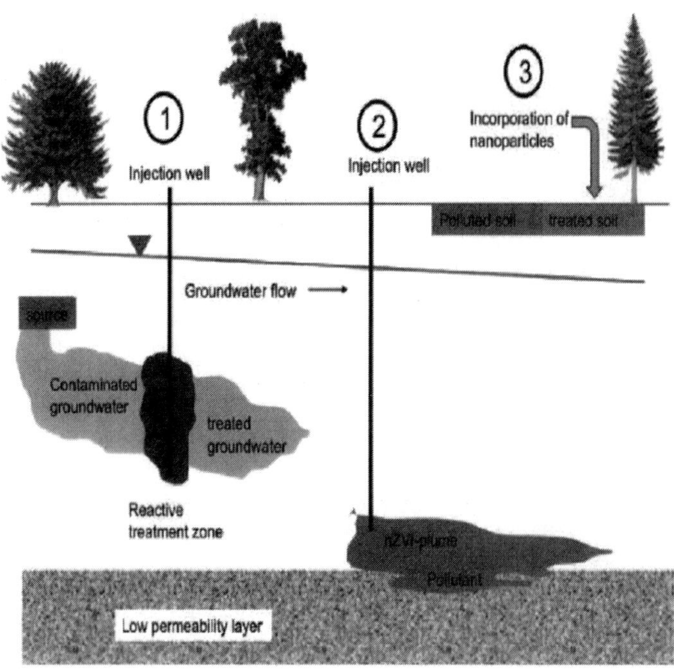

Fig. 1: In-situ technologies used to treat polluted groundwater and soil
(1) Injection of nZVI to form a reactive barrier (2) Injection of mobile nZVI to form an nZVI plume (3) Incorporation of NP into topsoil to adsorb or degrade pollutants

Most of the field application of nanoparticles for remediation at full scale have focused on nZVI and related products. Full scale commercial applications of nZVI in remediation are quickly becoming common, and there already are competitive markets among purveyors of nZVI materials and geotechnical service providers with various methods for deploying nZVI. The rapid emergence of remediation technologies based on nZVI has obscured and perhaps even exacerbated some pervasive misconceptions regarding the fundamental principles underlying this technology and the practical implications of its use in the environment. A common type of *in-situ* remediation method, currently used to clean up contaminated groundwater is the permeable reactive barrier (PRB). PRBs are treatment zones composed of materials that degrade or immobilize contaminants as the groundwater passes through the barrier. They can be installed as permanent, semi permanent or replaceable barriers within the flow path of a contaminant plume.

Site characterization and evaluation

Nanoremediation has site specific requirements that must be met in order for it to be effective. Adequate site characterization is essential, including information about site location, geologic conditions, and the concentration and types of contaminants. Geologic, hydro geologic, and subsurface conditions include composition of the soil matrix, porosity, hydraulic conductivity, ground water gradient and flow velocity, depth to water table, and geochemical properties (pH, ionic strength, dissolved oxygen and concentrations of nitrate, nitrite, and sulfate). All of these variables need to be evaluated before nano particles are injected to determine whether the particles can infiltrate the remediation source zone, and whether the conditions are favourable for reductive transformation of contaminants. The sorption or attachment of nanoparticles to soil and aquifer materials depends on the surface chemistry of soil and nanoparticles, ground water chemistry, and hydro dynamic conditions. The reactions between the contaminants and the nZVI depend on contact or probability of contact between the pollutant and nano particles (USEPA, 2008).

Potential applications of nanomaterials for remediation

Recently, a variety of nanomaterials for potential use to adsorb or destroy contaminants as part of either *in-situ* or *ex-situ* processes. Many different nanoscale materials have been explored for remediation, such as nanoscale zeolites, metal oxides, carbon nanotubes and fibers, enzymes, various noble metals [mainly as bimetallic nanoparticles (BNPs)] and titanium dioxide.

Iron oxides can strongly adsorb metals in soils. Adding nanoscale metal oxides to soils will thus immobilize soil metals (Schorr, 2007). A mixture of iron and iron oxide has also been shown to be effective in phosphate removal even more effective than higher-cost products such as activated alumina, while being active for even longer periods. Green rust is a very reactive iron oxide can be used to reduce Cr (VI) to Cr (III), which is not soluble and much less toxic than the mutagenic Cr (VI) (Rickerby and Morrison, 2007). Carbon-based nanomaterials, such as dendrimers and polymers, are also currently being explored for the removal of metals and organics from soils and groundwaters (Mueller and Nowack, 2009). Iron oxide minerals can be used to adsorb not only metals but also arsenic (Rickerby and Morrison, 2007). Arsenic in groundwaters in the Bengal region of Southeast Asia and elsewhere constitutes a major hazard to the health of millions of people who use these waters for drinking, cooking and irrigation (Smedley and Kinniburgh, 2002). Iron oxide NPs can bind arsenic 5–10 times more effectively than larger particles. In laboratory tests, more than 99% of the arsenic in water was bound by 12 nm diameter iron oxide nanoparticles.

A new study by Pan et al. (2010) found that nanoscale magnetite particles can be used successfully to immobilize phosphate in soil by adsorption. They compared the immobilization efficiency of micro and nanoscale particles stabilized by a coating of carboxymethyl cellulose and found that only nanoscale particles were able to penetrate the soil column. Non-stabilized 'nanomagnetite' could not pass through the soil column under gravity because it quickly agglomerated into microparticles. The coated NPs were more transportable due to their small size and the increased negative charge associated with the carboxymethyl groups. Consequently, a large proportion (72%) of the coated NPs could pass through the column due to their charge repulsion with the negatively charged soil particles. Transport over a certain distance is needed to achieve a good distribution of the reactive particles in the soil matrix and therefore an even adsorption capacity for phosphate. Much research has been conducted on nanosorbents, and although the results are promising, there are only a few commercial applications so far. The problems with using nanosorbents in soils are similar to those encountered with conventional adsorbents (O'Day and Vlassopoulos, 2010).

The use of nZVI in groundwater remediation is the most widely investigated environmental nanotechnological technique and has considerable potential benefits (Nowack, 2008). Recent research has suggested that as a remediation technique, nZVI have several advantages;

- Effective for the transformation of a large variety of environmental contaminants
- Inexpensive and
- Nontoxic to environment
- There are two ways to use nZVI in groundwater remediation;
- nZVI is injected to form a reactive barrier of iron particles
- nZVI is injected in surface-modified form (e.g. coated with polyelectrolytes, surfactants or cellulose/polysaccharides) to establish a plume of reactive iron, which destroys any organic contaminants within the aqueous phase.

The potential applications of nZVI for site remediation are illustrated in fig.2. Several studies have shown that nZVI as a reactive barrier is very effective in the reductive degradation of halogenated solvents, such as chlorinated methanes, brominated methanes, trihalomethanes, chlorinated ethenes, chlorinated benzenes and other polychlorinated hydrocarbons, in groundwater (Schorr, 2007). nZVI has also been shown to be effective against pesticides and dyes (Zhang, 2003). Efficient removal by nZVI of polycyclic aromatic hydrocarbons (PAHs) adsorbed to soils has been reported at room temperature (Chang et al., 2007), while under the same conditions only 38% of the polychlorinated biphenyls (PCBs) were destroyed because of the very strong sorption of PCBs to the soil matrix (Varanasi et al., 2007).

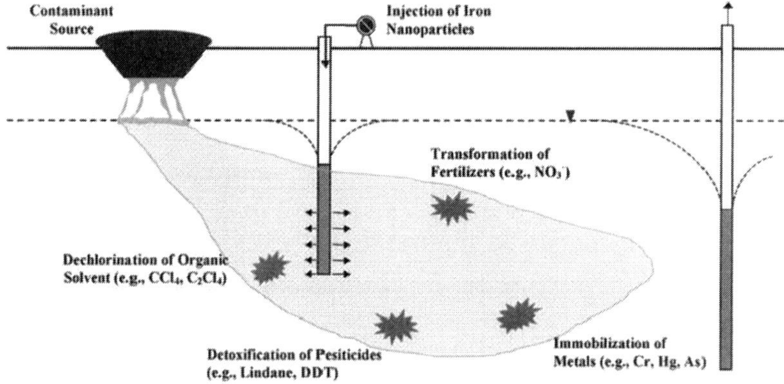

Fig. 2: Nanoscale iron particles for *in-situ* remediation

In an aqueous media, ZVI will react with dissolved oxygen, water and if present with other oxidants such as nitrate and possible contaminants. Due to the redox reactions taking place, the application of nZVI results in an increase of the pH and a decrease of the oxidation reduction potential (ORP). The consumption of oxygen further leads to an anaerobic environment while the reduction of water yields hydrogen. If chlorinated hydrocarbons are present, the following reaction takes place:

$$R - Cl + Fe^0 + H_2O \rightarrow R - H + Fe^{2+} + Cl^{-} + OH^-$$

There are two pathways for the degradation of chlorinated hydrocarbons by nZVI. The first is hydrogenolysis through sequential dehalogenation (PCE to TCE to DCE to VC to ethene). The second is beta elimination (TCE to chloroacetylene to acetylene to ethene). The latter occurs in 70-90 per cent of the reactions with nZVI primarily when the contaminant comes in direct contact with the nZVI. If nZVI reacts with ionic heavy metals such as Pb^{2+}, the following reaction takes place.

$Pb^{2+} + Fe^0 \rightarrow Pb^0 + Fe^{2+}$

Semiconductor photocatalysts such as titanium dioxide (TiO_2), zinc oxide (ZnO), iron oxide (Fe_2O_3) and tungsten oxide (WO_3) act as photocatalysts (Nagaveni et al., 2004). Due to their light absorbing capabilities, they are employed in a variety of applications. Both titanium dioxide and zinc oxide can be used as pigments to provide whiteness for substances, such as paint and paper. The ability of photocatalysts to absorb ultraviolet light makes them useful in sunscreen as well as cosmetics to provide opaqueness to the creams or lotions. These properties also can be exploited for antimicrobial coatings; the photocatalytic properties allow thin coatings to be self cleaning and to have disinfecting properties after exposure to UV radiation. Photocatalysts have the ability to oxidize organic pollutants into nontoxic materials. Traditionally, TiO_2 has been used in advanced photochemical oxidation (APO) processes for environmental remediation because of its low toxicity, high photoconductivity, high photostability, availability, and low cost (Nagaveni et al., 2004). Field and pilot scale applications have demonstrated the ability of TiO_2 to remediate fuel contaminated groundwater containing benzene, toluene, ethylbenzene, and xylene (BTEX) compounds. In addition, TCE, Methyl-*tert*-butyl ether (MTBE), chloroform, ethylbenzene, and nitrobenzene can be destroyed by semiconductor photocatalysis (Rajeshwar et al., 2001).

Recently, ZnO has been proposed as a dual function photocatalytic material. Kamat et al. (2002) reported that ZnO possessed both sensing and remediating capabilities for organic contaminants in water. The study concluded that the ZnO films showed a high degree of sensitivity on the order of 1 ppm to aromatic compounds such as chlorinated phenols. Under UV lighting, the films degraded the aromatic compounds.

Nanotubes are engineered molecules most frequently made from carbon. They are electrically insulating, highly electronegative and easily polymerized. Nanotubes have also been made from TiO_2 and have demonstrated potential for use as a photocatalytic degrader of chlorinated compounds (Chen et al., 2005). Bench scale research has shown TiO_2 nanotubes to be particularly effective at high temperature, capable of reducing contaminant chemicals by greater than 50 percent in three hours (Xu, 2005). Bench scale tests using ferritin, an iron storage protein, have indicated that it can reduce the toxicity of contaminants such as chromium and technetium in surface water and groundwater to facilitate remediation. Like TiO_2, ferritin is photocatalytic; in one bench scale project, the addition of visible light caused ferritin to reduce toxic, water soluble hexavalent chromium to the less toxic trivalent chromium, which is not water soluble and precipitates out of solution. Dendrimers are hyper branched, well organized polymer molecules made up of three components:

core, branches, and end groups. Dendrimer surfaces terminate in several functional groups that can be modified to enhance specific chemical activity. Fe^0/FeS nanocomposites, synthesized using dendrimers as templates, could be used to construct permeable reactive barriers for the remediation of contaminated groundwater. Bench scale research has indicated that dendrimers have flexible delivery options (Diallo et al., 2006).

Nanoscale calcium peroxide

Nanoscale calcium peroxide has recently been used for the clean-up of oil spills (Karn et al., 2009). The nano-sized calcium peroxide used as an oxidant in the remediation of soils containing various organic contaminants, such as gasoline, heating oil, methyl tertiary butyl ether (MTBE), ethylene glycol and solvents. Nanoscale calcium peroxide is claimed to be highly efficient in removing aromatics and is also used in enhanced bioremediation. The oxygen produced in the reaction of calcium peroxide with water leads to an aerobic environment that supports natural bioremediation by aerobic organisms present in the soil.

Conclusion

Nanoparticles can potentially be used for the remediation of soil and groundwater. The small particles are highly reactive and have great sorption capacity. Nanoremediation can reduce the overall costs of cleaning up large scale contaminated sites, reduce cleanup time, eliminate the need for treatment and disposal of contaminated dredged soil, and reduce some contaminant concentrations to near zero and it can be done *in-situ*. Nanotechnology based products and processes promise to be ecologically beneficial in terms of energy consumption, materials saving, replacements of toxic materials and resource preservation. In order to prevent any potential adverse environmental impacts, proper evaluation including full scale ecosystem wide studies of these nanoparticles needs to be addressed before this technique is used on a mass scale. However, technical challenges, such as the delivery of the particles to the target area, have to be solved. There are also concerns regarding the release of large quantities of manufactured nanoparticles into the soil prior to extensive human and ecological toxicity testing.

References

Banfield, J.F. and Zhang, H.Z. 2001. Nanoparticles in the environment. In: Banfield J.F and Navrotsky A (eds), Nanoparticles and the Environment. Reviews in Mineralogy and Geochemistry 44, Mineralogical Society of America, Chantilly, VA: pp. 1-58.

Chang, M.C., Shu, H. Y., Hsieh W.P. and Wang, M.C. 2007. Remediation of soil contaminated with pyreneusing ground nanoscale zero-valent iron. *J. Air Waste Manage. Assoc.*, **57:** 221-227.

Chen, Y., John, C., Stephen, H., Larry, S. and David, H. 2005. Preparation of a Novel TiO_2-Based p-n Junction Nanotube Photocatalyst. *Environ. Sci. Technol.*, **39(5):** 1201-1208.

Diallo, M., Hudrlik, P. and Hudrlik, A. (2006). Fe^0/FeS Dendrimer nanocomposites for reductive dehalogenation of chlorinated haliphatic compounds: synthesis, characterization and bench scale laboratory evaluation of materials performance. Available at: http://www.howard.edu.

Gillham, R.W. and O'Hannesin, S.F. 1994. Enhanced degradation of halogenated aliphatics by zero-valent iron. *Ground Water*, **32**: 958-967.

Hochella, M.F. and Madden, A.S. 2005. Earth's nano-compartment for toxic metals. *Elem.*, **1:** 199-203.

Hodson, M.E. 2010. The need for sustainable soil remediation. *Elem.*, **6:**363-368.

Kamat, P., Rebecca, H. and Roxana, N. 2002. A "Sense and Shoot" Approach for photocatalytic degradation of organic contaminants in water. *J. Phy. Chem.,* **106**(4):788-794.

Karn, B., Kuiken, T. and Otto, M. 2009. Nanotechnology and *in-situ* remediation: A review of the benefits and potential risks. Environ. *Health Perspect.*, **117**:1823-1831.

Li, X.Q., Elliott, D.W. and Zhang, W.X. 2006. Zero-valent iron nanoparticles for abatement of Environmental pollutants: Materials and engineering aspects. *Crit. Rev. olid State Mater.Sci.,* **31**: 111-122.

Masciangioli, T. and Zhang, W. 2003. Environmental nanotechnology: Potential and pitfalls. *Environ. Sci. Technol.,* **37**:102–108.

Mueller, N.C. and Nowack, B. 2009. Nanotechnology developments for the environment sector – report of the observatory nano EU FP7 project. Available at www.observatorynano.eu/project/document/2790.

Nagaveni, K., Sivalingam G., Hegde, M.S. and Giridhar, M. 2004. Photocatalytic degradation of organic compounds over combustion-synthesized nano-TiO_2. *Environ. Sci. Technol.,* **38(5):**1600-1604.

Nowack, B. 2008. Pollution prevention and treatment using nanotechnology. In: Krug H (ed) Nanotechnology. Wiley-VCS Verlag GmbH & Co,Weinheim, pp. 1-15.

Day, P.A. and Vlassopoulos, D. 2010. Mineralbased amendments for remediation. *Elem.*, **6:** 375-380.

Otto, M., Floyd, M.and Bajpai, S. 2008. Nanotechnology for site reme-diation. *Remediation*, **19(1):** 99–108.

Pan, G., Li, L., Zhao, D. and Chen, H. 2010. Immobilization of non-point phosphorus using stabilized magnetite nanoparticles with enhanced transportability and reactivity in soils. *Environ. Pollut.,* **158**: 35-40.

Rajeshwar, K., Chenthamarakshun, C.R., Scott, G. and Milijana, D. (2001). Titania-based heterogeneous photocatalysis. *Pure Appl. Chem.,* **73(12):** 1849-1860.

Rickerby, D. and Morrison, M. 2007. Report from the Workshop on Nanotechnologies for Enviromental Remediation, JRC Ispra. Available at www.nanowork.com/ nanotechnology/repoer/reportpdf/report101.pdf.

Schorr, J.R. 2007. Promise of Nanomaterials for Water Cleanup. Water Conditioning & Purification. Available at www.wcponline.com/pdf/0701Schorr.pdf.

Smedley, P.L. and Kinniburgh, D.G. 2002. A review of the source, behaviour and distribution of arsenic in natural waters. *Appl. Geochem.,* **17**: 517-568.

Som, C., Berges, M., Chaudhry, Q., Dusinska, M., Fernandes, T.F., Olsen, S.I. and Nowack, B. 2010. The importance of life cycle concepts for the development of safe nanoproducts. *Toxicol.*, **269**: 160-169.

USEPA. 2004. Cleaning up the nations waste sites: Markets and technology trends. EPA 542-R-04-015.http://www.epa.gov/tio/download/market/2004market.pdf

USEPA. 2007. Nanotechnology White Paper. Available at (www.epa.gov/osa/pdfs/nanotech/epa) nanotechnology whitepaper- 0207.pdf.

USEPA. 2008. Nano technology for site remediation: Fact Sheet. EPA 542-F-08-009. Washington, DC. 236-328.

Varanasi, P., Fullana, A. and Sidhu, S. 2007. Remediation of PCB contaminated soils using iron nano-particles. *Chemosphere*, **66:** 1031-1038.

Xu, Y. 2005. Removal of copper from contaminated soil by use of poly (amidoamine) dendrimers. *Environ. Sci. Technol.*, **39(7)**: 2369-2375.

Zhang, W. 2003. Nanoscale iron particles for environmental remediation: An overview. *J. Nanopart. Res.,* **5:** 323-332.

Section 5: Biosafety

32

Biosafety of Nanoparticles

S.K. Rajkishore, K.S. Subramanian, R. Sunitha and K. Gunasekaran

Introduction

Nanotoxicity is a serious concern to be addressed sooner than later as nano-based processes and products in various fields are being extensively exploited. Nanotoxicology is the science of engineered nanodevices and nanostructures that deals with their effects in living organisms (Oberdorster et al., 2005). The unique property of nanoparticles is in its very large surface area to mass ratio, which is considered to be a boon to create novel products. But this distinctive property also challenges the way we identify, understand and address potential risks caused by nanoparticles (Rajkishore et al., 2011). It is sufficient to alert us to the fact that some engineered nanoparticles do indeed behave differently to their more conventional counterparts and may present new and unusual risks. Kahru and Dubourguier (2010) assessed the existing literature on nanotoxicity to main food chain levels (bacteria, algae, crustaceans, ciliates, fish, yeasts and nematodes) and classified the NPs of Ag and ZnO as "extremely toxic", C60 fullerenes and CuO NP as "very toxic", SWCNTs and MWCNTs as "toxic" and TiO_2 NP as "harmful". In addition, a review carried out by Rajkishore et al. (2013) concluded that most of the nanoparticles which find its application in a wide spectrum of disciplines have tremendous potential to cause toxicity at various trophic levels in ecological pyramids. Such dataset is sufficient to alert us to the fact that some engineered nanoparticles do indeed behave differently to their more conventional counterparts and may present new and unusual risks.

Risk to biological organisms

Numerous investigations have documented that particle at nanoscale exhibit toxicity effects in a wide array of biological systems (Table 1). Literature survey reveals that even primitive organisms such as algae which have proved to tolerate adverse and fluctuating environmental conditions are prone to nanotoxicity. Further, there are pouring evidences from across the globe that documents toxicity to microbes, plants, fishes, rodents and humans as a result of nanoparticle interaction.

Table 1: Some published literature on nanotoxicity

Model system	Nanoparticle	Effects	References
Algae	TiO_2	Affected photosynthetic activity	Navarro et al. (2008a); Aruoja et al. (2009)
	CeO_2	Inhibited growth	Hoecke et al. (2009)
Microbes	ZVI	Cell disruption	Lee et al. (2008)
	CuO	Antibacterial activity	Mahapatra et al. (2008); Dimkpa et al. (2011)
	Ag	Increased cell permeability and ultimately cell death	Sondi and Salopek-Sondi (2004)
	ZnO	100% mortality	Jiang et al. (2009)
Plants	CNT	Programmed cell death	Shen et al. (2010)
	ZnO	Inhibited seed germination and root growth	Yang and Watts (2005); Lin and Xing et al. (2007)
Fish	Se	Hyper-accumulation	Li et al. (2008)
	ZVI	Morphologic changes	Li et al. (2009b)
Rodents	Al_2O_3	Inflammation, crossed blood brain barrier	Li et al. (2009a)
	TiO_2	Inflammation	Oberdorster et al. (1992) and Baggs et al. (1997)
	ZVI	Toxic to nerve cells	Phenrat et al. (2009)
	CNT	Death, necrosis, inflammation and cell injury	Lam et al. (2004); Shvedova et al. (2005)
	Au	Moved from mother's placenta to fetus	Warheit (2004)
Humans	Magnetite	Neurodegenerative diseases	Kirschvink et al. (1992); Dunn et al. (1995)
	CNT	Mesothelioma	Donaldson et al. (2006); Fisher et al. (2012)
	Au	Penetrated into the sperm head and tails causing fragmentation.	Wiwanitkit et al. (2007)

Science behind nanotoxicity

Scanning through the literature base in the field of nanotoxicology reveals the fact that the excessive generation of reactive oxygen species (ROS), resulting in oxidative stress is the prime mechanism that contributes for nanotoxicity in most of the living organisms (Oberdorster et al., 2005; Shvedova et al., 2010; Rajkishore et al., 2013). ROS play pivotal roles in the initiation of numerous signal transduction pathways that are linked to apoptosis, inflammation and proliferation (Shukla et al., 2003). The chemical reactions that contribute to the production of ROS after nanoparticle interaction are diagrammatically represented in Fig 1 as well as briefly discussed below.

a. The presence of transition metals or redox cycling organic chemicals on the nanoparticle surfaces can generate excess of ROS. Moreover, the transition metals can also generate hydroxyl radicals through the Fenton reaction (Nel et al., 2001). The Fenton chemistry is one of the mechanisms by which metal impurities like ferrous iron on the CNT surface can react with hydrogen peroxide and produce hydroxyl radical. For instance, the iron based nanoparticles are presumed to react with peroxides in the environment generating free radicals.

b. Nanoscale TiO_2 during UV exposure is found to generate ROS through the formation of electron-hole pairs as a result of photoactivation effect leading to oxidative stress and inflammation (Long et al., 2006). In simple terms, it can be explained that upon irradiation, the electrons in the valence band of nanoparticles are promoted to conduction band, leaving a hole. These holes at the valence band will have an oxidation potential of +2.6 V in comparison with normal hydrogen electrode and therefore can oxidize water or hydroxide into hydroxyl radicals.

c. In some cases, the surface of nanoparticles that possess discontinuous crystal planes or material defects creates active electronic state and facilitates in the production of ROS (Xia et al., 2009).

d. The particle dissolution (e.g., ZnO, CdSe, Cu) can produce free ions and this in turn may trigger the ROS production (Derfus et al., 2004; Meng et al., 2007).

Overall, the investigations with various biological model systems bring to light that depending upon the nature and type of nanoparticles, ROS are generated through different reactions and can ultimately result in cellular and tissue injury responses such as inflammation, apoptosis, necrosis, fibrosis, hypertrophy, metaplasia, and carcinogenesis (Nel et al., 2006).

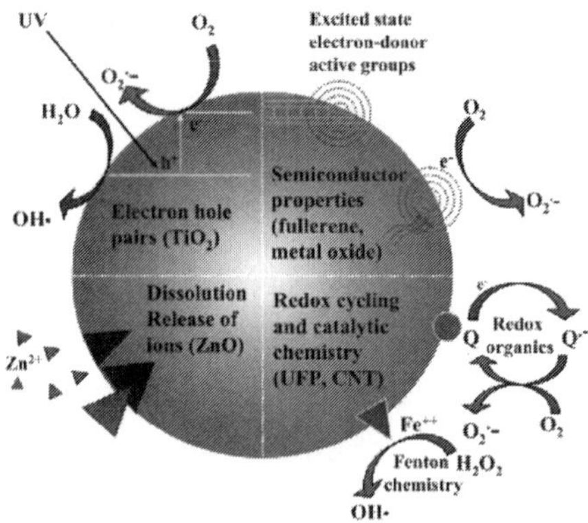

Fig 1: Nanoparticle and ROS production (Source: Li et al., 2008; Xia et al., 2009)

Key factors that affect nanotoxicity

1. Dose

It is quite interesting to note that nanotechnology challenges the age old saying "Dose makes the poison". This notion is most debated by several researchers in the field of nanotoxicology and further investigation is warranted. Toxic effects of nanoparticles do not always appear to correlate with particle mass dose. This is because a high concentration of nanoparticles may promote particle aggregation and could therefore reduce toxic responses compared to lower concentrations of the same particles (Buzea et al., 2007).

2. Surface Area

The size of the materials when reduced, contributes to changes in surface physical and chemical properties (Jones and Grainger, 2009). Therefore, the relative portion of surface atoms to bulk atoms is considerably different in nano-sized when compared to microsized particles of same chemistry. For instance, less than 1% of atoms of a microparticle occupy surface positions, while 10% of the atoms in a 10 nm particle reside on its surface.

3. Size

To understand the size dependent toxicity, the cytotoxicity studies involving different sized gold nanoparticles provides a great scope. Gold nanoclusters (1.4 nm) were observed to be toxic to cells owing to their specific interaction with major grooves of DNA, whereas smaller or larger gold particles did not behave in this way (Pan et al., 2007). The gold nanoparticles of 35 nm size were non-toxic to a murine macrophage-like cell line (Shukla et al., 2003). Overall, the literature survey reveals that the gold particles with a size of 13 nm and above, commonly typified as colloids, may be viewed as non-toxic (Jahnen-Dechent and Simon, 2008). On the other hand, gold particles below 2 nm have shown an unexpected degree of toxicity in different cell lines (Schmid, 2008). Nanoparticles with its tiny size get great chance to enter or cross biological barriers like blood brain barrier. In addition, other have observed that silica nanoparticles of 40–80 nm in diameter can enter the cell nucleus and localize to distinct subnuclear domains in the nucleoplasm, but do not colocalize with nucleoli. Further, these nanoparticles induced the formation of nucleoplasmic protein aggregates. In contrast, fine and course (0.5–2 μm) silica particles located exclusively in the cytoplasm (Chen and Mikecz, 2005).

4. Crystalline Structure

Different crystal forms are reported to cause varied levels of toxicity. This statement is supported with the findings of cytotoxic properties of titanium dioxide nanoparticles that positively correlate with their crystalline structure (Shvedova et al., 2010). Titania exists in a variety of crystal structures and the most researched forms are rutile, anatase, and brookite (Fadeel and Bennett, 2010). Investigations with titanium dioxide nanoparticles of size ranging between 3-10 nm, demonstrated that anatase titanium dioxide was 100 times more toxic than an equivalent sample of rutile titanium dioxide (Sayes et al, 2007). Furthermore, the crystal structure of titanium dioxide also dictates the mode of cell death. Anatase titanium dioxide nanoparticles, regardless of size, were reported to induce necrosis, whereas rutile titanium dioxide nanoparticles triggered apoptosis.

5. Surface Coating

The surface chemistry of nanoparticle also dictates the toxic responses in living organisms. When nanoparticles make contact with cells and a thorough understanding of its surface composition is vital to understand the interactions of nanoparticles within the biological systems (Jones and Grainger, 2009). In many cases, the contaminants on the surface of ENPs are found to cause toxicity. For example, the surface of CNTs when contaminated with ferrous iron can induce the production of ROS through Fenton's reactions inside biological system (Nel et al., 2001). The frequent problem at the nano-bio-interface is due to the possible adsorption of the biomolecules on the surface of ENPs which also contribute to the cellular responses, in particular immunological responses (Shevoda et al., 2010). Hence, it is also crucial to distinguish between undesirable cellular responses to nanoparticles themselves and residual materials associated with the nanoparticle such as surfactants, adsorbed biomaterials or transition metals as a product of the synthetic process.

6. Opsonization

The ENPs are often employed as composites and rarely used as sole active agent. In several cases, the ENPs are encapsulated within a host system or require functionalization of their external surface i.e. chemical modification through the use of tethering or coupling agents (Fadeel and Garcia-Bennett, 2010). The main idea behind these modifications is to enable the ENPs to interact in a suitable manner with the biological environment. These modifications will easily disperse in biological media or to protect the nanoparticle against degradation. In addition, the nanoparticles may also bind to proteins in biological fluids, which in turn could affect their biological performance. It is pointed out that adsorbed proteins could play a vital role in modulating the uptake and toxicity of nanoparticles (Dutta et al., 2007; Cedervall et al., 2007). For example, the contamination of gold nanoparticles with the endotoxin, lipopolysaccharide (LPS) results in the activation of dendritic cells and ultimately interferes with the assessment of biological (immuno-modulatory) effects of these nanoparticles (Vallhov et al., 2006). As a whole it is proposed that the opsonized proteins constitute a major element of the biological identity of the nanoparticle (Fadeel and Garcia-Bennett, 2010).

Bioassays to assess nanotoxicity

Fadeel and Garcia-Bennett (2010) reviewed the effectiveness and validity of assays for determining the toxicity and concluded that more than one assay may be required for nanotoxicity risk assessment. Therefore, three major categories of assays namely cytotoxic, genotoxic and alterations in gene expression assays are generally proposed to evaluate the toxicity of nanoparticles in *in vitro* system (Subbulakshmi, 2011).

1. Cytotoxic assays are used to detect the cell viability, plasma membrane integrity and cellular metabolism. This includes, Trypan Blue Exclusion Assay, In Vitro cell viability assay – WST 1, Lactase Dehydrogenase Assay (LDH) Assay.

2. Genotoxicity assays are used to study the DNA structure breakage, mutagenicity and chromosomal aberration. Example: Ames Assay, Comet Assay and 8-Oxo-dG Assay. In 8-Oxo-dG Assay, the 8-hydroxy-2-deoxy Guanosine (8-OH-dG) is a product of oxidative damage of DNA by reactive oxygen and nitrogen species and serves as an established marker of oxidative stress.

3. Gene expression assays (gene profiling) is employed to assess the alterations in the gene expression as a result of nanoparticle interaction.

Northern blot analysis, quantitative real-time polymerase chain reaction (qRT-PCR), PCR arrays and micro arrays are some of the gene expression assays.

Conclusion

The nanoparticles with the unique property of large surface area to mass ratio have a great potential to revolutionize a wide array of disciplines. Nevertheless, the increased chemical reactivity and biological activity as a result of large surface area is also a concern that needs to be addressed in the light of biosafety issues. Plethora of literature documents that nanoparticles causes toxicity in various biological systems. Despite such concerns, there is still scope to employ the nanoparticles, provided the key factors that contributes to toxicity is well understood. Scientists applying the nanotechnology concepts to their respective fields should also focus on biosafety tests. Such exercise helps to know the undesirable properties of different types of nanoparticles and the means to avoid them. Though there are set of assays to test the biosafety of nanoparticles, it is more important to undertake such investigations on case-to-case basis. Overall, the nanotechnology is recommended with a word of caution.

References

Aruoja, V., Dubourguier, H. C., Kasemets K. and Kahru, A. 2009. Toxicity of nanoparticles of Cuo, Zno and TiO_2 to Microalgae *Pseudokirchneriella subcapitata*. *Sci. Total Environ.* **407**: 1461–1468.

Baggs, R.B., Ferin, J. and Oberdorster, G. 1997. Regression of pulmonary lesions Produced by Inhaled titanium dioxide in rats. *Veterinary Pathology*, **34**: 592-597.

Buzea, C., Pacheco, I. and Robbie, K. 2007. Nanomaterials and nanoparticles: Sources and Toxicity. *Bio. Inter. Phases*. **2:** 17–71.

Cedervall, T., Lynch, I., Lindman, S., Berggard, T., Thulin, E., Nilsson, H., Dawson, K. A. and Linse, S. 2007. Understanding the Nanoparticles - Protein Corona Using Methods to Quantify Exchange Rates and Affinities of Proteins for Nanoparticles. *Proc. Natl. Acad. Sci.*, U. S. A. 104: 2050–2055.

Chen, M. and Mikecz, A. V. 2005. Formation of nucleoplasmic protein aggregates impairs nuclear function in response to SiO_2 nanoparticles. *Exp. Cell Res.*, **305**:51–62.

Derfus, A.M., Chan, W.C.W. and Bhatia, S.N. 2004. Probing the cytotoxicity of semiconductor quantum dots. *Nano Lett*. **4**: 11–18.

Dimkpa, C.O., Calder, A., Britt, D.W., McLean, J.E. and Anderson, A.J. 2011. Responses of a soil bacterium, *Pseudomonas chlororaphis O6* to commercial metal oxide nanoparticles compared with responses to metal ions. *Environmental Pollution*. **159**: 1749-1756.

Donaldson, K., Aitken, R., Tran, L., Stone, V., Duffin, R., Forrest, G. and Alexander, A. 2006. Carbon Nanotubes: A Review of their properties in relation to pulmonary toxicology at workplace. *Toxicological Sciences*. **92**: 5–22.

Dunn, J.R., Fuller, M., Zoeger, J., Dobson, J., Heller, F. and Hamnmann, J. 1995. Magnetic material in the human hippocampus. *Brain. Res. Bull*. **36**: 149–153.

Dutta, D., Sundaram, S. K., Teeguarden, J. G., Riley, B. J., Fifield, L. S., Jacobs, J. M., Addleman, S. R., Kaysen, G. A., Moudgil, B. M. and Weber, T. J. 2007. Adsorbed proteins influence the biological activity and molecular targeting of nanomaterials. *Toxicol. Sci.* **100**: 303–315.

Fadeel, B. and Bennett, A. E. G. 2010. Better safe than sorry: Understanding the toxicological properties of inorganic nanoparticles manufactured for biomedical applications. *Advanced Drug Delivery Reviews*. **62:** 362–374.

Fisher, C., Rider, A.E., Han, Z.J., Kumar, S., Levchenko, I. and Ostrikov, K. 2012. Applications and Nanotoxicity of Carbon Nanotubes and Graphene in Biomedicine. *Journal of Nanomaterials*. **2012**: 1-19.

Hoecke, K.V., Quik, J.T.K., Mankiewicz-Boczek, J. and Schamphelaere. K. 2009. Fate and effects of CeO_2 nanoparticles in aquatic ecotoxicology tests. *Environ. Sci. Technol.* **43**: 4537-4546.

Jahnen-Dechent, W. and Simon, U. 2008. Function follows form: shape complementarity and nanoparticle toxicity. *Nanomedicine.* **3**: 601-603.

Jiang, W., Mashayekhi, H. and Xing, B. 2009. Bacterial toxicity comparison between nano- and micro-scaled oxide particles. *Environ. Pollut.* **157:** 1619–1625.

Jones, C. and Grainger, D.W. 2009. *In vitro* assessments of nanomaterial toxicity. *Adv. Drug Deliv. Rev.* **61**: 438–456.

Kahru, A. and Dubourguier, H.C. 2010. From ecotoxicology to nanoecotoxicology. *Toxicology.* **269**: 105-119.

Kirschvink, J.L., Kirschvink, A. and Woodford, B.J. 1992. Magnetite biomineralization in the human Brain. *Proc. Natl. Acad. Sci.*, **89**: 7683–7687.

Lam, C .W., James, J.T., McCluskey, R. and Hunter, R.L. 2004. Pulmonary toxicity of single-wall carbon nanotubes in mice 7 and 90 days after intratracheal instillation. *Toxicol. Sci.* **77**: 126–134.

Lee, W., An, Y., Yoon, H. and Kweon. H. 2008. Toxicity and bioavailability of copper nanoparticles to the terrestrial plants Mung bean (*Phaseolus radiates*) and Wheat (*Triticum aestivum*): Plant uptake for water insoluble nanoparticles. *Environmental Toxicology and Chemistry.* **27**: 1915-1921.

Li, H., Zhang, J., Wang, T., Luo, W., Zhou, Q. and Jiang, G. 2008. Elemental selenium particles at nano-size (Nano-Se) are more toxic to medaka (*Oryzias latipes*) as a consequence of hyper-accumulation of selenium: A Comparison with Sodium Selenite. *Aquatic Toxicology,* **89**: 251–256.

Li, H., Zhou Wu, F. J., Wang, T. and Jiang, G. 2009b. Effects of waterborne nano-iron on medaka (*Oryzias latipes*): Antioxidant Enzymatic Activity, Lipid Peroxidation and Histopathology. *Ecotoxicol. Environ. Saf.* **72**: 684-692.

Li, X.B., Zheng, H., Zhang, Z.R., Li, M., Huang, Z.Y., Schluesener, H.J., Li, Y.Y. and Xu, S.Q. 2009a. Glia activation induced by peripheral administration of aluminum oxide nanoparticles in rat brains nanomedicine: Nanotechnology. *Biology and Medicine*. **5:** 473–479.

Lin, D. and Xing, B. 2007. Phytotoxicity of nanoparticles: Inhibition of seed germination and root Growth. *Environmental Pollution*. **150**: 243-250.

Long, T.C., Saleh, N., Tilton, R.D., Lowry, G.V. and Veronesi, B. 2006. Titanium dioxide (P25) produces reactive oxygen species in immortalized brain microglia (BV2): Implications for nanoparticle neurotoxicity. *Environ. Sci. Technol*. **40**: 4346-4352.

Manapatra, O., Bhagat, M., Gopalakrishnan, C., Arunachalam, K.D.2008. Ultrafine dispersed CuO nanoparticles and their antibacterial activity. *J. Exp. Nanosci.* **3**: 185-193.

Meng. H., Chen, Z., Xing, G., Yuan, H. and Chen, C. 2007. Ultrahigh reactivity provokes nanotoxicity: Explanation of oral toxicity of nano-copper particles. *Toxicol. Lett.* **175**: 102-110.

Navarro, E., Baun, A., Behra, R., Hartmann, N. B., Filser, J. and Miao, A. 2008a. Environmental behavior and ecotoxicity of engineered nanoparticles to algae, plants and fungi. *Ecotoxicology*. **17**: 372-386.

Nel, A., Xia, T., Madler, L. and Li, N. 2006. Toxic potential of materials at the nanolevel. *Science*. **311**: 622-627.

Nel, A.E., Diaz-Sanchez, D. and Li, N. 2001. The role of particulate pollutants in pulmonary inflammation and asthma: Evidence for the involvement of organic chemicals and oxidative stress. *Curr. Opin. Pulm. Med.* **7:** 20–26.

Oberdorster, G., Ferin, J., Gelein, R., Soderholm, S.C. and Finkelstein, J. 1992. Role of the alveolar macrophage in lung injury: Studies with ultrafine particles. *Environment Health Perspect*. **97**:193–197.

Oberdorster, G., Oberdorster, E. and Oberdorster, J. 2005. Nanotoxicology: An emerging discipline evolving from studies of ultrafine particles. *Environ. Health Perspect*. **113**: 823.

Pan, Y., Neuss, S., Leifert, A., Fischler, M. and Wen, F. 2007. Size-dependent cytotoxicity of gold nanoparticles. *Small.* **3**:1941–1949.

Phenrat, T., Long, T.C., Lowry, G.V. and Veronesi, B. 2009. Partial oxidation ("Aging") and surface modification decrease the toxicity of nano-sized zerovalent Iron. *Environ. Sci. Technol*. **43**: 195-200.

Rajkishore, S.K., Doraisamy, P., Maheswari, M. and Subramanian, K.S. 2011. Nanotoxicity. In: *Nano Agriculture: Principles and Practices*, K.S. Subramanian, A. Lakshmanan, N. Natarajan, K. Gunasekaran, C.R. Chinnamuthu, P. Latha and C. Sharmila Rahale (Eds.). Shri Garuda Graphics, Coimbatore, pp. 290-293.

Rajkishore, S.K., Subramanian, K.S., Natarajan, N. and Gunasekaran, K. 2013. Nanotoxicity at various trophic levels: A review. *The Bioscan.* **8(3):**.975-982.

Sayes, C. M., Reed, K. L. and Warheit, D. B. 2007. Assessing toxicity of fine and nanoparticles: Comparing *in vitro* measurements to *in vivo* pulmonary toxicity profiles. *Toxicol.Sci*. **97**:163–180.

Schmid, G. 2008. The relevance of shape and size of Au55 clusters. *Chem. Soc. Rev.* **37**: 1909–1930.

Shen, C.X., Zhang, Q.F., Li, J., Bi, F.C. and Yao, N. 2010. Induction of programmed cell death in *Arabidopsis* and rice by single-wall carbon nanotubes. *American Journal of Botany*. **97**: 1602–1609.

Shukla, A., Gulumian, M., Hei, T. K., Kamp, D., Rahman, Q. and Mossman, B.T. 2003. Multiple roles of oxidants in the pathogenesis of asbestos-induced diseases. *Free Radical Biology and Medicine*. **34**: 1117–1129.

Shvedova, A. A., Kisin, E. R., Mercer, R., Murray, A. R. and Johnson, V.J. 2005. Unusual inflammatory and fibrogenic pulmonary responses to single-walled carbon nanotubes in mice. *Am. J. Physiol. Lung Cell. Mol. Physiol.* **289**: 698–708.

Shvedova, A.A., Kagan, V. E. and Fadeel, B. 2010. Close encounters of the small kind: Adverse effects of man-made materials interfacing with the nano-cosmos of biological systems. *Annu. Rev. Pharmacol. Toxicol*. **50:** 63–88.

Sondi, I. and Salopek-Sondi, B. 2004. Silver nanoparticles as antimicrobial agent: A Case Study on *E.coli* as a Model for Gram-Negative Bacteria. *J. Colloid Interface Sci*. **275**: 177-182.

Subbulakshmi, V. 2011. Laboratory Protocols for Nanotoxicity Studies, In: *Nano Agriculture: Principles and Practices*, K.S. Subramanian, A. Lakshmanan, N. Natarajan, K. Gunasekaran, C.R. Chinnamuthu, P. Latha and C. Sharmila Rahale (Eds.), Shri Garuda Graphics, Coimbatore, pp. 247-252.

Vallhov, H., Qin, J., Johansson, S.M., Ahlborg, N., Muhammed, M.A., Scheynius, A. and Gabrielsson, S. 2006. The Importance of an endotoxin-free environment during the production of nanoparticles used in medical applications. *Nano Lett*. **6**: 1682–1686.

Warheit D. B., Laurence, B. R., Reed, K.L., Roach, D.H., Reynolds, G.A.M. and Webb, T.R. 2004. Comparative pulmonary toxicity assessment of single-wall carbon nanotubes in rats. *Toxicol. Sci*. **77**: 117–125.

Wiwanitkit, V., Sereemaspun, A. and Rojanathanes, R. 2007. Effect of gold nanoparticles on spermatozoa: The first world report. *Fertility and Sterility*. **11**: 132-134.

Xia, T., Li, N. and Nel, A.E. 2009. Potential health impact of nanoparticles. *Annu. Rev. Public. Health*. 30: 137-150.

Yang, L. and Watts, D.J. 2005. Particle surface characteristics may play an important role in phytotoxicity of alumina nanoparticles. *Toxicology Letters*. **158**: 122–132.

33

Assays for Nano-toxicity Assessment

Venkita Subbulakshmi

Introduction

The characterization of the potential health impact of engineered materials produced through nanotechnology is an emerging issue of considerable discussion and debate. The current focus on nanotoxicity fails to assess the risk in different local environments and populations. People in developing countries may be more prone to adverse effects of nanoparticles because of underlying health conditions and malnutrition. Moreover, genetic susceptibility to toxic effects varies in diverse ethnic groups and geographical areas. The scientific community needs to identify these information gaps before developing regulations and standard methodologies for nanotoxicity assessment.

The skin is the largest organ of the body and serves as a primary route of environmental and/or occupational exposure; it is one of the principal portals of entry by which environmental toxicants or nanomaterials can enter into the body. At present, there is no information on whether nanoparticles can be absorbed across the stratum corneum barrier or whether systemically administered particles can accumulate in dermal tissue. Skin is unique because it provides an environment within the avascular epidermis where particles could potentially lodge and not be susceptible to removal by phagocytosis. The ability for nanomaterials to traverse the skin is a primary determinant of their dermatotoxic potential.

One of the major decisions is to be faced in assessing the skin absorption and toxicity of nanomaterials is how to conduct the experiments. Should in vitro cell cultures, flow-through diffusion cells or cell lines be used? In vivo studies conducted in rat or preferably pig skin (since it is anatomically, physiologically, and biochemically similar to man) would be ideal. However, there are limitations in obtaining the quantity and quality of some nanomaterials to conduct in vivo studies. Therefore, in some cases it may be best to study their interactions in vitro in order to estimate the in vivo starting dose for toxicity. In vitro studies have shown that multi-walled carbon nanotubes (not derivatized or optimized for biological applications) are capable of localizing within and initiating an irritation response in human epidermal keratinocytes, which are a primary route of occupational exposure (Monteiro-Riviere et al., 2005).

A systematic tier approach can be implemented for evaluating nanomaterials. In vitro testing, followed by escalation to more complex testing models, may render useful information in the evaluation of these materials in lieu of chronic bioassays. Short-term mechanistic studies, in vitro studies, and ultra fine particle epidemiological studies can provide important enhancements to traditional toxicity assays. Integrating this information increases our confidence in the hazard identification of nanoparticles, and when coupled with exposure considerations, the risk assessment of nanomaterials.

Determination of the mode of action (MOA) for effects observed with nanomaterials will be a key issue in understanding data obtained from toxicological evaluations and its extrapolation for the determination of potential human health risk. Some of the key MOA considerations for nanoscale materials include: (1) do their unique physicochemical properties translate into unique MOAs?; (2) what are the best experimental strategies to obtain data that can identify key events with which to evaluate these MOAs?; (3) can evaluation of a core set of parameters and/or model materials be used to determine MOAs that can be applied to emerging nanomaterials?

Nanomaterials cannot simply be considered as a single homogeneous class. In addition, our current understanding of the MOA of nanomaterials is very limited, since research to date has only been conducted on a few example materials. There are a number of parameters that will be key to understanding the MOA for a given material, including the size/shape / aspect ratio, hardness / deformability, composition, surface area and surface chemistry, types of coatings / modifications and stability. Given that nanomaterials vary greatly based on these parameters, extrapolation of a MOA from one material to another will need to be made with extreme caution until more knowledge is gained on a broader class of materials.

A number of studies have documented *in vitro* and *in vivo* toxicity of exposure to nanoparticles. Evidence suggests they can induce DNA damage, reactive oxygen species, damage to cellular organelles and cell death. This necessitates the requirement of Biosafety studies to be performed while developing any products with nanoparticles.

Importance of invitro tests

The predictive value of in vitro cytotoxicity tests is based on the idea of 'basal' cytotoxicity – that toxic chemicals affect basic functions of cells which are common to all cells and that the toxicity can be measured by assessing cellular damage. The development of *in vitro* cytotoxicity assays has been driven by the need to limit animal experimentation whenever possible and to carry out tests with small quantities of compounds.

Invitro test models and assays

Test models

As nanoparticles are known to cause toxicity through inhalation, dermal integration and in some cases by metabolism, lung cell lines (MRC-5,), epithelial cell line (HeLa, A549) and liver cell line (HepG2) will be generally used. Depending on the applicability and area of environmental exposure biosafety studies can be supplemented with specific cell lines and tissue as the product warrants.

Assays: Cell health can be monitored by numerous methods. Plasma membrane integrity, DNA synthesis, DNA content, enzyme activity, presence of ATP, and cellular reducing conditions are known indicators of cell viability and cell death.

There are three major categories of assays which will help in evaluating the toxicity of nano particles in *in vitro* system.

1. Cytotoxic assays (which mainly focus on cell viability, plasma membrane integrity and cellular metabolism).
2. Genotoxicity assays (which study the DNA structure breakage, mutagenicity, Chromosomal aberration etc.).
3. Alteration in gene expression assays.

Cytotoxicity assays

Any cytotoxic agent acting through the cell impacts the basic cellular metabolism. To study the basic cellular metabolism, the response of cell to any toxic agent is measured by the enzymes present on the cellular surface. This, in turn, reflects the changes in cell proliferation. There are many assays to study the intactness of cell and cellular metabolism which are described below:

I. Trypan Blue Exclusion Assay

In this assay cells are treated with agents, trypsinized, and subsequently stained with trypan blue, a diazo dye which is taken up by dead cells, but excluded by viable cells. Unstained cells reflect the total number of viable cells recovered from a given dish. This method is advantageous because it conveys the actual number of viable cells and increases (cell proliferation) or decreases (cytotoxicity) in comparison to control, untreated cells.

II. *In Vitro* cell viability assay

Cell proliferation is quantified by WST-1/ MTT/ XTT reagent. This assay is by far the most sensitive and convenient method for quantifying cell proliferation and viability.The first assay type is the measurement of cellular metabolic activity. An early indication of cellular damage is a reduction in metabolic activity. Tests which can measure metabolic function measure cellular ATP levels or mitochondrial activity (via MTS metabolism). Mitochondrial activity is measured by this assay.

Principle: This is a colorimetric assay for the quantification of cell viability and proliferation. The enzyme succinate tetrazolium reductase cleaves tetrazolium salts (WST1) to formazan that is colored product and can be quantitatively measured in spectrophotometer at 450 nm. This enzyme belongs to the respiratory chain of the mitochondria and is active only in viable cells. The amount of formazan dye produced directly correlates the number of metabolically active cells.

III. Cytotoxicity assay (LDH Assay)

Another parameter often tested is the measurement of membrane integrity. The cell membrane forms a functional barrier around the cell, and traffic into and out of the cell is highly regulated by transporters, receptors and secretion pathways. When cells are damaged, they become 'leaky' and this forms the basis for the second type of assay. Membrane integrity is determined by measuring lactate dehyrogenase (LDH) in the extra cellular medium. This enzyme is normally present in the cytosol, and cannot be measured extracellularly unless cell damage has occurred. It has been shown that changes in metabolic activity are better indicators of early cell injury and that effects on membrane integrity are indicative of more serious injury, leading to cell death. The enzyme lactate dehydrogenase (LDH-L) is distributed in all cells. Several colorimetric LDH-L assay methods have been developed. Most of these assays are based on the coupling of the reduction of NAD and tetrazolium salts (INT). Nachlas, et al. (2006) described an LDH-L assay using phenazine methosulfate (PMS) as the intermediate electron carrier between NADH and INT. Allain et al, (2000) replaced PMS with the enzyme diaphorase.

Principle; The assay is based on the cleavage of a tetrazolium salt when LDH is present in the culture supernatant. The procedure involves incubating the cells with any cytotoxic agent in culture may result in cell death. An increase in the amount of dead or plasma membrane-damaged cells during the assay results in an increase of LDH in the culture supernatant.

LDH catalyzes the oxidation of lactate to pyruvate in the presence of NAD which is subsequently reduced to NADH. The formation of NADH is coupled with the reduction of INT to INTH catalyzed by the enzyme of diaphorase. INTH is bright red formazan which is measured photo metrically at 500 ± 5 nm. The color intensity is proportional to the LDH-L activity of the sample.

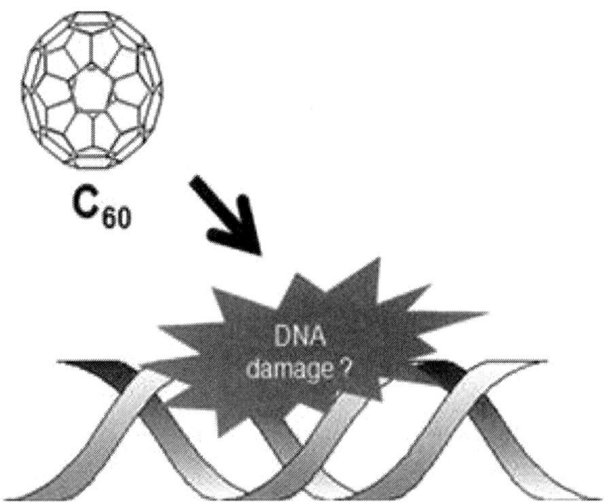

Genotoxicity Assays

Generation of DNA damage is considered to be an important initial event in carcinogenesis. A considerable battery of assays exists for the detection of different genotoxic effects of compounds in experimental systems, or for investigations of exposure to genotoxic agents in environmental or occupational settings. The nanoparticles have been shown to be generating free radicals and these are highly reactive with DNA. Oxidative damage in DNA has often been suggested as a contributing factor in the process of aging and the development of cancer. There are many assays which can be employed to study the genotoxic potential of the Nanoparticles, which are described below.

I. Determination of Gene Mutations Using the Ames Assay in *Salmonella typhimurium* and *Escherichia coli*

The reverse mutation (Ames) assay in *Salmonella typhimurium* employs bacteria deficient in DNA repair mechanisms that are unable to grow in the absence of histidine. Following exposure to compounds of interest, reversion to a histidine-positive phenotype (indicating a reverse mutation in the histidine locus) is established by counting colonies that have been grown in histidine-free media. Inclusion of an exogenous metabolizing system (Aroclor induced rat liver S9 microsomal fraction) allows for the detection of mutagens requiring metabolic activation to form DNA-reactive intermediates. In addition to several strains of *S. typhimurium* (e.g.TA98, TA100, TA102, TA1535, TA1537, TA1538), each allowing detection of different mutation types, this assay has been adapted in a strain of *Escherichia coli* (WP2*uvr*A) to identify base-pair substitutions based on reversion at the tryptophan locus. Given the possible differences in cellular uptake of particulates and genomic complexity between prokaryotes and eukaryotes, genotoxicity data for nanomaterials obtained from the Ames assay should be interpreted carefully. The Ames assay should not be considered as a stand alone assay for identifying genotoxicity elicited by nanomaterials in humans and other vertebrates, and should instead be supplemented with additional studies as described hereafter.

II. Cytogenetic Assessment of Chromosome Damage through Analysis of Chromosomal Aberration Induction and Micronuclei

In addition to determining mutations of a particular gene, it is important to evaluate effects on the number and integrity of chromosomes via karyotype analyses. Such analyses can be carried out directly via simple staining techniques (5% Giemsa) and microscopy, and entail evaluation of changes in the morphological appearance of chromosomes (chromosomal aberrations representing clastogenicity) and the presence of micronuclei. Protocols typically involve treatment of cells during S-phase (due to the sensitivity of cells at this point in the cell cycle) followed by treatment at predetermined intervals with a substance such as Colcemid® or colchicine that is capable of arresting the cells in metaphase. Alternatively, assessment of chromosomal breakage and chromosome loss events can be carried out by identifying the presence of micronuclei. Micronuclei are chromosomal fragments or whole chromosomes that are not incorporated into the nucleus of either daughter cell at anaphase, and are therefore bound by a membrane and remain in the cytoplasm through subsequent cell cycles. Micronucleus assays typically employ a cytokinesis-block technique in which cytochalasin B is used to inhibit cytokinesis, thus allowing micronuclei to be assessed in binucleated cells. This is important since micronuclei are most accurately quantified in bi nucleated cells that have undergone only one cell division. Karyotypic analyses as described above have been carried out for a number of nanomaterials.

III. Comet Assay

The single cell gel electrophoresis (comet) assay is technically simple, relatively fast, cheap, and DNA damage can be investigated in virtually all mammalian cell types without requirement for cell culture. The comet assay can be employed as a genotoxicity test in evaluating the genetic toxicology of environmental agents, encompassing both experimental animal models and biomonitoring. The simple version of the alkaline comet assay detects DNA migration caused by strand breaks, alkaline labile sites, and transient repair sites. The pH > 13 version is capable of detecting DNA single-strand breaks (SSB), alkali-labile sites (ALS), DNA-DNA/DNA-protein cross-linking, and SSB associated with incomplete excision repair sites. Relative to other genotoxicity tests, the advantages of the SCG assay include its demonstrated sensitivity for detecting low levels of DNA damage, requirement for small numbers of cells per sample, its flexibility, its low costs, its ease of application, and the short time needed to complete a study. The cells are embedded in agarose and lysed, generating nucleus-like structures in the gel (referred to as nucleoids). Following alkaline electrophoresis, the DNA strands migrate toward the anode, and the extent of migration depends on the number of SB in the nucleoids. The migration is visualized and scored in a fluorescence microscope after staining.

IV. 8-Oxo-dG Assay

Oxidative damage in DNA has often been suggested as a contributing factor in the process of aging and the development of cancer. One of the most studied and important lesions produced in DNA by reactive oxygen species is 7-hydro-8-Oxodeoxyguanosine (8-OxodG). Due to its mispairing with deoxyadenosine, 8-OxodG is mutagenic. Also, 8-OxodG was found to be highly miscoding during replication with purified DNA polymerases *in vitro*. 8-OxodG is not poorly repaired in the cells, causing a block of DNA replication. The production of 8-Hydroxyguanine (8-Oxo-dG) is almost exclusively elicited by oxidative stress. Polymerases preferentially insert adenine opposite 8-Oxo-dG. Therefore, oxidatively damaged adducts, without repair, are susceptible to G to T transitions.

In *Escherichia coli*, 8-OxodG is removed from DNA by either a specific glycosylase, known as formamidopyrimidine glycosylase (Fpg protein), in the initial step of base excision repair (BER), or by the UvrABC enzymes through nucleotide excision repair (NER). Cells deficient in either pathway do not show an appreciable defect in the repair of 8-oxodG, but the double mutants exhibit sensitivity to agents producing 8oxodG lesions. Eukaryotic cells may also employ multiple pathways for the removal of this common lesion. When the cells fail to remove the lesion it leads to misincorporation, mutagenicity and genotoxicity.

Quantification of 8- Oxo-dG

8-hydroxy-2-deoxy Guanosine (8-OH-dG) is a product of oxidative damage of DNA by reactive oxygen and nitrogen species and serves as an established marker of oxidative stress. Hydroxylation of guanosine occurs in response to both normal metabolic processes and a variety of environmental factors. Increased levels of 8-OH-dG are associated with the aging process as well as with a number of pathological conditions including cancer, diabetes, and hypertension. 8-OH-dG can be quantified in EIA which is a competitive assay which can be performed in cell culture, plasma, microbial lysates and other sample matrices. The EIA utilizes an anti-mouse IgG-coated plate and a tracer consisting of an 8-OH-dG-enzyme conjugate. This format has the advantage of providing low variability and increased sensitivity compared to assays that utilize an antigen-coated plate.

Assays for alteration in gene expression

Gene expression assays, i.e., gene profiling, are an important tool for screening different environmental particles, including nanoparticles. Techniques used to assess gene expression include: Northern blot analysis, quantitative real-time polymerase chain reaction (qRT-PCR), PCR arrays and micro arrays.

I. Real Time Polymerase Chain Reaction (RT-PCR)

Real-time PCR is a quantitative method for the determination of copy number of PCR templates, such as DNA or cDNA and consists of two types: probe-based and intercalator based. Probe-based real-time PCR, also known as Taqman PCR, requires a pair of PCR primers (as regular PCR does), and an additional fluorogenic oligonucleotide probe with both a reporter fluorescent dye and a quencher dye attached. The intercalator-based (SYBR Green) method requires a double-stranded DNA dye in the PCR reaction which binds to newly synthesized double-stranded DNA and renders fluorescence. Both methods require a special thermocycler equipped with a sensitive camera that monitors the fluorescence in each well of a 96-well plate at frequent intervals during the PCR reaction.

PCR arrays are important tools for analyzing the expression of a focused panel of genes. Each 96-well plate includes SYBR Green-optimized primer assays for a thoroughly researched panel of relevant, pathway-or disease-focused genes. In PCR arrays, 96 different gene-specific products are simultaneously amplified under uniform cycling conditions using specific master mix formulation and subsequently detected.

II. Micro array Analyses

Gene expression profiling by micro array analysis has enabled the measurement of mRNA levels of thousands of genes in a single RNA sample. In this technique, a glass slide or membrane is spotted or "arrayed" with DNA fragments or oligo nucleotides that represent specific gene coding regions. Purified RNA is then fluorescently or radioactively labeled and hybridized to the slide/membrane. After thorough washing, the raw data is obtained by laser scanning or auto radiographic imaging and subsequently entered into a database and analyzed by a number of statistical methods.

Closing notes

Biosafety is integral to modern biotechnology. The adoption of modern biotech products needs to be balanced with adequate biosafety safeguards. Scientific risk assessment and cost benefit analysis have to be done by case by case studies individually. Biosafety panel of tests, need based adoption in NP products is recommended. Also the participation of scientists from various fields is important. Dissemination of knowledge and information occupy the integral part of Biosafety.

References

Ames, B.N., Durston, W.E., Yamasaki, E., Lee, F.D. 1973. Carcinogens are mutagens: a simple test system combining liver homogenates for activation and bacteria for detection. *Proc Natl Acad Sci . USA* **70**:2281–2285.

Beckett, W. et.al 2005. Comparing inhaled ultrafine versus fine zinc oxide particles in healthy adults: A human inhalation study. *Am. J. Respir. Crit. Care Med.* **171**: 1129–1135.

Bermudez, E., Mangum, J. B.,Wong, B. A., Asgharian, B. Hext, P.M.,Warheit, D. B., and Everitt, J. I. 2004. Pulmonary responses of mice, rats, and hamsters to subchronic inhalation of ultrafine titanium dioxide particles. T*oxicol Sci.* **77**: 347–357.

Buege, J.A., Aust, S.D. 1978. Microsomal lipid peroxidation. *Methods Enzymol* **52**:302–310.

Castranova, V. 2011. Overview of Current Toxicological Knowledge of Engineered Nanoparticles *J. Occup. Environ. Med.* **53:**S14-S17.

Invernizzi N. 2011. Nanotechnology between the lab and the shop floor: what are the effects on labor? *J. Nanopart. Res.* **13(6):**2249-68.

Lanone, S., Boczkowski, J. 2006. Biomedical applications and potential health risks of nanomaterials: molecular mechanisms. *Curr Mol Med,* **6:**651–663.

Møller, P. 2006. The alkaline comet assay: towards validation in biomonitoring of DNA damaging exposures. *Basic Clin Pharmacol Toxicol.* **98**(4):336-45.

Monteiro-Riviere, N. A., Nemanich, R. J., Inman, A. O., Wang, Y. Y., and Riviere, J. E. 2005. Multi-walled carbon nanotube interactions with human epidermal keratinocytes. *Toxicol. Lett.* **155**, 377–384.

Monteiro-Riviere, N.A., Inman, A.O., Zhang, L.W. 2009. Limitations and relative utility of screening assays to assess engineered nanoparticle toxicity in a human cell line. *Toxicol Appl Pharmacol* **234**:222–235.

Nanotechnology research directions for societal needs to 2020: Retrospective and outlook. 2011;Available from:http://www.wtec.org/nano2/Nanotechnology_ Research_ Directions_to_2020/.

Nel, A., Xia, T., Madler, L., Li, N. 2006. Toxic potential of materials at the nanolevel. *Science* **311**:622–627.

Quinti, L., Weissleder, R., Tung, C.H. 2006. A fluorescent nanosensor for apoptotic cells. *Nano Lett.* **6:** 488–490.

Schellenberger, E.A., Reynolds, F., Weissleder, R., Josephson, L. 2004. Surface-functionalized nanoparticle library yields probes for apoptotic cells. *Chembiochem* **5**: 275–279.

Schulte, P.A., Geraci, C.L., Hodson, L., Zumwalde, R., Castranova, V., Kuempel, E.D., Methner MM, Hoover, M., Murashov, V. 2010. Nanotechnologies and nanomaterials in the occupational setting *Ital. J. Occup. Environ. Hyg.* **1:** 63-68.

Schulte, P.A., Salamanca-Buentello, F. 2007. Ethical and scientific issues of nanotechnology in the workplace Environ. *Health Perspect.* **115:** 5-12.

Tice RR, et.al. Single cell gel/comet assay: guidelines for in vitro and in vivo genetic toxicology testing. *Environ Mol Mutagen*. 2000; **35(3):**206-21.

Warheit, D.B., Hoke, R.A., Finlay, C., Donner, E.M., Reed, K.L., Sayes, C.M. 2007. Development of a base set of toxicity tests using ultrafine TiO2 particles as a component of nanoparticle risk management. *Toxicol Lett* **171**: 99–110.

34

Regulatory Framework for Nanomaterials

A. Lakshmanan, K.S. Subramanian and K.Gunasekaran

Introduction

Engineered nanoparticles are a group of products that are defined by their size, ranging from 1 to 100 nm. Due to the smaller size, the physio-chemical properties of nano particles become very different from those of the same material in a larger form. Total production of nano-materials is expected to increase considerably in the near future. Nanotoxicology is yet another emerging field of science dealing with the effects of nano-devices and engineered nano-particles on living organisms. It has been observed that the materials in nano-dimensions react totally different in comparison to their existence in micro or macro sizes. The production of engineered nanomaterials was 2000 tons in the year 2004 and it is expected to increase to 58,000 tons in 2020. Such increase is likely to impact the ecosystems. On the other hand, number of nano-based products in the world had increased from 50 in 2005 to 1100 in 2010 and likely to double in a year or two. Thus, there is a growing anxiety to gain insights into the mechanisms associated with the toxicity of nanoparticles and nano-agri products. We may not be facing a grey goo challenge in the foreseeable future, but managing the risk of products where nanostructure and chemistry conspire to create new properties is presenting sufficient challenges of its own.

In contrast to this increase in productivity, remarkably little is known on the hazards that nanoparticles may pose to the terrestrial environment. Limited information is available on the biosafety studies and risk assessment of engineered nanoparticles, though there are numerous projects on its application. Also, the current focus on nanotoxicity fails to assess the risk in different local environments and populations. Research on the hazards and effects that nanoparticles may exhibit, is still in its infancy when compared to the research on the toxicity of conventional chemicals, and the majority of the studies performed so far is focused on aquatic organisms. Surprisingly few data are available on effects of nanoparticles on soil fauna and their predators. People in developing countries may be more prone to adverse effects of nanoparticles because of underlying health conditions and malnutrition. Moreover, genetic susceptibility to toxic effects varies in diverse ethnic groups and geographical areas (Xue, et al., 2006). The scientific community needs to identify these information gaps before developing regulations and standard methodologies for nanotoxicity assessment. A number of studies have documented *in vitro* and *in vivo* toxicity of exposure to nanoparticles. Evidence suggests that they can induce DNA damage, reactive oxygen species, damage to cellular organelles and cell death. This necessitates the requirement of Biosafety studies to be performed while developing any products with nanoparticles to ensure that nanomaterial based industries emerge as tools for sustainability rather than environmental liabilities. Hence, developing toxico kinetics spectrum for various engineered nano particles is aimed in the proposed study. Moreover, current nanomaterial fabrication methods suggest a substantial potential for environmental and health effects that are not from the nanomaterials themselves, but from associated energy inputs, feedstocks, and wastes. Such "collateral damages" associated with nanomaterial production are suggested by the high embodied energy in many nanomaterials, the feedstocks of a non-nano nature that are known to be hazardous (Mills, 2006).

Nanotechnology initiatives in India

The emergence of nanotechnology in India has witnessed the engagement of a diverse set of players, each with their own agenda and role. Nanotechnology in India is a government led initiative. Industry participation has very recently originated. Nanotechnology R&D barring a few exceptions is largely being ensued at public funded universities as well as research institutes. Given the enabling nature of nanotechnology and ability to develop along with existing technologies, it has the potential to be utilized as a tool to address key development related challenges in diverse sectors like energy, water, agriculture, health, environment and the like. Enabling energy storage, production and conversion within renewable energy frameworks has been cited as the primary area where nanotechnology applications might aid developing countries. Nanotechnology interventions might be sought at specific junctures to improve quantity and quality of water and wastewater treatment systems. Enhancement of agricultural productivity has been identified as a critical area of nanotechnology application for attaining the Millennium Development Goals. In light of the developments worldwide hailing nanotechnology as a technology with the potential of addressing a number of developing country needs, India has sought to promote nanotechnology applications in sectors that are likely to have a wide impact, and influence the course of future development in the country. Sectors such as health, energy and environment have received greater attention by various technology departments in the government (DST, DBT and SERC). Department of Science and Technology (DST), the chief agency engaged in the development of nanotechnology, initiated India's principal programme, the Nanoscience and Technology Mission (NSTM) in 2007, with an allocation of Rupees 1000 crores for a period of five years. The five-year programme followed the flagship initiative, the Nanoscience and Technology Initiative (NSTI) that was in operation from 2001–06. Close to 200 projects have been undertaken in the NSTI and NSTM since 2002. The DST has also set up 'Centers of Excellence (CoE) for Nanoscience and Technology' established under the NSTI to undertake R&D to develop specific applications in a fixed period of time. On the whole the 19 CoE have been spread across 14 distinct institutions. These CoE have been set up primarily at those institutes that have either been engaging in nanotechnology based R&D prior to their establishment or have developed the resources to do so. Aside DST, several other agencies with diverse mandates are also actively engaged in supporting nanotechnology in the national arena. DBT (Department of Biotechnology) is supporting research in nanotechnology and the lifesciences. CSIR (Council of Scientific and Industrial Research), a network of 38 laboratories that engages in scientific and industrial R&D for socio-economic benefit has also commissioned R&D in nanotechnology in diverse areas. SERC (Science and Engineering Research Council) too has aided projects on nanotechnology. Support for these projects has been through its general R&D schemes for basic science and engineering science.

India's International Collaborations in the field of Nano technology

Several bilateral collaborations emerged in nanoscience and technology, as it was a part of nearly all the S&T agreements between India and other countries. Initiatives for joint R&D have figured prominently with Indian institutes engaging in projects of similar kind in the US, EU, Japan, Taiwan and Russia. The S&T Departments of Brazil, South Africa and India have embarked on a tri-lateral initiative to develope collaborative programmes in several common areas of interest, and nanotechnology being one of them. Other initiatives include Science and Technology Initiatives with Indian diaspora – Scientists and Technologists of Indian Origin Abroad (STIOs) for encouraging networking between Indian scientists and scientists and technologists of Indian origin that are based abroad.

The International Science and Technology Directorate (ISAD) of the CSIR that aims to strengthen cooperation between CSIR and international institutions has facilitated workshops and collaborative projects with international partners like South Africa, France, South Korea, China, Japan in the area of nanoscience and technology. Another forum for international collaboration is the Euro-India Net set up under the FP6 between EU and India to encourage collaborations between scientists from the two regions in the area of nanotechnology. A memorandum of understanding also has been signed between India and UNESCO to establish a Regional Centre for Education and Training in Biotechnology, where one of the focus areas is on nano-biotechnology (TERI, 2010).

Need to formulate national Nano technology strategy

Developing clear national strategies to engage with this emerging technology is imperative. It is imperative to link technology developments with social priorities and goals. The role of the state would be imperative in charting the trajectory of nanotechnology developments in developing countries. In general, in developing countries, the share of public investment in total nanotechnology R&D basket is relatively greater in comparison to private sector. Although the growing importance of private sector cannot be underestimated, but given that technology base for nanotechnology being in an embryonic phase, industry would not be able to sustain the research effort needed for the establishment of scientific and technological infrastructure. Therefore, there is a stronger case for the role of government support for fundamental research. The government has played a predominant role in research effort in nanotechnology in terms of funding, establishing the scientific and technological infrastructure and developing human skills and capacity.The dynamics of the sectors in terms of the nature and composition of research infrastructure and the level of technology production would determine to a great extent the role of agencies in support for research. For instance research support for application of nanotechnology in IT sector could be private agency driven. However, in the energy, health and other development sectors, the government should play an active role in providing research support for nanotechnology applications.

The role of the state is also of prime importance in defining regulatory objectives, developing the ambit and then selecting the tools from the toolkit that would best facilitate the achievement of the objectives. Normally, the government priority at a given point of time also influences regulatory choices made. However, the State should strike a balance between promoting a technology and regulating its risks. Furthermore, the State can play a crucial role in resisting pressures of globalisation in the context of technology development.

International trends in R&D lead to the development of certain products that can be classified as either high-end luxury goods or quality enhancement of products that benefit a small segment of society. The state particularly in the developing country context can set the agenda and resist the tendency to uncritically follow international trends in research that do not address their developmental needs.

Conceptualizing national capability in the field of Nano technology

A conceptual framework to assess national capability to respond to nanotechnology development needs to address the key opportunities and challenges created by this technology for developing countries in terms of the demands imposed on the science and technology infrastructure and by changing the nature of science and technology. Several issues emerge from the review of international developments in nanotechnology:

- There is a need for strong infrastructure to enable and stimulate R&D and commercialisation of nano products;
- Constraints and concerns among users must be addressed for successful deployment of technology;
- Appropriate strategies, policies and institutions are needed to engage with an emergent technology;
- Human resources with multidisciplinary perspectives is key for progress in nanotechnology;
- There is a need for addressing nanotechnology risks in the societal context;
- Regulatory oversight and preparedness for nanotechnology is necessary to channelize research efforts in a specific direction;
- Capacity building of regulatory and monitoring agencies;
- Transparency and public involvement in the design and implementation of regulatory structure in nanotechnology should be ensured.

Developing capabilities in emerging technologies thus would require: (a) skills of both scientific and non-scientific kind, including regulatory bodies, (b) a greater degree of linkages between various actors from academia, industry, policy makers would be necessary for successful market deployment of such technologies, (c) the interdisciplinary approaches in nanotechnology would demand a different R&D strategy as well as reorientation of science and technology activities in universities, research institutes, funding agencies and industry with a conducive institutional setting facilitating interactive learning would be essential to respond to and develop nanotechnology, (d) devising adaptive and responsive governance structures that can suitably regulate applications of nanotechnology in society, and (e) a flexible, dynamic policy environment that has the ability to create the conditions required for both knowledge generation and its effective utilization would form an important dimension guiding the process of development of capabilities (TERI, 2010).

Risks in nanotechnology and need for the regulatory frame work

Nanotechnology risks can be best understood in conjunction with its benefits. The complexity of the technology, the breadth of nanomaterials and applications, coupled with the possibility of its wide dissemination in the globalised world renders the technology unpredictable in many senses. The risks are heterogeneous as the field of nanotechnology itself and include environmental, health, occupational and socio economic risks.The unusual properties of nanomaterials that can enable rewarding applications for society might pose unknown or unforeseen environment, health and safety challenges. The pro-technology stance taken in several developed and developing countries at the cost of risk related research has led to an information gap around the impacts of nanomaterials. These aspects together with the commercialization and pervasiveness of nanoproducts serve to heighten the risk from this emergent technology. By virtue of their size, nanomaterials like other tiny particles might be able to enter the human body and those of other species imperceptibly through various pathways- inhalation, ingestion, dermal contact, etc. Early research also indicates that nanoparticles could reach various parts of the body where they may exert adverse effects. Nanoparticles, it is believed might be able to disrupt cellular, enzymatic and other organ related functions posing health hazards. On the other hand, nanoparticles might also be non-biodegradable and on disposal, these disposed materials might form a new class of non-biodegradable pollutant and pose a new threat to the environment (air, water, soil) and health. In light of these facts it appears that the greatest current risk is to the occupational health of the workers involved in the production, packaging or transport of the nanomaterials. With the increase in the application of nanomaterials in various products, the risk of the exposure of the consumers and the general public will also increase. Therefore it is crucial to examine and estimate the risk for regulating the production, use, consumption and disposal of these materials.

In this context, the use of life cycle analysis (LCA) has been encouraged as a first step to understand the risks from nanoapplications during their entire life cycle-from cradle to grave. Despite lack of substantial data on EHS impacts,

LCA can be vital in identifying and evaluating potential risks from nanoapplications and also serve to identify crucial knowledge gaps.Currently, there is clearly a lack of emphasis on risk related research in nanotechnology. Even where risk aspects are looked at, several gaps and challenges exist in the sphere of undertaking studies related to toxicity and EHS impacts.

These include lack of information on the nature and characteristics nanomaterials in applications, insufficient methods for detecting and measuring nanomaterials, inadequate breadth of risk related research. These issues must be addressed alongside conducting appropriate kinds of tests and interdisciplinary research to accurately elucidate the risks of nanomaterials and products. Providing adequate investment to fund robust risk research, including regulatory toxicology, is necessary to both address and anticipate risks accruing from nanotechnology.

Given that nanoproducts are already in the market, LCA studies and risk assessments must be emphasized in order to address the nature and extent of risks from these applications. Further, there is a need to understand the trade-offs that may have to be made between the risks of nanotechnology applications and the risk that emerges from lack of access to the technology.It is likely that nanotechnology would have major impacts on the existing social, economic and trade milieu as well as on global markets and commodity driven industries. History shows that past introductions of major technological interventions in the global community have resulted in socio- economic impacts in vulnerable communities. However, the nature and extent of such risks would depend on the nature and extent of commodity dependence of the respective countries. It would also depend upon whether nanotechnology application substitutes the use of the commodities or whether it complements the existing use of the commodities. Policies have to be implemented to reduce the vulnerability of commodity dependent developing countries in the context of emergence of socio- economic risks from nanotechnology applications.

Development of an adequate risk governance framework for addressing risks that surround nanotechnology is vital for the responsible development of nanotechnology that allows reaping benefits while minimizing risks.One of the key debates around nanotechnology regulation has been around whether there exists a need for a separate nanotechnology specific law to regulate the concerns around nanotechnology or not. A close look at the existing set of laws and rules in different areas of current and potential nanotechnology applications suggests that a nanotechnology specific legislation may not be necessary at this stage in India. Most of the challenges and concerns could be addressed by way of either intervention at the level of subordinate legislation or amendments in the existing instruments, or interventions at the level of implementation. Moreover, since nanotechnology being an emerging technology is still surrounded with uncertainty and lack of enough information about its impacts, instead of a definite nano-specific law, ensuring the capacity of existing regulations to address new risks as they become known would be the viable course forward. Applications of nanotechnology, owing to their very nature, interaction with other technologies, and extent are subject to several regulations already. However, most of these existing regulations require revision before they are able to regulate the risks associated with nanotechnology. In this context, precautionary principle, which has already been adopted in environmental regulation, should be extended to nanotechnology regulation, considering the uncertainties. However, the principle should be applied in a judicious manner taking a middle ground, reconciling the need to address risks and using the technology to address development needs.

Challenges in Nanotechnology governance

- One of the biggest challenges has been in terms of the interdisciplinary nature of nanotechnology per se and the scope of its applications. This has lead to significant overlaps in the areas for R&D support identified by different agencies.
- The gap between basic research and application is another challenge in nanotechnology, like several other technologies. There is poor lab-firm integration, which is compounded by the paucity of skilled manpower that could provide linkages between the technology and commercial domains.
- Being cost and risk intensive, and being dependent upon sophisticated and complex equipment, technical know-how and capacity, financial constraints often act as an impediment in this regard.
- The main challenges faced by regulatory institutions currently relate to the regulatory capacity, information asymmetry and absence of inter-agency coordination.
- Another challenge that nanotechnologists should address in their research is due priority to risk research. Currently, funding allocated for analysing risks from nanotechnology is low compared to the vast amounts invested for its commercial applications. On the other hand, other experts argue that research related to toxicity and risk assessment must be undertaken once the applicability of specific nano-applications are ascertained, especially when the prototype of the product has been developed and is available for field testing. This strategy might help enable a balanced approach between technology development and addressing risk issues. Moreover, it could also facilitate the judicious allotment of already constrained financial resources as well as prevent an overzealous focus

on risk issues especially in the nascent stages of product development that might in turn prevent the emergence of socially useful and significant applications (TERI, 2010).

Conclusion

Given that nanotechnology is comprehensive in its reach and interdisciplinary in nature, ensuring the accountability of actors involved in its application and regulation is essential. Its socially embedded nature entails that its credibility depends on fostering partnerships among the various stakeholders. This would ensure that scientific research does not overwhelm public perception and the social analysis of technology, an approach that would be particularly relevant in the case of nanotechnology governance. Currently, there exists a trust deficit between the optimism of the scientific community and the apprehensions expressed by public interest groups regarding nanotechnology's potential. Bridging this gap would be critical in determining the extent to which India can avail of international opportunities to enhance its capabilities on the development front. A comprehensive governance framework has to transcend all the vertical and horizontal levels involved in the development and application of technology. Horizontal level refers to distributed responsibilities between national government departments, statutory bodies, and other non-government groups in engaging with nanotechnology developments, capabilities, networking, environmental, health socio-economic and political impacts, and designing a regulatory framework. At a vertical level, the framework spans global, regional, national, state and local nanotechnology developments and policy making. Adopting such an approach would enable integration of relevant perspectives on issues, challenges, prospective emerging applications and factors that will be influential in determining the development pathways for nanotechnology.

Acknowledgement

The information on formulating national strategies, formulating risk governance and regulatory frame work of this article have been obtained from the briefing paper on 'Capability, Governance and Nanotechnology Developments: A Focus on India' by The Energy and Resource Institute and we extend our special thanks to the authors of this brief.

References

Mills, N.L. et al., Am. J. Respir. 2006. *Crit. Care Med.*, **173**:426-431.

Smyth, A.B., Talasila, P.C., Cameron, A.C. 1999. An ethanol biosensor can detect low-oxygen injury in modified atmosphere packages of fresh-cut produce. *Postharvest Biology and Technology*, **15(2):**127–134.

Starodub, N. F., Kanjuk, N. I., Kukla, A. L., Shirshov, Y.M. 1999. Multi-enzymatic electrochemical sensor: field measurements and their optimisation. *Analytica Chimica Acta,* **385(1-3):** 461-466.

Su, X.D., Li, S.F.Y., Kwang, J., Low, S. 2000. Piezoelectric quartz crystal based screening test for porcine reproductive and respiratory syndrome virus infection in pigs. *Analyst,* **125(4):** 725–730.

TERI. 2010. Nanotechnology development in India: building capability and governing the technology [TERI Briefing Paper], Supported by IDRC, Canada.

Tran-Minh, C., Pandey, P.C., Kumaran, S. 1990. Studies on acetylcholine sensor and its analytical applications based on the inhibition.

Xue, Z. G. et al., Zhong Nan Da Xue Xue Bao Yi Xue Ban, 2006. Biotoxicology and Biodynamics of Silica nanoparticle. *MEDLINE Journal,* **31**:6-8.